AF615971

INTEGER PROGRAMMING

SERIES IN DECISION AND CONTROL

Ronald A. Howard, Editor

INTEGER PROGRAMMING

by Robert Garfinkel and George L. Nemhauser

FINITE DIMENSIONAL LINEAR SPACES

by Roger W. Brockett

DYNAMIC PROBABILISTIC SYSTEMS
Volume I: MARKOV MODELS

by Ronald A. Howard

DYNAMIC PROBABILISTIC SYSTEMS
Volume II: SEMI-MARKOV AND DECISION PROCESSES

by Ronald A. Howard

OPTIMIZATION BY VECTOR SPACE METHODS

by David G. Luenberger

INTRODUCTION TO DYNAMIC PROGRAMMING

by George L. Nemhauser

OPTIMIZATION THEORY WITH APPLICATIONS

by Donald A. Pierre

ROBERT S. GARFINKEL,
Graduate School of Management
University of Rochester

GEORGE L. NEMHAUSER,
Department of Operations Research
Cornell University

Integer Programming

A WILEY-INTERSCIENCE PUBLICATION

JOHN WILEY & SONS

NEW YORK · LONDON · SYDNEY · TORONTO

Library of Congress Cataloging in Publication Data:

Garfinkel, Robert.
Integer programming.

(Series in decision and control)

Bibliography: p.
1. Integer programming. I. Nemhauser, George L., joint author. II. Title.

T57.7.G37 519.7'7 72-3881
ISBN 0-471-29195-1

Printed in the United States of America

10-9 8 7 6 5

To The Knicks

Preface

Integer programming deals with the class of optimization problems in which some or all of the variables are required to be integer. This book is intended to be a comprehensive one on the theory, methodology, and applications of integer programming and can be used both as a text and as a reference.

Since the pioneering work of Ralph Gomory in the late 1950s, integer programming has been one of the most exciting and rapidly developing areas of operations research. Very many real-world applications of integer programming models have been developed, and algorithms have been devised for general and specific purpose models. These algorithms vary widely in spirit and mathematical sophistication. One of our objectives is to unify these approaches as far as possible and to classify them into a few generic categories so that similarities and differences may be easily seen.

The book is designed to be a text for a one- or two-semester course, depending on the sophistication of the students. Numerous exercises, ranging from simple numerical calculations to research problems, are provided. Every technique that is developed is illustrated by at least one numerical example. A comprehensive set of notes is included at the end of every chapter. These notes give an historical summary of the development of the material of each chapter. More importantly, they supply a (hopefully complete) set of references to integer programming, including many works that are not described in the body of the text.

Chapter 1 introduces the reader to some methods of formulating and solving integer programming problems. In addition, a number of applications are presented. Chapter 2 is an introduction to those linear programming methods used in integer programming. A knowledge of some basic linear algebra is needed, and this is actually the only mathematical requirement for reading the entire text. Chapter 3 deals with integer programming and graphs. The first four sections are basic and should be included in any course. Sections 3.5–3.7 are special topics of interest. Chapter 4 on enumeration is one of the fundamental chapters and should be studied in its entirety. Chapter 5 on cutting planes is also essential. The only topics that could reasonably be omitted are the rather complex details of the proof of termination in Section 5.12 and the methods described in Sections 5.14–5.16 and 5.18. The knapsack problem, discussed in Chapter 6 is a specific integer programming model. However, the development of Chapter 7 is, to some extent, dependent on Sections 6.1 and 6.5–6.7. Section 6.12 is of importance in its own right.

Chapter 7 contains some of the most recent approaches to integer programming problems. The first six sections should be included in any course. The set covering and partitioning problems discussed in Chapter 8 are special models with a wide range of applications and relatively efficient algorithms. Theoretical and computational aspects of covering and partitioning are currently an active area for research. Chapter 9 on approximate methods is quite straightforward but very important for anyone interested in solving large problems. We would recommend including this chapter in a course involving the computational aspects of integer programming. Chapter 10 contains a variety of special topics involving nonlinearities. Chapter 11 contains computational experience. It is the only chapter written in the form of the notes; that is, it contains references to authors within the body of the text. Anyone interested in solving real-world problems is strongly advised to read this chapter.

The book, in draft form, has been used as a text for courses in integer programming at Cornell and Rochester, as well as at other universities. Our students and colleagues have been very helpful in finding errors and making suggestions. In particular, we would like to thank Egon Balas, Gerry Bennington, Mark Eisner, Mendu Rao, Don Ratliff, and David Rubin for their comments. Ivan Rosenberg, at the University of Rochester, read a number of the chapters in detail and provided many corrections. We are especially grateful to Les Trotter, at Cornell University, who read the manuscript in its entirety, and whose suggestions and corrections were invaluable. Finally, we are grateful for the efficient typing service provided by Cornell and Rochester. In particular, special thanks are due Debbie Cagey, Kathy King, and Lorraine Ziegenfuss, all of whom did a marvelous job in typing the bulk of the manuscript.

ROBERT S. GARFINKEL
GEORGE L. NEMHAUSER

Rochester, New York
Ithaca, New York
February 1972

Contents

INTEGER PROGRAMMING

1 Introduction

1.1. BACKGROUND

Integer programming is a branch of *mathematical programming*. A general mathematical programming problem can be stated abstractly as

$$\max f(x), \qquad x \in S \subseteq R^n \tag{1}$$

where R^n is the set of all n-dimensional vectors of real numbers and f is a real-valued function defined on S. The set S is called the *constraint set* and f is called the *objective function*.

Every $x \in S$ is called a *feasible solution* to (1). If there is an $x^o \in S$ satisfying

$$\infty > f(x^o) \geq f(x) \qquad \text{for all } x \in S$$

then x^o is called an *optimal solution* to (1). The objective in a mathematical programming problem is to establish whether an optimal solution exists and then to find one, or perhaps all, optimal solutions.

Although there is no generally agreed-upon definition, an *integer programming problem* is a mathematical programming problem in which

$$S \subseteq Z^n \subseteq R^n$$

where Z^n is the set of all n-dimensional integer vectors. A *mixed integer programming problem* is a mathematical programming problem in which at least one, but not all, of the components of $x \in S$ are required to be integers.

In an applied context, it is convenient to think of (1) as a model of decision making in which S represents the set of all permissible decisions and f assigns a utility or profit to each $x \in S$. Applications of this model abound in the real world and are relevant to various branches of engineering, business, and the physical and social sciences. Some of these applications are discussed in Sections 1.4 and 1.5 and briefly throughout the book.

Much of the work done in modeling and developing solution techniques for mathematical programming problems comes under the names *operations research* and *management science*. Stimulated by scientific applications to military planning problems during World War II, operations research and management science have now become disciplines in their own right.

Linear Programming Problems

Considerable success in analyzing (1) has been achieved for the *linear programming problem* (LP), where

$$f(x) = cx \tag{2}$$

and

$$S = \{x | Ax = b, x \geq 0\} \tag{3}$$

Here, A is an $m \times n$ matrix, b is an m-vector, c is an n-vector, and 0 is an n-vector of zeros. The set S is a *convex set*, since $x, y \in S$ imply $\alpha x + (1 - \alpha)y \in S$ for all $0 \leq \alpha \leq 1$ (see Exercise 1). A convex set defined by linear constraints is called a *polyhedron* or *polytope*.

The Use of Computers

It is very important to keep in mind that the aim of much of the research in mathematical programming is to develop a theory that leads to the construction of algorithms for use on modern, high-speed digital computers. An unsophisticated reader might be attracted to simple algorithms that can be used to solve small problems by hand. Instead, the following important aspects of algorithms should be considered:

1. Is finite termination guaranteed?
2. If so, is there an upper bound on the number of computations?
3. Has computational experience been promising with respect to the speed of the algorithm?
4. Are computer storage requirements reasonable?
5. If the algorithm does not terminate in a specified time, are feasible solutions generated?

Notation

In defining the linear programming problem, we referred to x as an n-vector, without specifying whether it was $n \times 1$ or $1 \times n$. However, from (3) it is clear that x is $n \times 1$, b is $m \times 1$, and 0 is $n \times 1$, since the various matrices must be conformable for matrix multiplication. From (2) it follows that c is $1 \times n$. In general, we will not use a transpose operation on a matrix, unless the possibility of confusion exists.

In comparing two n-dimensional vectors (n-vectors) a and b, we write $a \geq b$ to mean $a_j \geq b_j$, $j = 1, \ldots, n$. If $a \geq b$, and $a_j > b_j$ for some j, we write $a > b$. Similarly, for two sets A and B, $A \subseteq B$ means that A is a subset of B, while $A \subset B$ denotes A a proper subset of B. The number of elements (*cardinality*) of a finite set A is denoted by $|A|$.

1.2. DEFINING AN INTEGER LINEAR PROGRAMMING PROBLEM

Much of this book is devoted to analyzing the *integer linear programming* problem (ILP) in which $f(x)$ is given by (2) and

$$S = \{x | Ax = b, x \geq 0 \text{ integer}\} \tag{4}$$

In more standard form the ILP is written

$$\begin{aligned} &\max cx \\ &\quad Ax = b \\ &\quad\ \ x \geq 0 \text{ integer} \end{aligned} \tag{5}$$

In summation notation (5) is

$$\begin{aligned} &\max \sum_{j=1}^{n} c_j x_j \\ &\sum_{j=1}^{n} a_{ij} x_j = b_i, \qquad i = 1, \ldots, m \\ &\qquad x_j \geq 0 \text{ integer}, \qquad j = 1, \ldots, n \end{aligned} \tag{6}$$

All of the data in c, A, and b will always be assumed integer. This is actually equivalent to assuming the data rational, since multiplication of the objective function by any positive number, or any constraint by any nonzero number, does not change the problem (see Exercise 2).

The ILP can be written in any of a number of forms, and (5) and (6) have been arbitrarily chosen to be the ones generally used. Sometimes we will choose to minimize rather than maximize. This causes no difficulty, since for $x \in S$

$$-(\min - f(x)) = \max f(x)$$

Similarly, we could use the constraint form $\sum_{j=1}^{n} a_{ij} x_j \leq b_i$, or $\sum_{j=1}^{n} a_{ij} x_j \geq b_i$. Table 1 gives methods for converting from any of these constraint forms to any other.

Table 1

From \ To	$\le$	$\ge$	$=$
$\le$	———	$-\sum_j a_{ij}x_j \ge -b_i$	$\sum_j a_{ij}x_j + s_i = b_i$ $s_i \ge 0$
$\ge$	$-\sum_j a_{ij}x_j \le -b_i$	———	$\sum_j a_{ij}x_j - t_i = b_i$ $t_i \ge 0$
$=$	$\sum_j a_{ij}x_j \le b_i$ $-\sum_j a_{ij}x_j \le -b_i$	$\sum_j a_{ij}x_j \ge b_i$ $-\sum_j a_{ij}x_j \ge -b_i$	———

The new variables s_i and t_i introduced in Table 1 are known as *slack* and *surplus* variables, respectively. Finally, if $x_j \ge 0$ is not required, an appropriate transformation is $x_j = x_j' - x_j''$, $x_j', x_j'' \ge 0$, so that x_j is unrestricted in sign.

The LP obtained by dropping the integrality constraints from the ILP (5) will be referred to as the *corresponding LP*. We also refer to the LP as a *relaxation* of the ILP. In general, the problem

$$P1: \quad \max f(x), \qquad x \in S_1$$

is said to be a relaxation of the problem

$$P2: \quad \max f(x), \qquad x \in S_2$$

if $S_1 \supseteq S_2$. Similarly, $P2$ is said to be a *restriction* of $P1$. The concepts of relaxation and restriction will be used often throughout the text. Note that if x^o is an optimal solution to $P1$ and x^* is an optimal solution to $P2$, then $f(x^o) \ge f(x^*)$. Furthermore, if $x^o \in S_2$, then x^o is an optimal solution to $P2$ (see Exercise 3).

An important special case of the ILP (6) is the *binary* ILP, where $x_j \ge 0$ and integer is replaced by $x_j = 0, 1$. Although the binary ILP could be written in the form (6) by adding the constraints $x_j \le 1$, $j = 1, \ldots, n$, it will generally be written

$$\begin{aligned} &\max cx \\ &Ax = b \\ &x \text{ binary} \end{aligned} \tag{7}$$

There is an important class of binary ILP's in which $a_{ij} = 0, 1$ for all i and j, and $b_i = 1$ for all i. Special methods have been developed for these so-called *covering* and *matching* problems, and they will be discussed in Chapters 3 and 8.

A generalization of the ILP is the *mixed integer linear program* (MILP), where only some of the variables are constrained to be integer. It is written

$$\begin{aligned} \max\ & c_1x + c_2v \\ & A_1x + A_2v = b \\ & x \geq 0 \text{ integer} \\ & v \geq 0 \end{aligned} \tag{8}$$

where A_1 is $m \times n_1$ and A_2 is $m \times n_2$. If $n_1 = 0$, (8) is an LP, and if $n_2 = 0$, (8) is an ILP.

1.3. METHODS FOR SOLVING THE ILP

In this section, a brief introduction is given to two fundamental approaches for solving ILP's, *enumeration* and *cutting planes.* Since most algorithms for solving ILP's solve subproblems that are LP's, a summary of the basic methodology of linear programming will be presented in Chapter 2. From the relaxation concept introduced in Section 1.2, it follows that if the corresponding LP has an optimal solution x^o which is an integer vector, then x^o is a feasible solution, and thus an optimal solution, to the ILP. There is a class of ILP's (see Chapter 3) for which the corresponding LP always has an integer optimal solution. In this class of problems, the relaxation of the ILP to the corresponding LP does not change the essence of the problem.

In general, the corresponding LP does not have an optimal solution that is integer. Such is the case in the following simple example for which two solution techniques are sketched.

Example

$$\begin{aligned} \max\ & 2x_1 + x_2 \\ & x_1 + x_2 + x_3 = 5 \\ & -x_1 + x_2 + x_4 = 0 \\ & 6x_1 + 2x_2 + x_5 = 21 \\ & x_1, \ldots, x_5 \geq 0 \text{ integer} \end{aligned}$$

Since x_3, x_4, and x_5 are slack variables, the problem can be written

$$\begin{aligned} \max\ & 2x_1 + x_2 \\ & x_1 + x_2 \leq 5 \\ & -x_1 + x_2 \leq 0 \\ & 6x_1 + 2x_2 \leq 21 \\ & x_1, x_2 \geq 0 \text{ integer} \end{aligned} \tag{9}$$

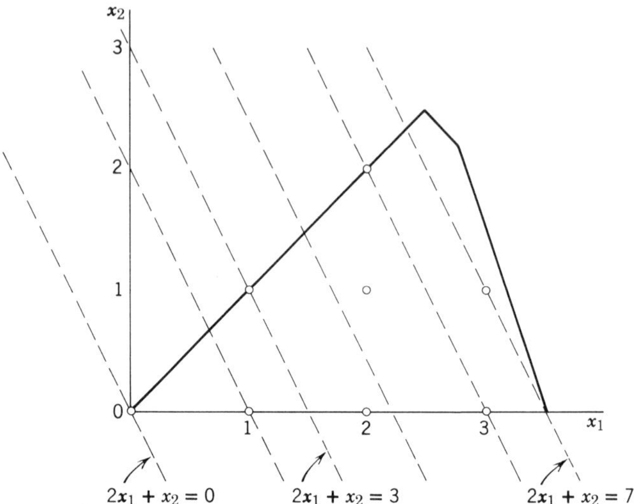

Figure 1

The feasible region of the corresponding LP is shown in Figure 1, along with a sequence of parallel lines representing constant values of the objective function. The circles represent the feasible solutions to (9). From Figure 1 it is clear that there are eight feasible solutions to (9) and that the optimal solution is $x_1 = 3$, $x_2 = 1$.

Enumeration

Without examining Figure 1, it is possible to get an upper bound on the number of feasible points. Note that the first constraint of (9), along with nonnegativity, implies $0 \leq x_1, x_2 \leq 5$. Also, the third constraint implies $0 \leq x_1 \leq 3$. Thus the constraints $0 \leq x_1 \leq 3$ and $0 \leq x_2 \leq 5$, along with the integrality requirements, limit the number of feasible points to not more than 24. For such a small problem, these 24 points could be *totally enumerated* to find that 16 are infeasible, 8 are feasible, and $(x_1, x_2) = (3, 1)$ is optimal.

Using a little more imagination, we might have added the first and second constraints to derive the valid constraint $2x_2 \leq 5$. Along with integrality, this implies $x_2 \leq 2$ and reduces the upper bound on the number of feasible points from 24 to 12.

Note that the feasible point $x_1 = 3$, $x_2 = 0$ yields an objective value of $f(x) = 6$. Thus every optimal solution to (9) must satisfy $2x_1 + x_2 \geq 6$.

Now $x_2 \leq 2$ implies $2x_1 \geq 4$ or $x_1 \geq 2$. Thus the candidates for optimality have been reduced to the set

$$Q = \{(x_1, x_2)|x_1 = 2, 3, x_2 = 0, 1, 2\}$$

Of these six points, (2, 0) and (2, 1) yield objective values smaller than 6. The optimal solution of the LP corresponding to (9) yields an objective value of $7\frac{3}{4}$. Since the objective value for an integer solution is an integer, it follows that $2x_1 + x_2 \leq 7$. This constraint is not satisfied by the point (3, 2). Now the candidates for optimality have been reduced to the set

$$Q = \{(2, 2), (3, 0), (3, 1)\}$$

The purpose of our rather naïve analysis of this example has been to motivate a class of systematic enumerative methods, which will be discussed in detail in Chapter 4.

Cutting Planes

Another class of approaches to the ILP is described in Chapter 5. It is based on the generation of a sequence of linear inequalities that "cut out" part of the feasible region of the corresponding LP while leaving the feasible region of the ILP intact. When enough of these cutting hyperplanes have been generated, the ILP will have the same optimal solution as the corresponding LP. In Figure 2, two cutting (hyper)planes are shown that accomplish this for the example (9). The remaining feasible region is shaded.

Suppose that the set $S = \{x|Ax = b, x \geq 0 \text{ integer}\}$ of feasible solutions to an ILP is bounded and therefore contains a finite number of points.

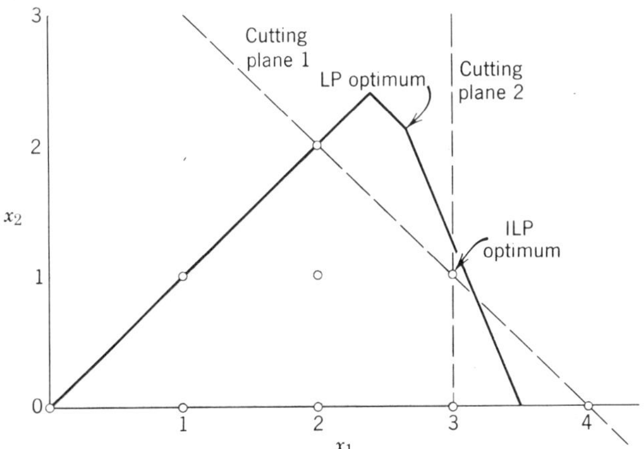

Figure 2

Define the set S^+ called the *convex hull* of S to be

$$S^+ = \{y | y = \sum \alpha_i x_i,\ \alpha_i \geq 0,\ \sum \alpha_i = 1,\ x_i \in S\}$$

Note that

$$S \subseteq S^+ \subseteq T = \{x | Ax = b,\ x \geq 0\}$$

where T is the set of feasible points to the corresponding LP.

For the example (9), the shaded region is S^+. The two cutting planes remove all of T–S^+ from T. In Chapter 5 we will see that a smaller set of cutting planes will generally suffice for solving the ILP. In fact, the problem of generating S^+ (see Chapter 7) is generally very difficult.

1.4. FORMULATING MODELS IN BINARY VARIABLES

In this section, it is shown how binary variables are used to model a variety of complex relationships. These include dichotomies, k-fold alternatives, and other nonlinearities. The models developed are ILP's, MILP's, or *integer nonlinear programming problems* (INLP's). Solution techniques for INLP's are discussed in Chapter 10. However, linear models are generally easier to solve and are therefore preferred when they can be achieved.

Dichotomies

Consider the problem

$$\max f(x), \qquad x \in S \tag{10}$$

and

$$g(x) \geq 0 \quad \text{or} \quad h(x) \geq 0 \quad \text{or both} \tag{11}$$

The problem, (10) and (11), is a mathematical programming problem not easily handled by any standard algorithm. However, the dichotomy (11) is equivalent to

$$\begin{aligned} g(x) &\geq \delta \underline{g} \\ h(x) &\geq (1 - \delta)\underline{h} \\ \delta &= 0, 1 \end{aligned} \tag{12}$$

where $\underline{g}$ and $\underline{h}$ are known finite lower bounds on g and h, respectively. To see this, observe that $\delta = 0$ yields $g(x) \geq 0$ and the trivially satisfied constraint $h(x) \geq \underline{h}$. Similarly, $\delta = 1$ gives $h(x) \geq 0$ and $g(x) \geq \underline{g}$.

If f, g, and h are linear, and $S = \{x | Ax = b, x \geq 0 \text{ integer}\}$, the resulting problem is an ILP. If, instead, $S = \{x | Ax = b, x \geq 0\}$, the problem has been transformed into an MILP.

k-Fold Alternatives

A generalization of the problem (10) and (11) is

$$\max f(x), \qquad x \in S \tag{13}$$

and at least k of the constraints

$$g_i(x) \geq 0, \qquad i = 1, \ldots, m \tag{14}$$

must hold, where $1 \leq k \leq m - 1$. The expression (14) can be replaced by the constraint set

$$g_i(x) \geq \delta_i \underline{g}_i, \qquad i = 1, \ldots, m \tag{15}$$

$$\sum_{i=1}^{m} \delta_i \leq m - k \tag{16}$$

$$\delta_i = 0, 1, \qquad i = 1, \ldots, m \tag{17}$$

where $\underline{g}_i$ is a known finite lower bound on g_i. Constraints (16) and (17) guarantee that at most $m - k$ of the constraints of (15) will be trivial. Thus at least k must be of the form (14). As before, the problem has been modeled as an INLP, MILP, or ILP depending on the form of f, g_i, and S.

Conditional Constraints

Consider the constraint

$$f(x) > 0 \Rightarrow g(x) \geq 0 \tag{18}$$

The constraint (18) is equivalent to the dichotomy

$$g(x) \geq 0 \qquad \text{or} \qquad f(x) \leq 0 \qquad \text{or both} \tag{19}$$

since $p \Rightarrow q$ is equivalent to (not p) or q or both. Then from (11) and (12) it follows that (19) can be transformed into

$$\begin{aligned} g(x) &\geq \delta \underline{g} \\ f(x) &\leq (1 - \delta)\bar{f} \\ \delta &= 0, 1 \end{aligned}$$

where $\bar{f}$ is a known finite upper bound on f.

Discrete Variables

In some applications variables are constrained to sets of discrete values. For instance, consider the constraint

$$x_j \in S_j = \{s_{1j}, \ldots, s_{pj}\} \tag{20}$$

Clearly, (20) is equivalent to the constraint set

$$\begin{aligned} x_j &= \sum_{k=1}^{p} s_{kj}\delta_{kj} \\ &\sum_{k=1}^{p} \delta_{kj} = 1 \\ &\delta_{kj} = 0, 1, \qquad k = 1, \ldots, p \end{aligned} \tag{21}$$

In particular, the transformation (21) works for the constraint $0 \leq x_j \leq u_j$, x_j integer, by defining $S_j = \{0, 1, \ldots, u_j\}$. However, in this simple case a transformation requiring fewer variables is

$$\begin{aligned} x_j &= \sum_{k=0}^{t_j} 2^k\delta_{kj} \leq u_j \\ &\delta_{kj} = 0, 1, \qquad k = 0, \ldots, t_j \end{aligned} \tag{22}$$

where $2^{t_j} \leq u_j < 2^{t_j+1}$.

Piecewise Linear Approximation of Nonlinear Functions

Consider the problem of approximating a nonlinear function $f_j(x_j)$, $0 \leq x_j \leq u_j$, as shown in Figure 3. The curve has been approximated by a

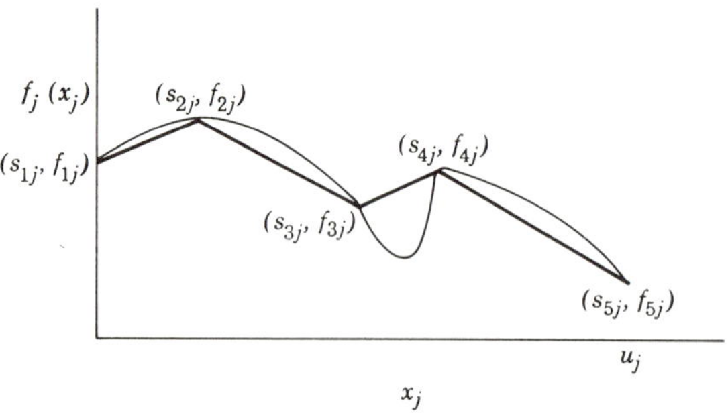

Figure 3

set of line segments; this approximation is known as "piecewise linearization." Once the grid $S_j = \{0 = s_{1j}, s_{2j}, \ldots, s_{p-1,\,j}, s_{pj} = u_j\}$ is chosen, the function $g_j(x_j)$ given by the set of line segments can be represented as follows. Let

$$\begin{aligned} x_j &= \sum_{k=1}^{p} \lambda_k s_{kj} \\ \text{and} \qquad g_j(x_j) &= \sum_{k=1}^{p} \lambda_k f_{kj} \\ \sum_{k=1}^{p} \lambda_k &= 1 \\ \lambda_k \geq 0, &\qquad k = 1, \ldots, p \end{aligned} \tag{23}$$

For a given x_j^*, $g_j(x_j^*)$ given by (23) is a point on the set of line segments if no more than two of the λ_k are positive and these are of the form λ_k, λ_{k+1}. This condition can be achieved by the constraints

$$\begin{aligned} &\lambda_1 \leq \delta_1 \\ &\lambda_k \leq \delta_{k-1} + \delta_k, \qquad k = 2, \ldots, p-1 \\ &\lambda_p \leq \delta_{p-1} \\ &\sum_{k=1}^{p-1} \delta_k = 1 \\ &\delta_k = 0, 1, \qquad k = 1, \ldots, p-1 \end{aligned} \tag{24}$$

From (24) it follows that for some t, $1 \leq t \leq p-1$, $\delta_t = 1$ and $\delta_k = 0$, $k \neq t$. Thus $\lambda_t \leq 1$, $\lambda_{t+1} \leq 1$, and $\lambda_k = 0$ otherwise.

1.5. APPLICATIONS

In this section some applications of ILP's, MILP's, and INLP's are presented. The models have been chosen because they represent applications that have widespread interest, or because they are prototypes for a large class of problems.

Capital Budgeting

A firm has n projects that it would like to undertake but, because of budget limitations, not all can be selected. In particular, project j has a present value of c_j, and requires an investment of a_{ij} dollars in the time period i, $i = 1, \ldots, m$. The capital available in time period i is b_i. The problem of maximizing the total present value subject to the budget constraints can be written as

$$\max \sum_{j=1}^{n} c_j x_j$$

$$\sum_{j=1}^{n} a_{ij} x_j \leq b_i, \qquad i = 1, \ldots, m \tag{25}$$

$$x_j = 0, 1, \qquad j = 1, \ldots, n$$

where $x_j = 1$ if project j is selected.

It is reasonable to assume that all of the data in (25) are nonnegative. Therefore (see Section 1.2) it can be assumed that the data consist of nonnegative integers. Given this restriction, the problem (25) for $m = 1$ is a *knapsack problem* (see Chapter 6).

Using the techniques of Section 1.4, the model (25) can be made considerably more realistic by including such factors as mutually exclusive and contingent projects (see Exercise 8).

Scheduling Jobs on Machines

Consider the problem of scheduling n jobs, $i = 1, \ldots, n$ on m machines, $k = 1, \ldots, m$. Assume, for simplicity, that each job requires exactly one operation on each machine and the operations on each job must be done in a specified order. In particular, define

$$r_{ijk} = \begin{cases} 1 & \text{if the } j\text{th operation of job } i \text{ requires machine } k \\ 0 & \text{otherwise} \end{cases}$$

Also, let the processing time of job i on machine k be p_{ik}. The decision variables are taken to be $x_{ik} \geq 0$, the starting time of job i on machine k. The schedule is assumed to begin at time zero.

To enforce the restriction that no two jobs may be on the same machine at the same time, we use the dichotomy

$$x_{rk} - x_{sk} \geq p_{sk} \tag{26}$$

or

$$x_{sk} - x_{rk} \geq p_{rk} \tag{27}$$

for all pairs of jobs r and s and all machines k. Clearly, it is not possible for (26) and (27) to hold simultaneously. As shown in Section 1.4, the dichotomy, (26) and (27), can be modeled in terms of binary variables. Let

$$\delta_{rsk} = \begin{cases} 1 & \text{if job } r \text{ precedes job } s \text{ (not necessarily directly) on machine } k \\ 0 & \text{otherwise} \end{cases}$$

Then (26) and (27) can be replaced by

$$\begin{aligned} x_{rk} - x_{sk} &\geq p_{sk} - M\delta_{rsk} \\ x_{sk} - x_{rk} &\geq p_{rk} - (1 - \delta_{rsk})M \\ \delta_{rsk} &= 0, 1 \end{aligned} \tag{28}$$

where M is sufficiently large. There are $m\binom{n}{2}$ constraint pairs of the form (28), and $m\binom{n}{2}$ binary δ variables.

Another set of constraints is required for each job i to guarantee that its operations are done in the correct order. Noting that the starting time of the jth operation of job i is $\sum_{k=1}^{m} r_{ijk}x_{ik}$, we maintain the correct order by requiring

$$\sum_{k=1}^{m} r_{ijk}(x_{ik} + p_{ik}) \leq \sum_{k=1}^{m} r_{i,j+1,k}x_{ik}, \qquad j = 1, \ldots, m - 1 \tag{29}$$

Thus there are $n(m - 1)$ constraints of the form (29).

There are many possible objective functions. A common one is to minimize the sum of the starting times of the last operation of each job, which is given by

$$\sum_{i=1}^{n} \sum_{k=1}^{m} r_{imk}x_{ik} \tag{30}$$

The problem of minimizing (30) subject to (28) and (29) is an MILP having mn continuous, nonnegative variables, $m\binom{n}{2}$ binary variables and $n(m - 1) + 2m\binom{n}{2}$ constraints. Clearly, unless m and n are very small, a very large MILP results. For instance, with $m = 4$ and $n = 10$, there are 40 continuous variables, 180 binary variables, and 390 constraints.

Political Districting

Here we investigate the topical question of devising optimal districting plans. Consider the fictional state shown in Figure 4. Let each of the numbered areas designate a population unit (possibly county or census tract). A population unit is indivisible in the sense that it must be included in exactly one voting district. Let the population of unit i be q_i, $i = 1, \ldots, m$. Also, let the number of districts required (number of representatives) be k.

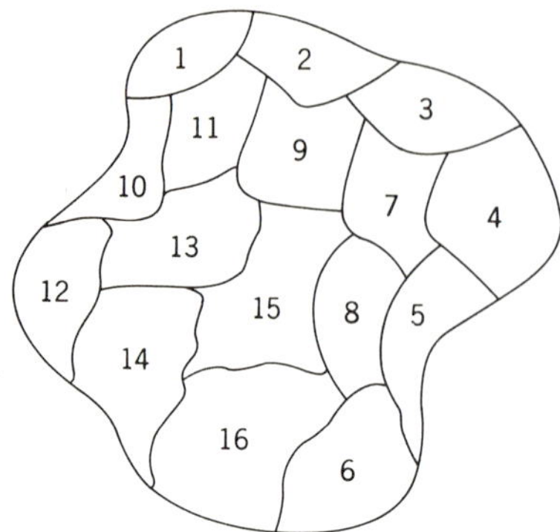

Figure 4

Ideally, from the "one man–one vote" principle, the population of each district should be

$$\frac{1}{k}\sum_{i=1}^{m} q_i = \bar{q}$$

This ideal cannot be achieved in general because of the indivisibility of the units.

Other criteria districts (sets of units) should satisfy include contiguity, compactness, and so forth. Suppose that there are n districts satisfying all such requirements. Let p_j be the population of district j, $j = 1, \ldots, n$, and

$$a_{ij} = \begin{cases} 1 & \text{if unit } i \text{ is in district } j \\ 0 & \text{otherwise} \end{cases}$$

Define a measure of unacceptability of district j to be $c_j = |p_j - \bar{q}|$. The variables are

$$x_j = \begin{cases} 1 & \text{if district } j \text{ is chosen} \\ 0 & \text{otherwise} \end{cases} \tag{31}$$

and a reasonable objective is to make the worst district as acceptable as possible. This objective can be stated mathematically as

$$\min \max_{j=1}^{n} c_j x_j \tag{32}$$

The "indivisibility" constraints are

$$\sum_{j=1}^{n} a_{ij} x_j = 1, \qquad i = 1, \ldots, m \tag{33}$$

In addition, exactly k districts are required so that

$$\sum_{j=1}^{n} x_j = k \tag{34}$$

The problem (31)–(34) is an INLP because of (32). However, it is easily converted into an ILP (see Section 1.7). If the objective were in summation form, the problem would be [except for the single constraint (34)] an ILP known as a *set partitioning* problem (see Chapter 8). This model has a wealth of other applications.

The Fixed Charge Problem

For some problems the cost of an activity is a nonlinear function of the activity level x_j, given by (see Figure 5)

$$f_j(x_j) = \begin{cases} d_j + c_j x_j & x_j > 0 \\ 0 & x_j = 0 \end{cases}$$

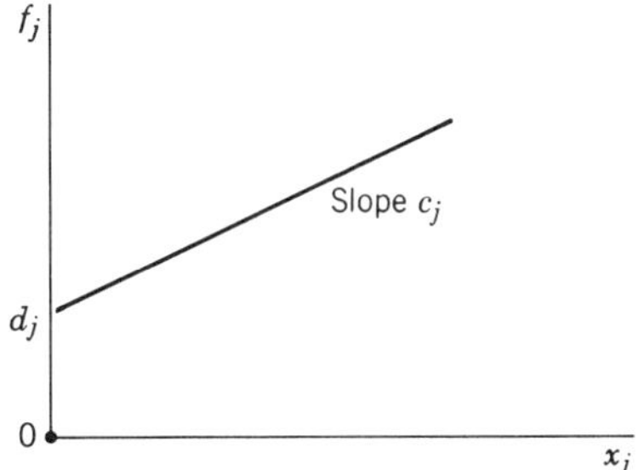

Figure 5

Assuming that $d_j > 0$ and f_j is to be minimized, $f_j(x_j)$ can be represented by

$$\begin{aligned} g_j(x_j, y_j) = c_j x_j &+ d_j y_j \\ x_j &\geq 0 \\ y_j &= 0, 1 \\ x_j(1 - y_j) &= 0 \end{aligned} \tag{35}$$

where y_j is an indicator of whether or not activity j is undertaken. The last constraint of (35) guarantees that $x_j > 0 \Rightarrow y_j = 1$. Minimizing g_j will insure $y_j = 0$ when $x_j = 0$.

The same result can be achieved if a finite upper bound u_j is known for x_j by replacing the last constraint of (35) by the linear constraint

$$x_j \leq u_j y_j \tag{36}$$

If $y_j = 1$, (36) is trivially satisfied. However, $y_j = 0 \Rightarrow x_j = 0$. Again when $x_j = 0$, minimization implies that $y_j = 0$. Thus, assuming $d_j > 0$ for all j, a fixed charge problem with linear constraints can be written as the MILP,

$$\begin{aligned}
\min \sum_{j=1}^{n} (c_j x_j + d_j y_j) & & \\
\sum_{j=1}^{n} a_{ij} x_j = b_i, & \quad i = 1, \ldots, m \\
x_j - u_j y_j \leq 0, & \quad j = 1, \ldots, n \\
x_j \geq 0,\ y_j = 0, 1, & \quad j = 1, \ldots, n
\end{aligned} \tag{37}$$

A particular fixed charge problem is the *plant* (warehouse) *location problem*. There are n customers, the jth one requiring b_j units of a particular commodity. There are m locations in which plants may operate to satisfy the demands. There is a fixed charge of d_i for opening plant i; the cost per unit of supplying customer j from plant i is c_{ij}. The capacity of plant i is h_i. The problem is

$$\begin{aligned}
\min \sum_{i=1}^{m} \left(\sum_{j=1}^{n} c_{ij} x_{ij} + d_i y_i \right) & & \\
\sum_{i=1}^{m} x_{ij} = b_j, & \quad j = 1, \ldots, n \\
\sum_{j=1}^{n} x_{ij} - h_i y_i \leq 0, & \quad i = 1, \ldots, m \\
x_{ij} \geq 0,\ y_i = 0, 1 & \quad \text{all } i \text{ and } j
\end{aligned} \tag{38}$$

The Quadratic Assignment Problem

The quadratic assignment problem can also be thought of as a location problem. There is a set of n locations in which m plants must be constructed, where $m < n$. The unit cost of shipping from location i to location j is c_{ij}. Also, the volume required to be shipped from plant k to plant l is d_{kl}. Let

$$x_{ik} = \begin{cases} 1 & \text{if plant } k \text{ is assigned to location } i \\ 0 & \text{otherwise} \end{cases} \tag{39}$$

Then

$$\sum_{i=1}^{n} x_{ik} = 1, \quad k = 1, \ldots, m \tag{40}$$

since each plant must have exactly one location. Also,

$$\sum_{k=1}^{m} x_{ik} \leq 1, \quad i = 1, \ldots, n \tag{41}$$

since each location can have at most one plant. If plant k is assigned to location i and plant l to location j, the shipping cost is $c_{ij}d_{kl} + c_{ji}d_{lk}$. Therefore, the total shipping cost is

$$\sum_{\substack{i,j=1\\i\neq j}}^{n} \sum_{\substack{k,l=1\\k\neq l}}^{m} c_{ij}d_{kl}x_{ik}x_{jl} \tag{42}$$

Thus the quadratic assignment problem is to minimize (42) subject to (39)–(41). The traveling salesman problem (Section 10.5) is a special case of the quadratic assignment problem (see Exercise 11).

Other Applications

Many other problems of practical interest in engineering, economics, business, and so forth, can be formulated as integer programs. Some brief descriptions and references are given in Section 1.7. However, it is important to realize that although almost any combinatorial optimization problem can be formulated as an integer program, the viability of the model will depend on the problem size and structure.

1.6. EXERCISES

1. Prove that the set $S = \{x | Ax = b, x \geq 0\}$ is convex.

2. Show that an LP with rational data can be converted into an LP with integer data, such that the sets of feasible (and therefore optimal) solutions to the two problems are identical. Can the same be said for LP's with real data?

3. Let P_1 be the problem $\max f(x)$, $x \in S_1$ and P_2 be $\max f(x)$, $x \in S_2$, where $S_1 \supseteq S_2$. If x^o is an optimal solution to P_1, show

(a) $f(x^o) \geq f(x)$ all $x \in S_2$.

(b) If $x^o \in S_2$, then x^o is an optimal solution to P_2.

4. Express the dichotomy

$$g(x) \geq 0 \quad \text{or} \quad h(x) \geq 0 \quad \text{but not both}$$

in binary variables.

5. Model in binary variables the constraint

$$(f(x) \geq 0 \quad \text{or} \quad g(x) \geq 0 \quad \text{or both}) \Rightarrow h(x) \geq 0$$

6. Is (12) equivalent to (11) if $\underline{g}$ or $\underline{h}$ are minus infinity?

7. Rewrite the problems

$$P1: \quad \max_{x \in S} f(x), \qquad P2: \quad \max_{x \in S} g(x)$$

where $g(x) \geq f(x)$ for all $x \in S$, in a form which shows that P_2 is a relaxation of P_1.

8. Show how the capital budgeting model (25) can be modified to incorporate constraints of the form: Project j may be undertaken only if project k is (is not) undertaken.

9. Modify the job-scheduling model to allow job i to require $Q_i < m$ operations.

(a) Assume that each operation of a given job is on a distinct machine.

(b) More than one operation on a given job may be on the same machine.

10. Show that the function $f_j(x_j)$ of the fixed charge problem is concave.

11. Consider the problem of the salesman who must visit n cities and return home. He wishes to minimize the total distance of his trip. This problem is called the *traveling salesman problem.* Show that it is actually a special case of the quadratic assignment problem.

1.7. NOTES

1.1. There are a number of survey papers on integer programming. A quite comprehensive one with an extensive bibliography is Balinski and Spielberg (1969). Geoffrion and Marsten (1972), which has been reprinted in Geoffrion (1972), emphasize the unification of the various approaches and also report recent computational experience. Other surveys are Balinski (1965), which has been reprinted in Kuhn (1970) and in Dantzig and Veinott (1968); Balinski (1967); Balinski (1970a); Beale (1965); and Dantzig (1957). Hu (1971) gives a survey and bibliography for discrete optimization.

The first book on integer programming was Hu (1969). Others are Greenberg (1971) and Plane and McMillan (1971). Saaty (1970) deals with the general problem of integer optimization.

1.4. This section is based on Dantzig (1960a), which is also reprinted in Chapter 26 of Dantzig (1963). Transforming in the opposite direction, Raghavachari (1969) has shown that a binary ILP can be transformed into a convex, quadratic maximization problem with continuous variables and linear constraints.

1.5. The capital budgeting model is discussed in Weingartner (1963) and (1966) and Unger (1970). A related problem is considered by Manne (1971).

There is a vast literature on integer programming and branch and bound formulations and techniques for job-machine scheduling. A comprehensive general reference is Conway, Maxwell, and Miller (1967). The model of this section is based on Manne (1960), which has been reprinted in Muth and Thompson (1963). Other models and methods are given in Bowman (1959); Wagner (1959); Wagner, Giglio, and Glaser (1964); Giglio and Wagner (1964); Ignall and Schrage (1965); Lomnicki (1965); Palmer (1965); McMahon and Burton (1967); Greenberg (1968); Balas (1969b) and (1970b); Gupta and Walvekar (1969); Gupta (1970); Charlton and Death (1970); Kirby and Scobey (1970); von Lanzenauer (1970); Bozoki and Richard (1970); Ashour (1970a) and (1970b); Arthanari and Ramamurthy (1970); and Florian, Trepant, and McMahon (1971).

The political districting model of this section is from Garfinkel and Nemhauser (1970). The model is one of a class of INLP's known as *bottleneck problems* (see Section 10.4). The conversion of a bottleneck INLP to an ILP is Exercise 17, Chapter 10. Other political districting models are found in Hess et al. (1965) and Wagner (1968).

An algorithm for the general linear fixed charge problem that uses the concavity of the objective function (see Exercise 10) is given in Murty (1968). There is a huge literature on the facilities location problem and the related fixed cost transportation problem. Of historical interest is Hirsh and Dantzig (1968), which is a reprint of a 1954 unpublished RAND memorandum. Other studies include Baumol and Wolfe (1958); Balinski (1961); Kuhn and Baumol (1962); Kuehn and Hamburger (1963); Manne (1964); Armour and Buffa (1965); Efroymson and Ray (1966); Feldman, Lehrer, and Ray (1966); Cooper and Drebes (1967); Gray (1967); Davis and Ray (1969); Denzler (1969); Jones and Soland (1969); Spielberg (1969a) and (1969b); Drysdale and Sandiford (1969); Shannon and Ignizio (1970); Steinberg (1970); Revelle, Marks, and Liebman (1970); and Roveda and Schmid (1971).

Koopmans and Beckmann (1957) appear to have formulated the quadratic assignment problem. Their model includes a fixed charge of assigning plant i to location j. Other studies of the quadratic assignment problem include Gilmore (1962); Lawler (1963); Land and Doig (1965); Carlson and Nemhauser (1966); Gavett and Plyter (1966); Hillier and Conners (1967); Greenberg (1969c); Graves and Whinston (1970), which is also reprinted in Abadie (1970); and Pierce and Crowston (1971). Carlson and Nemhauser (1966) give an application to classroom and examination scheduling. References for the related traveling salesman problem are given in Section 10.6.

The optimal allocation of resources in the critical path method for planning project schedules has been treated by integer programming methods. Studies include Pritsker, Watters, and Wolfe (1969); Crowston (1970); Teruo

(1970); Schrage (1970); Mason and Moodie (1971); and Davis and Heidorn (1971).

The problem of optimal assembly line balancing has been modeled and treated by integer programming—see Salveson (1955); Bowman (1960); Roberts and Villa (1970); and Thangavelu and Shetty (1971). A survey of integer programming applications in business is given by Cushing (1970).

Integer programming treatments of problems related to the determination of size and schedules for transportation fleets have been considered by D'Esopo and Lefkowitz (1964); Cristofides and Eilon (1969); Martin-Lof (1970); and Appelgren (1971). Little (1966) gives an integer programming model for the synchronization of traffic lights. Applications of integer programming in optimal power generation are developed by Dakin (1961) and Gately (1970). Childress (1969) gives several applications of integer programming in the chemical and oil industries. Mizukami (1968) considers a model for maximizing systems reliability.

A military application concerning the allocation of missiles to targets is developed by Miercort and Soland (1971). Another model with military applications concerns the optimal removal and/or addition of arcs in a flow network—see Hammer (1969a); Bellmore, Bennington, and Lubore (1970); and Ghare, Montgomery, and Turner (1971). This problem is mentioned in Chapter 8, along with many other applications of set covering and partitioning models.

Some other applications are given by Steinmann and Schwinn (1969); White and Francis (1971); Eilon and Cristofides (1971); Shaftel (1971); and Baugh, Ibaraki, and Muroga (1971).

Woolsey (1972) discusses some of the difficulties encountered in applying integer programming in the real world.

2 Linear Programming

2.1. INTRODUCTION

The LP is

$$\begin{aligned} \max x_0 &= cx \\ Ax &= b \\ x &\geq 0 \end{aligned}$$

given the data $c = (c_1, \ldots, c_n)$, $b = (b_1, \ldots, b_m)$, and $A = (a_1, \ldots, a_n)$, where $a_j = (a_{1j}, \ldots, a_{mj})$.

Undoubtedly, the linear programming model is one of the most important models in operations research and management science. Many real problems can be stated or approximated as LP's; optimal solutions are relatively easy to calculate; computer programs for solving large problems are readily available; and various subsidiary questions, such as sensitivity of the optimal solution to data changes, can be analyzed. Furthermore, linear programming algorithms are frequently used as subroutines for solving more difficult optimization problems in nonlinear and integer linear programming.

Because of the intimate theoretical and computational relationships between linear and integer linear programming, we need to survey the basic results of linear programming. The reader who is well grounded in linear programming might read this chapter quickly just to become familiar with our notation. The reader who is ignorant of linear programming will, most likely, find this chapter insufficient.

We restrict our presentation to stating and explaining the essential results. Some theorems are given without proof.

A vector x satisfying $Ax = b$ is called a *solution* to the LP. In addition, if $x \geq 0$, the solution is said to be *feasible*. Given an LP, there are three mutually exclusive and collectively exhaustive possibilities:

1. There is no feasible solution. In this case the problem is said to be *infeasible*.
2. There are vectors x and y such that $cy > 0$ and $x + \alpha y$ is a feasible solution for all nonnegative scalars α. In this case x_0 can be made arbitrarily large, and the problem is said to be *unbounded*.
3. There is a feasible solution x^o with $\infty > cx^o \geq cx$ for all feasible solutions x. In this case x^o is called an *optimal* solution.

Case 3 is the most important one, and we study it first. It is assumed, without loss of generality, that the rank of A is m; that is, the rows of A are linearly independent (see Exercise 17).

2.2. BASIC SOLUTIONS, EXTREME POINTS, AND OPTIMAL SOLUTIONS

Suppose that the columns of A are permuted so that $A = (B, N)$, where B is an $m \times m$ nonsingular matrix; that is det $B \neq 0$, where det B denotes the determinant of B. The matrix B is called a *basis matrix* for the LP. Let $x = (x_B, x_N)$, where x_B is the vector of *basic* variables associated with the columns of B and x_N is the vector of *nonbasic* variables associated with the columns of N. Then $Ax = b$ can be written as

$$Bx_B + Nx_N = b \tag{1}$$

Since B is nonsingular, B^{-1} exists and we can solve (1) for x_B in terms of x_N to obtain

$$x_B = B^{-1}b - B^{-1}Nx_N \tag{2}$$

The particular solution to (2)

$$x_B = B^{-1}b, \qquad x_N = 0 \tag{3}$$

is called a *basic solution*. If, in addition, $x_B \geq 0$, the solution (3) is called a *basic feasible* solution.

A fundamental theorem of linear programming is:

Theorem 1: *If an LP has an optimal solution, it has a basic optimal solution.*

Example

$$
\begin{aligned}
\max x_0 = 2x_1 + x_2 & \\
x_1 + x_2 + x_3 \qquad\qquad & = 5 \\
-x_1 + x_2 \qquad + x_4 \qquad & = 0 \\
6x_1 + 2x_2 \qquad\qquad + x_5 & = 21 \\
x_1, \ldots, x_5 & \geq 0
\end{aligned}
$$

This example will be used throughout this chapter to illustrate the various concepts and computational techniques.

Clearly there are at most $\binom{n}{m}$ bases. In the example, 6 of the 10 possible bases yield basic feasible solutions. The bases and corresponding basic feasible solutions are

$$B_1 = \begin{bmatrix} 1 & 1 & 0 \\ -1 & 1 & 1 \\ 6 & 2 & 0 \end{bmatrix} \qquad x_{B_1} = \begin{bmatrix} x_1 \\ x_2 \\ x_4 \end{bmatrix} = \begin{bmatrix} \frac{11}{4} \\ \frac{9}{4} \\ \frac{1}{2} \end{bmatrix}$$

$$B_2 = \begin{bmatrix} 1 & 1 & 0 \\ -1 & 1 & 0 \\ 6 & 2 & 1 \end{bmatrix} \qquad x_{B_2} = \begin{bmatrix} x_1 \\ x_2 \\ x_5 \end{bmatrix} = \begin{bmatrix} \frac{5}{2} \\ \frac{5}{2} \\ 1 \end{bmatrix}$$

$$B_3 = \begin{bmatrix} 1 & 1 & 0 \\ -1 & 0 & 1 \\ 6 & 0 & 0 \end{bmatrix} \qquad x_{B_3} = \begin{bmatrix} x_1 \\ x_3 \\ x_4 \end{bmatrix} = \begin{bmatrix} \frac{7}{2} \\ \frac{3}{2} \\ \frac{7}{2} \end{bmatrix}$$

$$B_4 = \begin{bmatrix} 1 & 1 & 0 \\ -1 & 0 & 0 \\ 6 & 0 & 1 \end{bmatrix} \qquad x_{B_4} = \begin{bmatrix} x_1 \\ x_3 \\ x_5 \end{bmatrix} = \begin{bmatrix} 0 \\ 5 \\ 21 \end{bmatrix}$$

$$B_5 = \begin{bmatrix} 1 & 1 & 0 \\ 1 & 0 & 0 \\ 2 & 0 & 1 \end{bmatrix} \qquad x_{B_5} = \begin{bmatrix} x_2 \\ x_3 \\ x_5 \end{bmatrix} = \begin{bmatrix} 0 \\ 5 \\ 21 \end{bmatrix}$$

$$B_6 = \begin{bmatrix} 1 & 0 & 0 \\ 0 & 1 & 0 \\ 0 & 0 & 1 \end{bmatrix} \qquad x_{B_6} = \begin{bmatrix} x_3 \\ x_4 \\ x_5 \end{bmatrix} = \begin{bmatrix} 5 \\ 0 \\ 21 \end{bmatrix}$$

The constraint set is graphed in (x_1, x_2) space in Figure 1. Note that x_{B_4}, x_{B_5}, and x_{B_6} correspond to point P_1, x_{B_2} to point P_2, x_{B_1} to P_3, and x_{B_3} to P_4. Points P_1, P_2, P_3, and P_4 are called the extreme points of the convex set

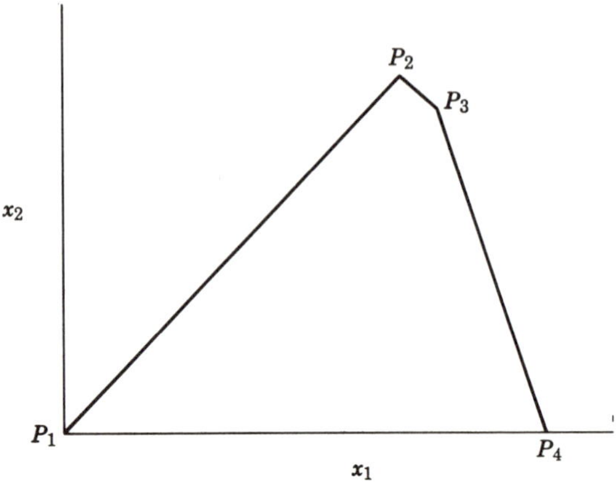

Figure 1

$S = \{x | Ax = b, x \geq 0\}$. A point $x \in S$ is an *extreme point* of the convex set S if there do not exist distinct $y_1, y_2 \in S$ and a scalar α, $0 < \alpha < 1$, such that $x = \alpha y_1 + (1 - \alpha)y_2$. From Figure 1, it is easy to see that for any objective function vector c, some extreme point must be optimal. This property is true for any LP with an optimal solution. Thus, in looking for an optimal solution, we only need to search the extreme points. The algebraic characterization of the extreme points is that a feasible solution corresponds to an extreme point if and only if it is basic. This characterization justifies Theorem 1, which says that the search for an optimal solution can be confined to basic feasible solutions.

The correspondence between basic feasible solutions and extreme points is, in general, many-to-one; this is illustrated in the example with three basic feasible solutions corresponding to the same extreme point. In each of these three solutions one basic variable is zero. A basic solution having at least one basic variable equal to zero is called *degenerate*. Degeneracy causes some complications in linear programming that will be discussed in Section 2.6.

2.3. IMPROVED BASIC FEASIBLE SOLUTIONS AND OPTIMALITY CONDITIONS

To find an optimal solution, we begin with any basic feasible solution and then attempt to find one with a higher value of x_0. Suppose that we have a basic feasible solution given by (2) and (3).

Let $c = (c_B, c_N)$, and thus

$$x_0 = c_B x_B + c_N x_N \tag{4}$$

Using (2) to eliminate x_B from (4) yields

$$x_0 = c_B B^{-1} b - (c_B B^{-1} N - c_N) x_N \tag{5}$$

Substituting $x_N = 0$ in (5) yields $x_0 = c_B B^{-1} b$ as the value of the basic solution given by (3). It is convenient to write (2) and (5) as

$$\begin{bmatrix} x_0 \\ x_B \end{bmatrix} = \begin{bmatrix} c_B B^{-1} b \\ B^{-1} b \end{bmatrix} - \begin{bmatrix} c_B B^{-1} N - c_N \\ B^{-1} N \end{bmatrix} x_N$$

and to introduce some new notation. Let $x_0 \equiv x_{B_0}$, $x_B \equiv (x_{B_1}, \ldots, x_{B_m})$, and R be the index set of the columns in N. Also let

$$y_0 \equiv \begin{bmatrix} c_B B^{-1} b \\ B^{-1} b \end{bmatrix} \equiv \begin{bmatrix} y_{00} \\ y_{10} \\ \vdots \\ y_{m0} \end{bmatrix}$$

and if $j \in R$ so that a_j is a column of N,

$$y_j \equiv \begin{bmatrix} c_B B^{-1} a_j - c_j \\ B^{-1} a_j \end{bmatrix} \equiv \begin{bmatrix} y_{0j} \\ y_{1j} \\ \vdots \\ y_{mj} \end{bmatrix}$$

Then (2) and (5) can be written as

$$x_{B_i} = y_{i0} - \sum_{j \in R} y_{ij} x_j, \qquad i = 0, 1, \ldots, m \tag{6}$$

The solution given by (3) is obtained by setting $x_j = 0$ for all $j \in R$ in (6).

Assume that x_B is nondegenerate and $y_{0j} < 0$ for some $j \in R$, say $j = k$. Then by increasing x_k and keeping all other nonbasic variables fixed at zero, x_0 increases linearly with slope $-y_{0k}$. Also, x_{B_i} is a linear function of x_k with slope $-y_{ik}$. If $y_{ik} > 0$, then $x_{B_i} \geq 0$ as long as $x_k \leq y_{i0}/y_{ik} \equiv \theta_{ik}$. When $x_k = \theta_{ik}$, $x_{B_i} = 0$.

Let x_{B_r} be any basic variable with

$$0 < \theta_{rk} = \min_{i=1,\ldots,m} (\theta_{ik}, y_{ik} > 0)$$

Then by keeping all nonbasic variables equal to zero, except x_k, and setting

$$\begin{aligned} x_k &= \theta_{rk} \\ x_{B_i} &= y_{i0} - y_{ik}\theta_{rk}, \qquad i = 0, 1, \ldots, m \end{aligned}$$

a new basic feasible solution is obtained with $x_k > 0$, $x_{B_r} = 0$, and $x_0 = y_{00} - y_{0k}\theta_{rk}$. Since $\theta_{rk} > 0$ and $y_{0k} < 0$, x_0 is increased.

Finding the new basic feasible solution corresponds to solving the rth equation of (6) for x_k to obtain

$$x_k = \frac{y_{r0}}{y_{rk}} - \sum_{j \in R - \{k\}} \frac{y_{rj}}{y_{rk}} x_j - \frac{1}{y_{rk}} x_{B_r} \tag{7}$$

and then eliminating x_k from the other equations of (6) to obtain for $i \neq r$

$$x_{B_i} = y_{i0} - \frac{y_{ik}y_{r0}}{y_{rk}} - \sum_{j \in R - \{k\}} \left(y_{ij} - \frac{y_{ik}y_{rj}}{y_{rk}}\right) x_j + \frac{y_{ik}}{y_{rk}} x_{B_r} \tag{8}$$

and then setting $x_{B_r} = 0$ and $x_j = 0$ for all $j \in R - \{k\}$. Thus, whenever there is a nondegenerate basic feasible solution with $y_{0j} < 0$, for some $j \in R$, say $j = k$, and $y_{ik} > 0$ for at least one i, a better basic feasible solution can be found by exchanging one column of N for one of B.

On the other hand, suppose that B is a basis matrix which yields a basic feasible solution x° with $y_{0j} \geq 0$ for all $j \in R$. From (6) the objective function can be written as

$$x_0^\circ = c_B B^{-1} b - \sum_{j \in R} y_{0j} x_j^\circ \tag{9}$$

Since x_B° does not appear in (9), $y_{0j} \geq 0$, and $x_j^\circ \geq 0$ for all $j \in R$, $c_B B^{-1} b$ is an upper bound on x_0. Because $x_B^\circ = B^{-1}b$, $x_N^\circ = 0$ is feasible and achieves this upper bound, it is optimal. We have proved

Theorem 2: *The basic solution to* (6) *is optimal if*

(i) $y_{i0} \geq 0$, $i = 1, \ldots, m$ *(feasibility).*
(ii) $y_{0j} \geq 0$, *for all* $j \in R$.

2.4. THE SIMPLEX ALGORITHM

Finding improved basic feasible solutions, as described above, is the essence of the simplex algorithm.

Algorithm

STEP 1: (Initialization.) Begin with a basic feasible solution (a method for finding an initial solution is given in Section 2.5). Go to Step 2.

STEP 2: (Test for optimality.) Is $y_{0j} \geq 0$ for every $j \in R$? If so, the current solution is optimal. If not, go to Step 3.

STEP 3: (Choose an entering variable.) Select a variable x_k, $k \in R$, with $y_{0k} < 0$ to enter the basis. (A common rule is to choose x_k if $y_{0k} = \min_{j \in R} y_{0j}$.) Go to Step 4.

STEP 4: (Choose a departing variable.) Select a variable x_{B_r} with $\theta_{rk} = \min_{i=1,\ldots,m} (y_{i0}/y_{ik}, y_{ik} > 0)$ to leave the basis. If there is a tie, break it arbitrarily. (The case $y_{ik} \leq 0$, $i = 1, \ldots, m$ will be discussed later.) Go to Step 5.

STEP 5: (Pivoting.) Solve (6) for x_k and x_{B_i}, $i \neq r$, in terms of x_j, $j \in R - \{k\}$ and x_{B_r} as in (7) and (8). Set $x_j = 0$, $j \in R - \{k\}$ and $x_{B_r} = 0$ to obtain a new basic feasible solution. Go to Step 2.

Each cycle between Steps 2 to 5 is called a simplex iteration, or simply an *iteration*. In the absence of degeneracy, each iteration must produce a new basic feasible solution, since x_0 is increased at each iteration. Consequently, if there is an optimal solution, it must be achieved in at most $\binom{n}{m}$ iterations. We have just proved

Theorem 3: *Given an LP with an optimal solution and with no degenerate basic feasible solutions, the simplex algorithm finds an optimal solution in at most* $\binom{n}{m}$ *iterations.*

Simplex Tableaus

Calculations in the simplex algorithm are in tables called *simplex tableaus.*

Table 1

		Nonbasic Variables		
		$\cdots$	$-x_j$ $\cdots$	$-x_k$ $\cdots$
	x_0	y_{00}	y_{0j}	y_{0k}
	$\vdots$	$\vdots$	$\vdots$	$\vdots$
Basic Variables	x_{B_i}	y_{i0}	y_{ij}	y_{ik}
	$\vdots$	$\vdots$	$\vdots$	$\vdots$
	x_{B_r}	y_{r0}	y_{rj}	y_{rk}
	$\vdots$	$\vdots$	$\vdots$	$\vdots$
	x_{B_m}	y_{m0}	y_{mj}	y_{mk}

Table 1 represents (6), with the ith row of the table corresponding to the ith equation.

Suppose at some iteration we want to make x_k basic and x_{B_r} nonbasic. This change of basis is accomplished in tabular form using the following rules:

1. Divide the rth row by y_{rk} (y_{rk} is called the *pivot element* for the iteration).
2. Multiply the new rth row by y_{ik} and subtract it from the ith row, $i = 0, 1, \ldots, m, i \neq r$.
3. Replace the old kth column by its negative divided by y_{rk}, except for the element at the intersection of row r and column k. This element (y_{rk}) is replaced by $1/y_{rk}$.

The transformed tableau is shown in Table 2.

Table 2

	$\cdots - x_j$		$\cdots - x_{B_r} \cdots$
x_0	$y_{00} - (y_{0k}y_{r0}/y_{rk})$	$y_{0j} - (y_{0k}y_{rj}/y_{rk})$	$-y_{0k}/y_{rk}$
$\vdots$	$\vdots$	$\vdots$	$\vdots$
x_{B_i}	$y_{i0} - (y_{ik}y_{r0}/y_{rk})$	$y_{ij} - (y_{ik}y_{rj}/y_{rk})$	$-y_{ik}/y_{rk}$
$\vdots$	$\vdots$	$\vdots$	$\vdots$
x_k	y_{r0}/y_{rk}	y_{rj}/y_{rk}	$1/y_{rk}$
$\vdots$	$\vdots$	$\vdots$	$\vdots$

Carrying out this transformation is equivalent to putting (6) in the form of (7) and (8).

Example

For the example given in Section 2.2, an obvious basic feasible solution is shown in Table 3. The arrows indicate variables entering and leaving the basis.

Table 3

		$\downarrow$ $-x_1$	$-x_2$
x_0	0	-2	-1
x_3	5	1	1
x_4	0	-1	1
$\leftarrow x_5$	21	6	2

Even though this basic feasible solution is degenerate, we will attempt to solve the problem using the simplex algorithm. Subsequent tableaus are shown in Tables 4 and 5.

Table 4

		$-x_5$	$-x_2$ ↓
x_0	7	$\frac{1}{3}$	$-\frac{1}{3}$
← x_3	$\frac{3}{2}$	$-\frac{1}{6}$	$\frac{2}{3}$
x_4	$\frac{7}{2}$	$\frac{1}{6}$	$\frac{4}{3}$
x_1	$\frac{7}{2}$	$\frac{1}{6}$	$\frac{1}{3}$

Table 5

		$-x_5$	$-x_3$
x_0	$\frac{31}{4}$	$\frac{1}{4}$	$\frac{1}{2}$
x_2	$\frac{9}{4}$	$-\frac{1}{4}$	$\frac{3}{2}$
x_4	$\frac{1}{2}$	$\frac{1}{2}$	-2
x_1	$\frac{11}{4}$	$\frac{1}{4}$	$-\frac{1}{2}$

In Table 3, x_1 becomes basic and x_5 nonbasic, with the results shown in Table 4. In Table 4, x_2 becomes basic and x_3 nonbasic. The optimal solution is shown in Table 5, $x_3^0 = x_5^0 = 0$, $x_1^0 = \frac{11}{4}$, $x_2^0 = \frac{9}{4}$, $x_4^0 = \frac{1}{2}$, and $x_0^0 = \frac{31}{4}$. In the example, beginning with a degenerate basic feasible solution did not cause any difficulty.

Other Tableaus

The tabular representation of (6) given in Table 1 is one of three commonly used schema. We have chosen it, because it is the most compact.

Closely related to the representation given in Table 1 is the one of Table 1a. In Table 1a, there is a row for each variable (nonbasic as well as basic). The equations are

$$x_i = y_{i0} - \sum_{j \in R} y_{ij} x_j \qquad \text{all } i$$

Table 1a

			Nonbasic Variables	
			$\cdots - x_j \cdots$	$- x_k \cdots$
	x_0	y_{00}	y_{0j}	y_{0k}
All	$\vdots$	$\vdots$	$\vdots$	$\vdots$
Variables	x_j	0	-1	0
	$\vdots$	$\vdots$	$\vdots$	$\vdots$
	x_{Bi}	y_{i0}	y_{ij}	y_{ik}
	$\vdots$	$\vdots$	$\vdots$	$\vdots$

and row 0 is for x_0, row 1 for x_1, and so forth. If x_i is basic, row i corresponds to some row in Table 1. If x_i is nonbasic, the equation is merely $x_i = x_i$, that is, $y_{i0} = 0$, $y_{ij} = 0$ for $j \neq i$, and $y_{ii} = -1$. Adding these trivial equations facilitates some proofs because each variable retains a fixed row position.

The tableau that is, perhaps, most widely used in elementary expositions represents the equations

$$0 = y_{i0} - \sum_{j=1}^{n} y_{ij}x_j, \qquad i = 0, 1, \ldots, m$$

and is shown in Table 1b. There is a row for each basic variable and a column for each variable (basic as well as nonbasic). In particular, if x_{B_i} is the basic variable in row i, $y_{ij} = 0$ if x_j is basic and $j \neq B_i$, and $y_{iB_i} = 1$. Thus it is possible to use the convention of column 1 for x_1, column 2 for x_2, and so forth. In this representation each variable retains a fixed column position, and there are at least m unit-vectors in the tableau.

Table 1b

			All Variables	
			$-x_j \cdots$	$-x_{B_i} \cdots$
	x_0	y_{00}	y_{0j}	0
	$\vdots$	$\vdots$	$\vdots$	$\vdots$
Basic	x_{B_i}	y_{i0}	y_{ij}	1
Variables	$\vdots$	$\vdots$	$\vdots$	$\vdots$
	x_{B_r}	y_{r0}	y_{rj}	0
	$\vdots$	$\vdots$	$\vdots$	$\vdots$

Unbounded Solutions

A change of basis in the simplex algorithm, in the absence of degeneracy, corresponds to moving from one extreme point to an adjacent extreme point along the edge of the convex set joining the two extreme points. In the example shown in Tables 3–5 we moved from P_1 to P_4 to P_3 of Figure 1.

Given an extreme point, if x_k, $k \in R$, is increased to $\theta > 0$, then x_B is increased by $\theta(-y_{1k}, \ldots, -y_{mk})$. Suppose that $y_{ik} \leq 0$, $i = 1, \ldots, m$; then, as x_k increases, all of the basic variables are nondecreasing. Thus $x_B = (y_{10}, \ldots, y_{m0}) - \theta(y_{1k}, \ldots, y_{mk})$, $x_k = \theta$ and $x_j = 0$, otherwise is a feasible solution for any $\theta \geq 0$. This infinite collection of feasible solutions for all $\theta \geq 0$, is called an *extreme ray*. (Generally a ray is defined to contain the origin, but our definition will be more convenient in later developments.) If $y_{0k} < 0$, the objective function increases along the extreme ray, thus

x_0 can be made arbitrarily large and the problem is unbounded. An unbounded solution can always be identified by the simplex algorithm because of

Theorem 4: *If an LP is unbounded, there exists a basic feasible solution* x_B *and a vector* y_k *with* $y_{0k} < 0$ *and* $y_{ik} \leq 0$, $i = 1, \ldots, m$.

Example

$$\begin{aligned} \max x_0 = x_1 & \\ x_1 + x_2 - x_3 &= 1 \\ x_1, x_2, x_3 &\geq 0 \end{aligned}$$

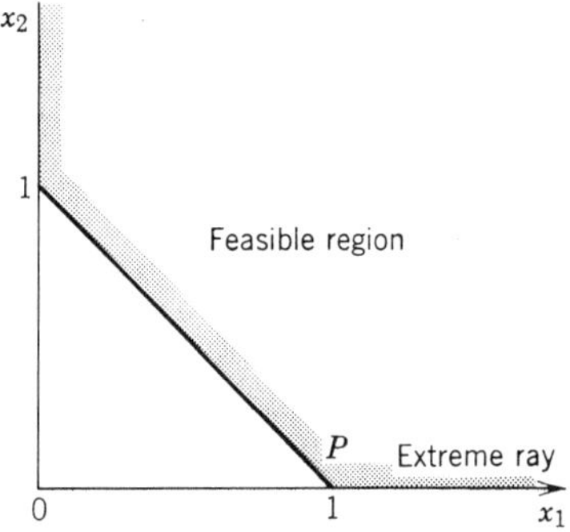

Figure 2

The extreme point P (see Figure 2) corresponds to the basic feasible solution $x_1 = 1$, $x_2 = x_3 = 0$, and $x_0 = 1$. The tableau representation at P is given in Table 6.

Table 6

		$-x_2$	$-x_3$
x_0	1	1	-1
x_1	1	1	-1

Since $y_{03} = -1$, x_0 increases as x_3 increases. But, since $y_{13} = -1$, x_1 increases as well. The extreme ray is given by $x_1 = 1 + \theta$, $x_2 = 0$, and $x_3 = \theta$. As θ becomes arbitrarily large, $x_0 = 1 + \theta$ becomes arbitrarily large, and the problem is unbounded.

2.5. FINDING A FIRST BASIC FEASIBLE SOLUTION. THE TWO-PHASE METHOD

To start the simplex algorithm a basic feasible solution is required. However it can be shown that

Theorem 5: *An LP has a feasible solution if and only if it has a basic feasible solution.*

In some problems a basic feasible solution can be found by inspection. Suppose $b \geq 0$ and A contains at least m unit vectors. Then the columns of A can be permuted to obtain the matrix (A', I). Writing (1) as

$$Ix_B + A'x_N = b$$

it is clear that $x_B = b$ and $x_N = 0$ is a basic feasible solution. This case occurs in the common situation in which the original problem is

$$\begin{aligned} \max x_0 &= cx \\ Ax &\leq b \geq 0 \\ x &\geq 0 \end{aligned}$$

since the equality version is

$$\begin{aligned} \max x_0 &= cx \\ Ax + Ix_s &= b \\ x, x_s &\geq 0 \end{aligned}$$

We put $x_s = x_B$ and $x = x_N$.

Assume now that the constraints are written as $Ax = b \geq 0$ and A does not contain an identity matrix. Also, without loss of generality, A can be partitioned as

$$A = \left[\begin{array}{c|c} A_1 & I_t \\ \hline A_2 & 0 \end{array}\right]$$

where I_t is an identity matrix of order t (t may be zero). Thus the constraints can be written as

$$\begin{aligned} A_1x_N + I_tx_t &= b_1 \\ A_2x_N \qquad &= b_2 \end{aligned} \tag{10}$$

Now consider a related LP with an $(m - t)$ dimensional vector $x_\alpha \geq 0$, of *artificial variables*, and the constraint set

$$\begin{aligned} A_1x_N + I_tx_t \qquad\quad &= b_1 \\ A_2x_N + \qquad I_{m-t}x_\alpha &= b_2 \end{aligned} \tag{11}$$

The constraints (11) have an obvious basic feasible solution $x_t = b_1$, $x_\alpha = b_2$, and $x_N = 0$. Furthermore, (10) has a basic feasible solution if and only if (11) has a basic feasible solution with $x_\alpha = 0$. Our objective is, beginning with the basic feasible solution $x_t = b_1$, $x_\alpha = b_2$, and $x_N = 0$, to find a basic feasible solution with $x_\alpha = 0$ or to show that none exists. This can be accomplished by solving the LP

$$\begin{aligned} \max z_0 = \qquad\qquad\qquad\qquad\qquad & -\mathbf{1}x_\alpha \\ x_0 - c_Nx_N - c_tx_t \qquad\qquad &= 0 \\ A_1x_N + I_tx_t \qquad\quad &= b_1 \\ A_2x_N + \qquad I_{m-t}x_\alpha &= b_2 \\ x_N, x_t, x_\alpha &\geq 0 \end{aligned} \tag{12}$$

where $\mathbf{1}$ is the sum vector $(1, \ldots, 1)$. The variable x_0 always remains basic. The initial basic feasible solution to (12) is $x_t = b_1 - A_1x_N$, $x_\alpha = b_2 - A_2x_N$, $z_0 = -\mathbf{1}(b_2 - A_2x_N)$, $x_0 = c_tb_1 + (c_N - c_tA_1)x_N$ and $x_N = 0$.

For any feasible solution to (12), $x_\alpha > 0$ implies $z_0 < 0$, and $x_\alpha = 0$ implies $z_0 = 0$. Thus, if the optimal value of z_0 is negative, (11) is infeasible. On the other hand, any basic feasible solution to (12) with $z_0 = 0$ provides a basic feasible solution to (11). This is obvious if such a basic feasible solution contains no artificial variables; see Exercise 13 for the case in which the basic feasible solution is degenerate and contains artificial variables. Once such a solution is found, the z_0 row and the artificial variables are dropped. (Actually, the artificial variables can be dropped as they leave the basis.) We then continue with the maximization of x_0; this method is called the *two-phase method.* In *phase* 1, we maximize z_0 to find a basic feasible solution to the original problem. In *phase* 2, we maximize x_0. An infeasible problem will terminate at the end of phase 1.

Example

$$\begin{aligned} \max x_0 = -5x_1 \qquad\quad & -21x_3 \\ x_1 - x_2 + 6x_3 - x_4 \qquad &= 2 \\ x_1 + x_2 + 2x_3 \qquad - x_5 &= 1 \\ x_1, \ldots, x_5 &\geq 0 \end{aligned}$$

The problem is solved by the two-phase method in Tables 7–10. The artificial variables are x_6 and x_7. The objective row $z_0 = -x_6 - x_7$ is rewritten by substituting for x_6 and x_7, to obtain

$$\begin{aligned} z_0 &= -2 + x_1 - x_2 + 6x_3 - x_4 - 1 + x_1 + x_2 + 2x_3 - x_5 \\ &= -3 + 2x_1 + 8x_3 - x_4 - x_5 \end{aligned}$$

Phase 2 begins with Table 9, and an optimal solution is given in Table 10.

Table 7

		$-x_1$	$-x_2$	↓ $-x_3$	$-x_4$	$-x_5$
z_0	-3	-2	0	-8	1	1
x_0	0	5	0	21	0	0
← x_6	2	1	-1	6	-1	0
x_7	1	1	1	2	0	-1

Table 8

		$-x_1$	↓ $-x_2$	$-x_4$	$-x_5$
z_0	$-\frac{1}{3}$	$-\frac{2}{3}$	$-\frac{4}{3}$	$-\frac{1}{3}$	1
x_0	-7	$\frac{3}{2}$	$\frac{7}{2}$	$\frac{7}{2}$	0
x_3	$\frac{1}{3}$	$\frac{1}{6}$	$-\frac{1}{6}$	$-\frac{1}{6}$	0
← x_7	$\frac{1}{3}$	$\frac{2}{3}$	$\frac{4}{3}$	$\frac{1}{3}$	-1

Table 9

		↓ $-x_1$	$-x_4$	$-x_5$
x_0	$-\frac{63}{8}$	$-\frac{1}{4}$	$\frac{21}{8}$	$\frac{21}{8}$
x_3	$\frac{3}{8}$	$\frac{1}{4}$	$-\frac{1}{8}$	$-\frac{1}{8}$
← x_2	$\frac{1}{4}$	$\frac{1}{2}$	$\frac{1}{4}$	$-\frac{3}{4}$

Table 10

		$-x_2$	$-x_4$	$-x_5$
x_0	$-\frac{31}{4}$	$\frac{1}{2}$	$\frac{11}{4}$	$\frac{9}{4}$
x_3	$\frac{1}{4}$	$-\frac{1}{2}$	$-\frac{1}{4}$	$\frac{1}{4}$
x_1	$\frac{1}{2}$	2	$\frac{1}{2}$	$-\frac{3}{2}$

2.6. DEGENERACY AND CYCLING

Suppose that there is a degenerate basic feasible solution such that, for some r and k, $y_{r0} = 0$, $y_{0k} < 0$, and $y_{rk} > 0$. By introducing x_k into the basis,

x_{B_r} (or some other basic variable equal to zero) must leave the basis and, although a new basis is obtained, the solution and value of the objective function remain the same. The new basic solution $x_{B_i} = y_{i0}$, $i \neq r$, $x_k = 0$, $x_j = 0$, otherwise is also degenerate. Examples have been constructed in which a finite sequence of degenerate bases obtained by the simplex algorithm yield a cycle; that is, the sequence $\{B_1, \ldots, B_{k-1}, B_k\}$ occurs with $B_k = B_1$.

This unhappy phenomenon is called *cycling*. A wealth of empirical evidence indicates that cycling does not occur in practical problems. However, the simplex algorithm can be modified so that cycling will never occur. Although the procedure used to protect against cycling is not of practical importance in linear programming, it is used in some integer programming algorithms. Consequently, we will cover it in detail. The procedure involves specializing Step 4 of the simplex algorithm. Recall that, in Step 4, a variable is chosen to leave the basis. When there is more than one candidate, an arbitrary choice is made. By giving a specific rule for breaking ties, the possibility of cycling can be eliminated. Before giving the rule, some new definitions are introduced.

A vector $v \neq 0$ is called *lexicographically positive* if its first nonzero component is positive. We use the notation $v \overset{L}{>} 0$ to denote v lexicographically positive. For example, $v = (0, 0, 7, -6, 4) \overset{L}{>} 0$. A vector v is said to be *lexicographically greater* than a vector u if $v - u \overset{L}{>} 0$. A sequence of vectors $\{v^t\}$ is called *lexicographically increasing* if $v^{t+1} - v^t \overset{L}{>} 0$ for all t. If $-v \overset{L}{>} 0$, v is called *lexicographically negative* (denoted $v \overset{L}{<} 0$) and if $v - u \overset{L}{<} 0$, v is *lexicographically smaller* than u. The notation $v \overset{L}{\geq} 0$ means that $v = 0$ or $v \overset{L}{>} 0$.

Table 11

										$-x_j$	
x_0	y_{00}	0	$\cdots$	0	$\cdots$	0		$\cdots$	y_{0j}	$\cdots$
x_{B1}	y_{10}	1	$\cdots$	0	$\cdots$	0		$\cdots$	y_{1j}	$\cdots$
$\vdots$	$\vdots$	$\vdots$		$\vdots$		$\vdots$			$\vdots$	
x_{Bi}	y_{i0}	0	$\cdots$	1	$\cdots$	0		$\cdots$	y_{ij}	$\cdots$
$\vdots$	$\vdots$	$\vdots$		$\vdots$		$\vdots$			$\vdots$	
x_{Bm}	y_{m0}	0	$\cdots$	0	$\cdots$	1		$\cdots$	y_{mj}	$\cdots$

The changes required in the simplex algorithm to avoid cycling involve a modification of Table 1. In the tableau corresponding to the initial basic feasible solution, an mth order identity matrix is inserted between the solution column and the columns corresponding to the nonbasic variables (see Table 11). The m unit columns are not used to determine the variable to enter the basis. However, at each iteration, they are transformed using the standard rules given in the simplex algorithm. As will be shown, the only purpose of these columns is to identify the variable to leave the basis when the choice given in Step 4 of the simplex algorithm is not unique.

Let $v_0 = (y_{00}, 0, \ldots, 0)$ and $v_i = (y_{i0}, 0, \ldots, 1, \ldots, 0)$, $i = 1, \ldots, m$, be the coefficients in the first $m + 1$ columns of the ith row in Table 11. Since $y_{i0} \geq 0$, $i = 1, \ldots, m$, it follows that $v_i \overset{L}{>} 0$, $i = 1, \ldots, m$.

Suppose that $y_{0k} < 0$ and x_k is chosen to enter the basis. Let $S_k = \{i | i \geq 1, y_{ik} > 0\}$ and $u_i = v_i / y_{ik}$ for all $i \in S_k$. Assume that u_r is the lexicographically smallest of these vectors, that is,

$$u_r = \underset{i \in S_k}{\text{lexmin}}\, u_i \tag{13}$$

The vector u_r must be unique, since the v_i's are linearly independent. Choose x_{B_r} to leave the basis. Note that x_{B_r} chosen by (13) satisfies the rule of the simplex algorithm, since

$$u_r = \underset{i \in S_k}{\text{lexmin}}\, u_i \Rightarrow \frac{y_{r0}}{y_{rk}} = \min_{i \in S_k} \frac{y_{i0}}{y_{ik}}$$

Pivoting on row r and column k yields

$$\hat{v}_r = \frac{v_r}{y_{rk}}$$

$$\hat{v}_i = v_i - \frac{y_{ik}}{y_{rk}} v_r = v_i - y_{ik}\hat{v}_r, \qquad i \neq r$$

Since $v_r \overset{L}{>} 0$ and $y_{rk} > 0$, $\hat{v}_r \overset{L}{>} 0$. If $y_{ik} \leq 0$, then $-y_{ik}\hat{v}_r \overset{L}{\geq} 0$ and $\hat{v}_i \overset{L}{>} 0$. If $y_{ik} > 0$, we have

$$u_i - u_r = \frac{v_i}{y_{ik}} - \hat{v}_r \overset{L}{>} 0 \tag{14}$$

from (13). Multiplying (14) by $y_{ik} > 0$ yields $\hat{v}_i \overset{L}{>} 0$.

Thus $\hat{v}_i$, $i = 1, \ldots, m$, are lexicographically positive. Furthermore, $\hat{v}_i$, $i = 1, \ldots, m$, are linearly independent, since adding multiples of one vector to another cannot destroy independence. By induction, under the pivot row selection rule just given, $\hat{v}_i$, $i = 1, \ldots, m$, remain linearly independent and lexicographically positive at each iteration.

At the qth iteration the tableau is completely determined by the basis B_q. Thus if tableaus q and t are such that $v_0^q \neq v_0^t$, then $B_q \neq B_t$. Note that if the sequence $\{v_0^t\}$ is lexicographically increasing, no t_1 and t_2 ($t_1 \neq t_2$) exist such that $v_0^{t_1} = v_0^{t_2}$. However,

$$v_0^{t+1} = v_0^t - \frac{y_{0k}^t}{y_{rk}^t} v_r^t = v_0^t - y_{0k}^t v_r^{t+1}$$

so that $v_r^{t+1} \overset{L}{>} 0$ and $y_{0k}^t < 0$ imply $v_0^{t+1} \overset{L}{>} v_0^t$.

Example

Consider the tableau in Table 12. Given that x_5 enters the basis, the

Table 12

					$-x_4$	$-x_5$ ↓
x_0	0	0	0	0	2	-1
x_1	0	1	0	0	1	1
x_2	0	0	1	0	2	4
$\leftarrow x_3$	0	0	0	1	-1	3

simplex algorithm would permit the removal of either x_1, or x_2, or x_3. However $u_3 = (0, 0, 0, \frac{1}{3}) \overset{L}{<} u_2 = (0, 0, \frac{1}{4}, 0) \overset{L}{<} u_1 = (0, 1, 0, 0)$. Thus x_3 is chosen as the departing variable. Note that in the transformed tableau (Table 13), $\hat{v}_i \overset{L}{>} 0$ and $\hat{v}_0 - v_0 \overset{L}{>} 0$. The reader can verify that if x_1 or x_2 were chosen to depart, then $\hat{v}_3 \overset{L}{<} 0$.

Table 13

					$-x_4$	$-x_3$
x_0	0	0	0	$\frac{1}{3}$	$\frac{5}{3}$	$\frac{1}{3}$
x_1	0	1	0	$-\frac{1}{3}$	$\frac{4}{3}$	$-\frac{1}{3}$
x_2	0	0	1	$-\frac{4}{3}$	$\frac{10}{3}$	$-\frac{4}{3}$
x_5	0	0	0	$\frac{1}{3}$	$-\frac{1}{3}$	$\frac{1}{3}$

2.7. THE DETERMINANT AND INVERSE OF THE BASIS MATRIX

Sometimes it is useful to know B^{-1} and the absolute value of the determinant of the basis matrix denoted by $|\det B| = D$. We assume that $A = (A', I)$.

Given $B = (a_{B_1}, \ldots, a_{B_m})$, B^{-1} can be obtained from the tableau. Specifically $B^{-1} = (b_1^*, \ldots, b_m^*)$ where, if the variable s_k corresponding to the kth unit vector e_k is nonbasic, then $b_k^* = (y_{1k}, \ldots, y_{mk})$. This result follows, since $(y_{1k}, \ldots, y_{mk}) = B^{-1}e_k$. If $s_k = x_{B_j}$, then $b_k^* = e_j$, as can be seen from the rules of matrix multiplication.

The basis matrix in Table 5 is

$$B = \begin{bmatrix} 1 & 0 & 1 \\ 1 & 1 & -1 \\ 2 & 0 & 6 \end{bmatrix}$$

Its inverse may be read directly from Table 5 as

$$B^{-1} = \begin{bmatrix} \frac{3}{2} & 0 & -\frac{1}{4} \\ -2 & 1 & \frac{1}{2} \\ -\frac{1}{2} & 0 & \frac{1}{4} \end{bmatrix}$$

We can write $B^{-1} = B^{+}/\det B$, where B^{+} is the *adjoint* matrix of B. It is important to note that if B is an integer matrix (a matrix having all of its elements integers), then B^{+} is also an integer matrix. Consequently, if $D = 1$, then B^{-1} must be an integer matrix. Thus, if b is an integer vector, then $x_B = B^{-1}b$ is an integer vector.

For relating linear programming solutions to integer programming solutions, the case $D = 1$ is of utmost importance. The difficulty in solving ILP's with many algorithms seems to grow with D. Loosely speaking, as D gets larger, x_B becomes more noninteger and more work needs to be done in modifying x_B to an integer solution.

For each simplex tableau, the determinant of the basis matrix can easily be calculated, as shown in

Theorem 6: *Let the tableaus be indexed* $i = 0, 1, 2, \ldots$ *and let* B_i *be the basis matrix for the ith tableau. Then*

$$D_p = D_0 \prod_{i=0}^{p} y_{r_i k_i}$$

where $y_{r_0 k_0} = 1$ *and* $y_{r_i k_i}$ *is the pivot element in going from the* $(i - 1)$*st to the ith tableau.*

PROOF: The theorem is trivially true for $p = 0$. Assume it true for $p - 1$; we must then prove that

$$D_p = y_{r_p k_p} D_{p-1}$$

Assume, only for simplicity of exposition, that there is an identity matrix in the first tableau. Therefore, B_{p-1}^{-1} appears in the $p - 1$st tableau. In transforming from the $p - 1$st to the pth tableau, the r_pth row of B_{p-1}^{-1} is divided by $y_{r_p k_p}$ and multiples of the r_pth row are added to other rows. By applying the two elementary rules for manipulating determinants:

1. If the ith row is multiplied by θ, the value of the determinant is multiplied by θ,
2. Adding a multiple of one row to another row does not change the value of the determinant,

we obtain

$$\det B_p^{-1} = \frac{\det B_{p-1}^{-1}}{y_{r_p k_p}}$$

But also for A_1 and A_2 square,

$$\det (A_1 A_2) = (\det A_1)(\det A_2)$$

Thus

$$(\det B)(\det B^{-1}) = \det I = 1$$

or

$$\det B = \frac{1}{\det B^{-1}} \qquad \text{and} \qquad D_p = y_{r_p k_p} D_{p-1} \qquad \blacksquare$$

We generally begin simplex iterations with $B_0 = I$. In this case,

$$D_p = \prod_{i=1}^{p} y_{r_i k_i}$$

As a simple example, in Table 3 we began a problem with $B_0 = I$ and executed two iterations with pivot elements of 6 and $\frac{2}{3}$. Thus the determinant of the basis matrix in Table 5,

$$\begin{bmatrix} 1 & 0 & 1 \\ 1 & 1 & -1 \\ 2 & 0 & 6 \end{bmatrix}$$

is $6(\frac{2}{3}) = 4$.

2.8. A MODIFIED SIMPLEX ALGORITHM FOR BOUNDED VARIABLES

Bounding constraints $0 < l_j \leq x_j \leq u_j$ arise quite naturally in LP's and in a widely used linear programming approach to ILP's. Although bounding constraints can be included in $Ax = b$, it is inefficient to do so, since the critical factor in the size of LP's that can be solved by the simplex algorithm is the number of rows (dimension of the basis matrix).

The lower bound constraints $l_j \leq x_j$ are simply disposed of by substitution. Let $x_j' = x_j - l_j$ and replace $x_j \geq l_j$ by $x_j' \geq 0$. The interesting result is that, with a slight modification of the simplex algorithm, upper bound constraints can be handled without a corresponding increase in the size of the basis. For notational simplicity, we consider a problem in which all of the variables are upper bounded

$$\begin{aligned} \max\ & cx \\ & Ax = b \\ & 0 \le x \le u \end{aligned} \tag{15}$$

A solution is called basic to (15) if each nonbasic variable is equal to *zero or* u_j. In other words, in the upper bounded algorithm one or more of the nonbasic variables x_j, $j \in R$, may be at their upper bounds.

Let $R_1 = \{j \mid j \in R, x_j = 0\}$ and $R_2 = \{j \mid j \in R, x_j = u_j\}$. In the equations

$$x_{B_i} = y_{i0} - \sum_{j \in R_1} y_{ij} x_j - \sum_{j \in R_2} y_{ij} x_j, \qquad i = 0, 1, \ldots, m$$

the current value of x_{B_i} is

$$z_{i0} = y_{i0} - \sum_{j \in R_2} y_{ij} u_j, \qquad i = 0, 1, \ldots, m$$

In the tableau for the upper bounded algorithm, z_{i0} appears in the solution column.

Proceeding as in Section 2.3, it is easy to show that for $j \in R_2$, $y_{0j} > 0$ implies that x_0 can be increased by decreasing x_j (making x_j basic). Furthermore, analogous to Theorem 2, we have

Theorem 7: *Given a basis matrix B, a sufficient condition for optimality is*

(i) $0 \le z_{i0} \le u_{B_i}$, $i = 1, \ldots, m$ (*feasibility*).
(ii) $y_{0j} \ge 0$ *for all* $j \in R_1$.
(iii) $y_{0j} \le 0$ *for all* $j \in R_2$.

To incorporate these ideas into the simplex algorithm, we keep a record of whether a nonbasic variable is zero or at its upper bound. In testing for optimality and choosing a variable to enter the basis, the opposite sign convention on y_{0j} is used for a nonbasic variable at its upper bound ($y_{0j} \le 0$ for optimality, $y_{0j} > 0$ for a candidate to enter the basis). Also, the rule for choosing the departing variable must be modified to yield a first basic variable that reaches zero or its upper bound. Suppose that x_k, $k \in R_1$ is the entering variable. A first variable to reach zero is given as previously by x_{B_r} determined from

$$\theta_{rk} = \min_{i=1,\ldots,m} \left(\frac{z_{i0}}{y_{ik}}, y_{ik} > 0 \right)$$

A first variable to reach its upper bound (excluding x_k) is given by x_{B_s} determined from

$$\alpha_{sk} = \min_{i=1,\ldots,m} \left(\frac{z_{i0} - u_{B_i}}{y_{ik}}, \ y_{ik} < 0 \right)$$

Also, $x_k \leq u_k$ must be considered. Thus the departing variable is determined from

$$\min \{\theta_{rk}, \alpha_{sk}, u_k\}$$

We will not worry about cycling and therefore assume that ties can be broken arbitrarily. If $\theta_{rk}(\alpha_{sk})$ is the minimum, x_{B_r} (x_{B_s}) departs and a simplex iteration is executed. If u_k is the minimum, x_k is both the entering and departing variable, and there is no change of basis. However, the change of x_k from zero to u_k must be recorded, and the solution column must be updated to

$$\hat{z}_{i0} = z_{i0} - y_{ik}u_k, \qquad i = 0, 1, \ldots, m$$

The case in which x_k, $k \in R_2$ is the entering variable yields similar results. A first variable to reach zero (excluding x_k) is given by x_{B_r} determined from

$$\bar{\theta}_{rk} = \max_{i=1,\ldots,m} \left(\frac{z_{i0}}{y_{ik}}, \ y_{ik} < 0 \right)$$

Here $\bar{\theta}_{rk} \leq 0$ represents the change allowed in x_k, so if $u_k < -\bar{\theta}_{rk}$, x_k reaches zero before x_{B_r}. A first variable to reach its upper bound is x_{B_s} determined from

$$\bar{\alpha}_{sk} = \max_{i=1,\ldots,m} \left(\frac{z_{i0} - u_{B_i}}{y_{ik}}, \ y_{ik} > 0 \right)$$

Thus, in this case, the departing variable is determined from

$$\max \{\bar{\theta}_{rk}, \bar{\alpha}_{sk}, -u_k\}$$

Algorithm

STEP 1: (Initialization.) Begin with a basic feasible solution as defined above (all nonbasic variables are either zero or at their upper bounds).

STEP 2: (Test for optimality.) If $y_{0j} \geq 0$ for all $j \in R_1$, and $y_{0j} \leq 0$ for all $j \in R_2$, the current solution is optimal. Otherwise go to Step 3.

STEP 3: (Choose an entering variable.) Select a variable x_k to enter the basis, $k \in R_1$ with $y_{0k} < 0$, or $k \in R_2$ with $y_{0k} > 0$. A common rule is to choose x_k if

$$|y_{0k}| = \max \left\{ \max_{j \in R_2} y_{0j}, \left| \min_{j \in R_1} y_{0j} \right| \right\}$$

If $k \in R_1$, go to Step 4. Otherwise go to Step 5.

STEP 4: (Choose a departing variable.) If $u_k \leq \min\{\theta_{rk}, \alpha_{sk}\}$, go to Step 6. Otherwise select x_{B_r} (x_{B_s}) to leave the basis if $\theta_{rk} \leq (\geq)\, \alpha_{sk}$. If x_{B_r} is selected, go to Step 8; otherwise go to Step 9.

STEP 5: (Choose a departing variable.) If $-u_k \geq \max\{\bar{\theta}_{rk}, \bar{\alpha}_{sk}\}$, go to Step 7. Otherwise select x_{B_r} (x_{B_s}) to leave the basis if $\bar{\theta}_{rk} \geq (\leq)\, \bar{\alpha}_{sk}$. If x_{B_r} is selected, go to Step 10; otherwise go to Step 11.

STEP 6: (Pivoting.) Record that $x_k = u_k$. Put $\hat{z}_{i0} = z_{i0} - y_{ik}u_k$, $i = 0, 1, \ldots, m$. Go to Step 2.

STEP 7: (Pivoting.) Record that x_k is zero. Put $\hat{z}_{i0} = z_{i0} + y_{ik}u_k$, $i = 0, 1, \ldots, m$. Go to Step 2.

STEP 8: (Pivoting.) Execute an ordinary simplex iteration with x_k entering and x_{B_r} departing [see (7) and (8)]. Record that x_{B_r} is zero. Go to Step 2.

STEP 9: (Pivoting.) Enter x_k and remove x_{B_s}. All columns are transformed by the usual simplex rules [(7) and (8)], except for the solution column, which is transformed by

$$\hat{z}_{s0} = \frac{z_{s0} - u_{B_s}}{y_{sk}}$$

$$\hat{z}_{i0} = z_{i0} - \frac{y_{ik}}{y_{sk}}(z_{s0} - u_{B_s}), \qquad i \neq s$$

Record that $x_{B_s} = u_{B_s}$. Go to Step 2.

STEP 10: (Pivoting.) Enter x_k and remove x_{B_r}. All columns are transformed by (7) and (8), except for the solution column, which is transformed by

$$\hat{z}_{r0} = u_k + \frac{z_{r0}}{y_{rk}}$$

$$\hat{z}_{i0} = z_{i0} - \frac{y_{ik}z_{r0}}{y_{rk}}, \qquad i \neq r$$

Record that x_{B_r} is zero. Go to Step 2.

STEP 11: (Pivoting.) Enter x_k and remove x_{B_s}. All columns are transformed by (7) and (8), except for the solution column, which is transformed by

$$\hat{z}_{s0} = u_k + \frac{z_{s0} - u_{B_s}}{y_{sk}}$$

$$\hat{z}_{i0} = z_{i0} - \frac{y_{ik}}{y_{sk}}(z_{s0} - u_{B_s}), \qquad i \neq s$$

Record that $x_{B_s} = u_{B_s}$. Go to Step 2.

Example

Consider the example given in Section 2.2 with the additional constraints $x_1 \leq 3$ and $x_2 \leq 2$. The initial solution shown in Table 14 is identical to the solution in Table 3. Nonbasic variables at their upper bounds are indicated with asterisks.

Table 14

		↑↓ $-x_1$	$-x_2$
x_0	0	-2	-1
x_3	5	1	1
x_4	0	-1	1
x_5	21	6	2

We choose x_1 as the entering variable and compute $\theta_{51} = \frac{7}{2}$. Since none of the basic variables is upper bounded, it is unnecessary to compute α_{s1}. Since $\theta_{51} > u_1 = 3$, x_1 is also the departing variable and is set to its upper bound. The tableau is transformed using Step 6 to obtain Table 15.

Table 15

		* $-x_1$	↓ $-x_2$	
x_0	6	-2	-1	
x_3	2	1	1	$x_1 = 3$
x_4	3	-1	1	
← x_5	3	6	2	

The entering variable is x_2 and the departing variable x_5, with $\theta_{52} = \frac{3}{2} < u_2 = 2$. An ordinary simplex iteration is executed (Step 8); the result is Table 16. Now, since x_1 is starred and $y_{01} > 0$, x_1 is chosen to enter the

Table 16

		* ↓ $-x_1$	$-x_5$	
x_0	$\frac{15}{2}$	1	$\frac{1}{2}$	
x_3	$\frac{1}{2}$	-2	$-\frac{1}{2}$	$x_1 = 3$
x_4	$\frac{3}{2}$	-4	$-\frac{1}{2}$	
← x_2	$\frac{3}{2}$	3	$\frac{1}{2}$	

basis. As x_1 decreases, x_3 and x_4 decrease and x_2 increases. Following Step 5, $\bar{\theta}_{31} = -\frac{1}{4} > -\frac{3}{8}$ and $\bar{\alpha}_{21} = (\frac{3}{2} - 2)/3 = -\frac{1}{6} > \bar{\theta}_{31}$. Because $\bar{\alpha}_{21} > -u_1$,

x_2 departs and is put to its upper bound. Step 11 is used for the transformation. This yields the optimal tableau of Table 17 and the optimal solution $x_0^0 = \frac{23}{3}$, $x_1^0 = \frac{17}{6}$, $x_2^0 = 2$, $x_3^0 = \frac{1}{6}$, $x_4^0 = \frac{5}{6}$, $x_5^0 = 0$.

Table 17

		* $-x_2$	$-x_5$	
x_0	$\frac{23}{3}$	$-\frac{1}{3}$	$\frac{1}{3}$	
x_3	$\frac{1}{6}$	$\frac{2}{3}$	$-\frac{1}{6}$	$x_2 = 2$
x_4	$\frac{5}{6}$	$\frac{4}{3}$	$\frac{1}{6}$	
x_1	$\frac{17}{6}$	$\frac{1}{3}$	$\frac{1}{6}$	

2.9. DUALITY

Corresponding to an LP,

$$\begin{aligned} \max x_0 &= cx \\ Ax &\le b \\ x &\ge 0 \end{aligned} \tag{16}$$

called the *primal* problem, there is a related LP,

$$\begin{aligned} \min v_0 &= vb \\ vA &\ge c \\ v &\ge 0 \end{aligned} \tag{17}$$

called the *dual* problem, where v is an m-dimensional row vector. It is easily shown that the dual of the dual is the primal, so that either problem may be called the primal or the dual. We will refer to (16) as the primal.

The example of Section 2.2 with the slack variables x_3, x_4, and x_5 removed is

$$\begin{aligned} \max x_0 &= 2x_1 + x_2 \\ x_1 + x_2 &\le 5 \\ -x_1 + x_2 &\le 0 \\ 6x_1 + 2x_2 &\le 21 \\ x_1, x_2 &\ge 0 \end{aligned} \tag{18}$$

The dual of (18) is

$$\begin{aligned} \min v_0 = 5v_1 \qquad &+ 21v_3 \\ v_1 - v_2 + &\ 6v_3 \geq 2 \\ v_1 + v_2 + &\ 2v_3 \geq 1 \\ v_1, v_2,\ & v_3 \geq 0 \end{aligned} \tag{19}$$

The optimal solution to (18), including the slack variables, is given in Table 5 as $x_0^o = \frac{31}{4}$, $x_1^o = \frac{11}{4}$, $x_2^o = \frac{9}{4}$, $x_4^o = \frac{1}{2}$, $x_3^o = x_5^o = 0$. The addition of slack variables to (19) yields

$$\begin{aligned} \min v_0 = 5v_1 \qquad &+ 21v_3 \\ v_1 - v_2 + &\ 6v_3 - v_4 \qquad = 2 \\ v_1 + v_2 + &\ 2v_3 \qquad - v_5 = 1 \\ & v_1, \ldots, v_5 \geq 0 \end{aligned} \tag{20}$$

Problem (20) was solved in Section 2.5; see Tables 7–10 and make the identification $v_i = x_i$ for $i \neq 0$ and $v_0 = -x_0$. The optimal solution to (20) is given in Table 10 as $v_0^o = \frac{31}{4}$, $v_1^o = \frac{1}{2}$, $v_3^o = \frac{1}{4}$, $v_2^o = v_4^o = v_5^o = 0$. Note that $v_0^o = x_0^c$; that is, the optimal primal and dual objective functions are equal. This is always true when the primal (dual) problem has an optimal solution as asserted by

Theorem 8: *Exactly one of the following four relationships between the primal and dual problems must hold:*

1. *Both the primal and dual problems have optimal solutions and* $v_0^o = x_0^o$.
2. *The primal is unbounded and the dual is infeasible.*
3. *The dual is unbounded and the primal is infeasible.*
4. *The primal and the dual are both infeasible.*

PROOF: The existence of case 4 is illustrated by the following pair of infeasible problems:

$$\begin{aligned} \max \quad & x_2 \\ x_1 \quad & \leq -1 \\ -x_2 & \leq 1 \\ x_1, x_2 & \geq 0 \end{aligned} \qquad\qquad \begin{aligned} \min -v_1 &+ v_2 \\ v_1 \qquad & \geq 0 \\ - v_2 & \geq 1 \\ v_1, v_2 & \geq 0 \end{aligned}$$

For any pair of primal and dual feasible solutions,

$$Ax \leq b,\ v \geq 0 \Rightarrow vAx \leq vb \tag{21}$$

Also

$$vA \geq c,\ x \geq 0 \Rightarrow vAx \geq cx \tag{22}$$

It follows from (21) and (22) that $vb \geq cx$. Therefore, if the primal (dual) is unbounded, the dual (primal) can have no feasible solution. Furthermore, if the primal has an optimal solution x^o, for all dual feasible solutions,

$$v_0 = vb \geq cx^o = x_0^o$$

Similarly, if the dual has an optimal solution v^o, for all primal feasible solutions,

$$x_0 = cx \leq v^o b = v_0^o$$

Thus to prove case 1, we give a dual feasible solution v^o with $v_0^o = x_0^o$. Since the primal has an optimal solution, there exists a basis matrix B having $y_{i0} \geq 0$, $i = 1, \ldots, m$ and $y_{0j} \geq 0$ for all $j \in R$ (see Exercise 19). The dual solution to be considered is $v^o = c_B B^{-1}$. We have

$$v^o b = c_B B^{-1} b = c_B x_B = x_0^o$$

and

$$v^o A = c_B B^{-1} A = c_B B^{-1}(B, N) = (c_B, c_B B^{-1} N) \geq (c_B, c_N)$$

since $y_{0j} \geq 0$ for all $j \in R$ is equivalent to $c_B B^{-1} N - c_N \geq 0$. To show $v^o \geq 0$, note that

$$v_i^o = c_B B^{-1} e_i = c_B B^{-1} e_i - c_{n+i}$$

since c_{n+i} (the objective coefficient for the ith slack variable) is zero. If the ith slack variable, x_{n+i}^o, is nonbasic,

$$c_B B^{-1} e_i - c_{n+i} = y_{0,n+i} \geq 0$$

by hypothesis. If x_{n+i}^o is basic, $B^{-1} e_i = e_j$ and $c_B e_j = 0$, since the ith column of B^{-1} is e_j and the jth element of c_B is c_{n+i} which is zero. ∎

The proof just given also establishes *complementary slackness*, $v_i^o x_{n+i}^o = 0$, $i = 1, \ldots, m$, since if $v_i^o = c_B B^{-1} e_i - c_{n+i} > 0$, then x_{n+i}^o must be nonbasic. Because of the symmetry between the primal and dual problems, a similar argument establishes the analogous complementary slackness condition $x_j^o v_{m+j}^o = 0$, $j = 1, \ldots, n$, where v_{m+j} is the surplus variable for the jth dual constraint. Thus, if the slack variable of the ith primal constraint is positive in an optimal solution, there exists an optimal dual solution with the ith dual variable equal to zero and conversely; and if the surplus variable of the jth dual constraint is positive in an optimal solution, there exists an optimal primal solution with the jth primal variable equal to zero and conversely. Stated more compactly, if $x = (x_r, x_s)$ and $v = (v_s, v_r)$, where the subscripts r and s refer to regular and slack (surplus) variables, respectively, then $v^o x^o = 0$.

The results given above are illustrated by the pair of problems (18) and (19). The inverse of the optimal primal basis matrix is

$$B^{-1} = \begin{bmatrix} \frac{3}{2} & 0 & -\frac{1}{4} \\ -2 & 1 & \frac{1}{2} \\ -\frac{1}{2} & 0 & \frac{1}{4} \end{bmatrix}$$

Thus an optimal dual solution is

$$v^o = (v_1^o, v_2^o, v_3^o) = c_B B^{-1} = (1 \quad 0 \quad 2) \begin{bmatrix} \frac{3}{2} & 0 & -\frac{1}{4} \\ -2 & 1 & \frac{1}{2} \\ -\frac{1}{2} & 0 & \frac{1}{4} \end{bmatrix} = (\tfrac{1}{2} \quad 0 \quad \tfrac{1}{4})$$

An optimal dual solution can also be obtained directly from Table 5: $x_1^o, x_2^o, x_4^o > 0 \Rightarrow v_4^o, v_5^o, v_2^o = 0$, $v_1^o = y_{03} = \frac{1}{2}$ and $v_3^o = y_{05} = \frac{1}{4}$. Similarly, from Table 10, an optimal primal solution can be obtained. Note that the pair of solutions $x^o = (x_1^o, x_2^o, x_3^o, x_4^o, x_5^o) = (\frac{11}{4}, \frac{9}{4}, 0, \frac{1}{2}, 0)$ and $v^o = (v_4^o, v_5^o, v_1^o, v_2^o, v_3^o) = (0, 0, \frac{1}{2}, 0, \frac{1}{4})$ satisfy the complementary slackness condition $v^o x^o = 0$.

Having concluded that by solving the primal (dual) we also solve the dual (primal), we can apply the simplex method to either one. In fact, as we shall see in Section 2.10, it is possible to begin with a dual feasible solution and solve the dual using the same tableau that we have used in the (primal) simplex method. The algorithm is called the *dual simplex* method and will be useful for several reasons. First, if there is an obvious dual feasible solution, but not a primal feasible solution, we can avoid phase 1 calculations. Second, when we have an optimal solution (i.e., primal and dual feasible) and then add another primal constraint, as is done in some integer programming algorithms, dual feasibility (but not necessarily primal feasibility) will be maintained. Finally, there are integer programming algorithms that operate exclusively on dual feasible solutions.

2.10. THE DUAL SIMPLEX ALGORITHM

In proving Theorem 8, we showed that if B is any primal basis matrix with $y_{0j} \geq 0$ for all $j \in R$, then

$$v_i = \begin{cases} 0 & \text{if } x_{n+i} \text{ is basic} \\ y_{0,n+i} & \text{if } x_{n+i} \text{ is nonbasic} \end{cases} \qquad i = 1, \ldots, m$$

$$v_{m+j} = \begin{cases} 0 & \text{if } x_j \text{ is basic} \\ y_{0j} & \text{if } x_j \text{ is nonbasic} \end{cases} \qquad j = 1, \ldots, n$$

is a dual feasible solution. If, in addition, $y_{i0} \geq 0$, $i = 1, \ldots, m$, the solution is primal feasible and optimal.

Suppose that the dual feasible solution is not primal feasible; in particular, assume that $y_{r0} < 0$. By dropping x_{B_r} from the basis and introducing x_k, where k satisfies

$$\frac{y_{0k}}{y_{rk}} = \max_{j \in R} \left[\frac{y_{0j}}{y_{rj}}, y_{rj} < 0 \right]$$

a new basic solution is obtained. It has the following two properties:

1. It is dual feasible.
2. In the absence of dual degeneracy ($y_{0j} > 0$ for all $j \in R$), the dual objective function decreases.

These results correspond directly to those obtained for the primal simplex method and will not be proved. Using them for making changes of bases yields the dual simplex algorithm, which is finite in the absence of dual degeneracy.

Algorithm

STEP 1: (Initialization.) Begin with a basis B such that $y_{0j} \geq 0$ for all $j \in R$. Go to Step 2.

STEP 2: (Test for optimality.) Is $y_{i0} \geq 0$ for $i = 1, \ldots, m$? If so, the present basic solution is optimal. If not, go to Step 3.

STEP 3: (Choose a departing variable.) Select a variable x_{B_r} with $y_{r0} < 0$ to leave the basis (a common rule is to choose x_{B_r} if $y_{r0} = \min y_{i0}$, $i = 1, \ldots, m$). Go to Step 4.

STEP 4: (Choose an entering variable.) Select a variable x_k with $y_{0k}/y_{rk} = \max_{j \in R} [y_{0j}/y_{rj}, y_{rj} < 0]$ to enter the basis. If there is a tie, break it arbitrarily. The case $y_{rj} \geq 0$ for all $j \in R$ corresponds to an unbounded dual solution, and therefore the primal is infeasible. Go to Step 5.

STEP 5: (Pivoting.) Execute a (dual) simplex iteration by making x_k basic and x_{B_r} nonbasic. Go to Step 2.

Eliminating Cycling

The possibility of cycling can be eliminated by beginning with and maintaining $y_j \overset{L}{>} 0$, for all $j \in R$. Assuming that the initial tableau has $y_j \overset{L}{>} 0$ for all $j \in R$, lexicographically positive columns can be maintained by specializing Step 4. Let $\alpha_j = y_j/y_{rj}$ for all $j \in R_r = \{j \mid j \in R, y_{rj} < 0\}$. Select x_k to enter the basis, where k is given by

$$\alpha_k = \operatorname*{lexmax}_{j \in R_r} \alpha_j \tag{23}$$

that is,

$$\alpha_k \overset{L}{\geq} \alpha_j \qquad \text{for all } j \in R_r$$

Note that k will be unique unless some other column is a scalar multiple of column k (see Exercise 15).

To prove that lexicographically positive columns are maintained, we have $\hat{y}_{B_r} = (-y_{0k}/y_{rk}, \ldots, 1/y_{rk}, \ldots, -y_{mk}/y_{rk})$. Now $y_k \overset{L}{>} 0$ implies that $\gamma_k = (y_{0k}, \ldots, y_{r-1,k}) \overset{L}{>} 0$, since $y_{rk} < 0$. But $\gamma_k \overset{L}{>} 0$ and $y_{rk} < 0$ imply that $\hat{\gamma}_k = (-y_{0k}/y_{rk}, \ldots, -y_{r-1,k}/y_{rk}) \overset{L}{>} 0$. Since $\hat{\gamma}_k$ is a vector consisting of the first r components of $\hat{y}_{B_r}$, it follows that $\hat{y}_{B_r} \overset{L}{>} 0$.

To prove $\hat{y}_j \overset{L}{>} 0$ for $j \neq 0, B_r$, three cases are considered:

1. $y_{rj} = 0$. Then $\hat{y}_j = y_j \overset{L}{>} 0$.
2. $y_{rj} > 0$. Let $\gamma_j(\hat{\gamma}_j)$ be a vector consisting of the first r components of $y_j(\hat{y}_j)$. Then $\hat{\gamma}_j = \gamma_j + y_{rj}\hat{\gamma}_k$ and $\gamma_j \overset{L}{\geq} 0$, $\hat{\gamma}_k \overset{L}{>} 0$, and $y_{rj} > 0$ imply $\hat{\gamma}_j \overset{L}{>} 0$. It follows that $\hat{y}_j \overset{L}{>} 0$.
3. $y_{rj} < 0$. Note that $\alpha_k \overset{L}{\geq} \alpha_j$ implies $\gamma_j \overset{L}{\geq} -y_{rj}\hat{\gamma}_k$, which in turn implies $\hat{\gamma}_j \overset{L}{\geq} 0$. However, even if $\hat{\gamma}_j = 0$, the $(r+1)$st component of $\hat{y}_j$ is $y_{rj}/y_{rk} > 0$. It follows that $\hat{y}_j \overset{L}{>} 0$.

Under these conditions cycling cannot occur because the solution column is lexicographically decreasing. This is true, since the vector consisting of the first r components of $\hat{y}_0 - y_0$ is $\hat{\gamma}_0 - \gamma_0 = y_{r0}\hat{\gamma}_k$ and $y_{r0} < 0$, $\hat{\gamma}_k \overset{L}{>} 0$, imply that $y_{r0}\hat{\gamma}_k \overset{L}{<} 0$.

Initial Solutions

If the primal problem is

$$\begin{aligned} &\max cx \\ &Ax - Ix_s = b \\ &x, x_s \geq 0 \end{aligned}$$

and at least one component of b is positive, no obvious primal basic feasible solution is available. However, if $c \leq 0$, there is an obvious dual feasible solution. Take $B = -I$ as the primal basis. The primal basic solution (not feasible) is $x_s = -b$ and $x = 0$. However,

$$y_{0j} = c_B B^{-1} a_j - c_j = -c_j \geq 0$$

since $c_B = 0$. Thus a basic feasible dual solution is $v_{m+j} = y_{0j}$, $j = 1, \ldots, n$ and $v_i = 0$, $i = 1, \ldots, m$. When no obvious basic dual feasible solution is known, a phase 1, similar to the one described for the primal simplex method can be used to obtain one.

Example

We return to the problem of Section 2.5:

$$
\begin{aligned}
\max x_0 = -5x_1 - 21x_3 & \\
x_1 - x_2 + 6x_3 - x_4 &= 2 \\
x_1 + x_2 + 2x_3 - x_5 &= 1 \\
x_1, \ldots, x_5 &\geq 0
\end{aligned}
$$

Since $c_j \leq 0$ for all j, a first dual feasible solution is obtained with $B = -I$, as shown in Table 18.

Table 18

		$-x_1$	$-x_2$	↓ $-x_3$
x_0	0	5	0	21
$\leftarrow x_4$	-2	-1	1	-6
x_5	-1	-1	-1	-2

Since $y_{40} = \min y_{i0} = -2$, x_4 is chosen to leave the basis. We have $y_{03}/y_{43} = -\frac{21}{6} > y_{01}/y_{41} = -5$, so that x_3 is chosen to enter the basis. The resulting tableau is shown in Table 19.

Table 19

		↓ $-x_1$	$-x_2$	$-x_4$
x_0	-7	$\frac{3}{2}$	$\frac{7}{2}$	$\frac{7}{2}$
x_3	$\frac{1}{3}$	$\frac{1}{6}$	$-\frac{1}{6}$	$-\frac{1}{6}$
$\leftarrow x_5$	$-\frac{1}{3}$	$-\frac{2}{3}$	$-\frac{4}{3}$	$-\frac{1}{3}$

Now we choose x_5 to leave the basis and x_1 becomes basic, yielding the optimal solution given in Table 20.

Table 20

		$-x_5$	$-x_2$	$-x_4$
x_0	$-\frac{31}{4}$	$\frac{9}{4}$	$\frac{1}{2}$	$\frac{11}{4}$
x_3	$\frac{1}{4}$	$\frac{1}{4}$	$-\frac{1}{2}$	$-\frac{1}{4}$
x_1	$\frac{1}{2}$	$-\frac{3}{2}$	2	$\frac{1}{2}$

2.11. ADDING VARIABLES AND CONSTRAINTS

Suppose, after having obtained an optimal solution to an LP, we consider the addition of a new variable x_p with objective coefficient c_p and constraint coefficients $(a_{1p}, \ldots, a_{mp})$. By setting $x_p = 0$, we retain a primal basic feasible solution. The vector y_p is given by

$$y_p = \begin{pmatrix} c_B B^{-1} a_p - c_p \\ B^{-1} a_p \end{pmatrix}$$

If $y_{0p} = c_B B^{-1} a_p - c_p \geq 0$, the present solution is still optimal. However, if $y_{0p} < 0$, the present solution may not be optimal. We introduce x_p into the basis and continue to execute *primal* simplex iterations.

Of much greater interest in integer programming is the addition of a new constraint after an optimal solution has been achieved. Suppose we introduce the constraint

$$\sum_{j=1}^{n} a_{pj} x_j + s_p = b_p, \qquad s_p \geq 0 \tag{24}$$

By eliminating the present basic variables from (24) and then solving for s_p, we obtain

$$s_p = y_{p0} - \sum_{j \in R} y_{pj} x_j$$

If $y_{p0} \geq 0$, the present solution $x_{B_i} = y_{i0}$, $i = 0, 1, \ldots, m$ and $s_p = y_{p0}$, is primal feasible and optimal, since the y_{0j} are not affected by the new constraint.

The more interesting case is $y_{p0} < 0$. Now the basic solution $x_{B_i} = y_{i0}$, $i = 0, 1, \ldots, m$, and $s_p = y_{p0}$ is not primal feasible. But, because we retain $y_{0j} \geq 0$ for all $j \in R$, the basic solution is dual feasible and we can continue the problem by removing s_p from the basis and executing dual simplex iterations.

If the original problem, together with (24), has a feasible solution, it has an optimal solution with $s_p = 0$. To show this, let $(x^{\text{o}}, s_p^{\text{o}})$ be the infeasible solution obtained after introducing the new constraint and (x^*, s_p^*) be the optimal solution obtained by the dual simplex algorithm. Let

$$\lambda = \frac{s_p^*}{s_p^* - s_p^{\text{o}}}$$

and $(x^+, s_p^+) = \lambda(x^{\text{o}}, s_p^{\text{o}}) + (1 - \lambda)(x^*, s_p^*)$. The solution (x^+, s_p^+) satisfies the original equations, $s_p^+ = 0$ and $cx^* \leq cx^+$. Thus $(x^+, 0)$ is an optimal solution.

Example

Suppose that in the optimal solution of Table 5, we add the constraint

$$x_5 \geq 1 \qquad \text{or} \qquad s_1 = -1 + x_5$$

Since $x_5 = 0$ in Table 5, the new basic solution $x_{B_i} = y_{i0}$, $i = 0, 1, \ldots, m$, and $s_1 = -1$, shown in Table 21, is not primal feasible. But, since it is dual feasible, we can reoptimize using the dual simplex algorithm. The slack

Table 21

		$\downarrow$ $-x_5$	$-x_3$
x_0	$\frac{31}{4}$	$\frac{1}{4}$	$\frac{1}{2}$
x_2	$\frac{9}{4}$	$-\frac{1}{4}$	$\frac{3}{2}$
x_4	$\frac{1}{2}$	$\frac{1}{2}$	-2
x_1	$\frac{11}{4}$	$\frac{1}{4}$	$-\frac{1}{2}$
$\leftarrow s_1$	-1	-1	0

variable s_1 leaves the basis and x_5 enters as shown in Table 22. The solution given in Table 22 is optimal, although, in general, it may take several iterations to restore primal feasibility.

Table 22

		$-s_1$	$-x_3$
x_0	$\frac{30}{4}$	$\frac{1}{4}$	$\frac{1}{2}$
x_2	$\frac{10}{4}$	$-\frac{1}{4}$	$\frac{3}{2}$
x_4	0	$\frac{1}{2}$	-2
x_1	$\frac{10}{4}$	$\frac{1}{4}$	$-\frac{1}{2}$
x_5	1	-1	0

Upper Bound Constraints on Basic Variables

Assume that there is an optimal solution with $x_{B_r} = y_{r0}$ and the constraint $x_{B_r} \leq u_r < y_{r0}$ is added. The constraint can be handled by using a dual version of the upper bounded algorithm (Section 2.8). Writing the constraint as an equality yields

$$s = u_r - x_{B_r} \tag{25}$$

Using (6) to eliminate x_{B_r} from (25) yields

$$s = u_r - y_{r0} - \sum_{j \in R} (-y_{rj})x_j \tag{26}$$

Since for all $j \in R$ the coefficients in (26) are the negative of the coefficients in the rth row of (6), pivoting on (26) (increasing s to zero) with column k chosen by the dual simplex criterion,

$$\begin{aligned} \frac{y_{0k}}{-y_{rk}} &= \max_{j \in R} \left(\frac{y_{0j}}{-y_{rj}}, y_{rj} > 0 \right) \\ &= -\min_{j \in R} \left(\frac{y_{0j}}{y_{rj}}, y_{rj} > 0 \right) = -\frac{y_{0k}}{y_{rk}} \end{aligned}$$

corresponds to pivoting on (6) for $i = r$ and column k. However, x_{B_r} is only reduced to u_r (not to zero). The tableau, except for the solution column, is transformed as usual. For the solution column,

$$\hat{y}_{r0} = \frac{y_{r0} - u_r}{y_{rk}}$$
$$\hat{y}_{i0} = y_{i0} - \frac{(y_{r0} - u_r)y_{ik}}{y_{rk}}, \qquad i \neq r \tag{27}$$

The solution given by (27) is dual feasible, since for $j \in R - \{k\}$, $\hat{y}_{0j} \geq 0$, $x_{B_r} = u_j$ and $\hat{y}_{0B_r} \leq 0$.

Dual simplex iterations are then continued until primal feasibility is restored. Because of the similarity with the algorithm given in Section 2.8, the details of the selection and pivoting rules will not be elaborated.

Example

The constraint $x_1 \leq 2$ is added to Table 5. To reduce x_1, it is necessary to make x_5 positive. Pivoting on x_1 (reducing it to 2) and x_5 yields Table 23.

Table 23

		*	↓	
		$-x_1$	$-x_3$	
x_0	7	-1	1	
x_2	3	1	1	$x_1 = 2$
$\leftarrow x_4$	-1	-2	-1	
x_5	3	4	-2	

The departing variable in Table 23 is x_4 and x_3 enters. Pivoting yields the optimal solution shown in Table 24.

Table 24

		*		
		$-x_1$	$-x_4$	
x_0	6	-3	1	
x_2	2	-1	1	$x_1 = 2$
x_3	1	2	-1	
x_5	5	8	-2	

2.12. EXERCISES

1. For the LP

$$\begin{aligned} \max x_0 &= x_1 + 2x_2 \\ x_1 + x_2 &\le 4 \\ -x_1 + x_2 &\le 4 \\ x_1 &\le 1 \\ x_1 - x_2 &\le 0 \\ x_1, x_2 &\ge 0 \end{aligned}$$

(a) Solve graphically.
(b) Enumerate all basic feasible solutions and show the correspondence between basic feasible solutions and extreme points.
(c) Solve by the primal simplex method.
(d) Solve by the primal upper bounded algorithm.
(e) Solve beginning with and maintaining lexicographically positive rows.
(f) Give the dual and its solution.

2. Prove Theorem 1.

3. Rank the following vectors in lexicographically increasing order:

$$a_1 = (2, -7, 4),\ a_2 = (-1, 3, 5),\ a_3 = (0, 1, 0),\ a_4 = (0, 0, 0)$$

4. Find the determinant of the optimal basis matrix in Exercise 1 by direct calculation and by application of Theorem 6.

5. Is condition (ii) of Theorem 2 necessary for optimality? Prove the necessity or give a counterexample.

6. Give the dual of

$$\begin{aligned} &\max cx \\ &Ax = b \end{aligned}$$

7. Prove (Farkas' lemma) if, for every solution of $Aw \ge 0$, it is also true that $bw \ge 0$, there exists $x \ge 0$ such that $xA = b$.

8. Solve

$$\begin{aligned} \max x_0 &= x_1 + 2x_2 + 3x_3 \\ x_1 \qquad\quad + x_3 &\le 5 \\ x_1 + 2x_2 \qquad\quad &\ge 8 \\ x_1 + x_2 + 2x_3 &= 10 \\ x_1, x_2, x_3 &\ge 0 \end{aligned}$$

using the primal two-phase method.

9. Solve

$$\begin{aligned} \max \quad -8x_1 - 4x_2 - 3x_3 & \\ 2x_1 - x_2 + x_3 + x_4 &\geq 1 \\ x_1 + x_2 \qquad - x_4 &\geq 2 \\ x_1, \ldots, x_4 &\geq 0 \end{aligned}$$

using the dual simplex algorithm.

10. Add the constraint $x_1 + x_2 \leq 6$ to the optimal solution of Exercise 1 and solve the new problem.

11. Add the constraint $x_1 \leq 1$ to the optimal solution of Exercise 1 and solve the new problem.

12. Add the constraint $x_1 \geq 2$ to the optimal solution of Exercise 1 and solve the new problem.

13. If phase 1 of the two-phase method ends with $z_0 = 0$ and at least one artificial variable in the basis, show how to construct a basic feasible solution to the original problem.

14. Justify the dual simplex pivoting rules.

15. Prove that α_k given by (23) is unique if the columns of A are pairwise independent.

16. Describe a phase 1 method for obtaining a dual feasible basic solution.

17. What happens in the simplex method when rank of $A < m$?

18. Give details of the selection and pivoting rules for the upper bounded, dual simplex algorithm described briefly in Section 2.11.

19. Show that if an LP has an optimal solution, there exists a basic feasible solution having $y_{0j} \geq 0$ for all $j \in R$.

2.13. NOTES

The material of Chapter 2 can be found in much greater detail in a variety of linear programming books. Some standard texts are Gale (1960), Charnes and Cooper (1961), Hadley (1962), Dantzig (1963), Gass (1964), and Simonnard (1966).

3 Integer Programming and Graphs

3.1. INTRODUCTION

A significant class of linear and integer linear programs is associated with mathematical objects called *graphs*. A pictorial representation of a graph is given in Figure 1. The circles correspond to elements of a set, and the lines indicate relationships between pairs of elements. In Section 3.2, precise definitions and properties of graphs will be given.

The results of Chapter 2 can be strengthened considerably for LP's

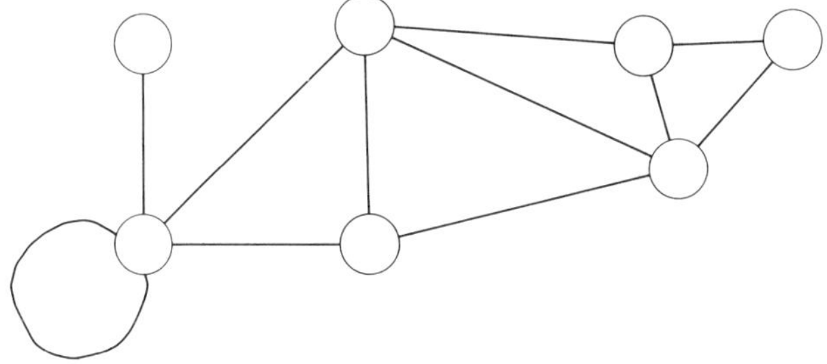

Figure 1

defined on graphs. An elegant duality theory for LP's on graphs exists that has led to very efficient specialized algorithms.

Our interest is in ILP's defined on graphs. One very significant result, to be given in Section 3.4, is that many ILP's on graphs are really LP's, in the sense that every basic solution to the corresponding LP is integer.

The well-known *assignment problem* belongs to this class. Suppose that there are m jobs and m men; each man is to be assigned a different job, and the value of assigning man i to job j is c_{ij}. The problem of maximizing total value is the ILP,

$$\max \sum_{i=1}^{m} \sum_{j=1}^{m} c_{ij}x_{ij}$$

$$\sum_{i=1}^{m} x_{ij} = 1, \qquad j = 1, \ldots, m$$

$$\sum_{j=1}^{m} x_{ij} = 1, \qquad i = 1, \ldots, m$$

$$x_{ij} = 0, 1, \qquad i, j = 1, \ldots, m$$

where $x_{ij} = 1$ if man i is assigned job j, and $x_{ij} = 0$ otherwise. In Section 3.4 a graphical interpretation of the assignment problem will be given, where it will be shown that the integrality constraints $x_{ij} = 0, 1$ are superfluous.

Another very important result (Section 3.6) is that there is a class of ILP's on graphs whose corresponding LP's have basic solutions which are "almost" integer. In particular, this class has basic feasible solutions whose components are integer or have the value one-half. Problems in this class can be solved much more efficiently than arbitrary ILP's.

3.2. BASIC DEFINITIONS AND ELEMENTARY PROPERTIES OF GRAPHS

Let $V = \{i | i = 1, \ldots, m\}$ be an arbitrary finite set and S be the set of all (distinct) *unordered* pairs (i, j) of elements of V, that is,

$$S = \{(i, j) | i \in V, j \in V\}$$

where (i, j) and (j, i) represent the same element. The pair $G = (V, E)$, $E \subseteq S$ is called a (undirected) *graph*; the elements of V are called *vertices* and the elements of E are called *edges* of the graph.

Although a graph is an abstract object, it has a convenient pictorial representation as shown in Figure 1, in which the circles represent the vertices and the lines represent the edges. Figure 1, with names given to the vertices and edges is repeated as Figure 2. Note that $V = \{1, 2, \ldots, 7\}$ and $E = \{e_1, e_2, \ldots, e_{10}, e_{11}\} = \{(1, 2), (2, 3), \ldots, (6, 7), (2, 2)\}$.

Graphs can be used to model a wide variety of real-world problems. An obvious application, which will be pursued in Section 3.4, is to represent transportation networks by graphs. For instance, in a road network, the vertices can represent intersections and the edges can represent streets.

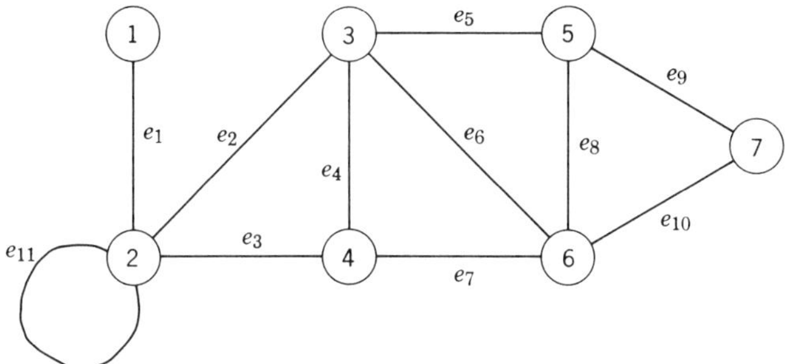

Figure 2

Directed Graphs

Let Q be the set of all *ordered* pairs of elements of V, that is,

$$Q = \{(i, j) | i \in V, j \in V\}$$

Note that, in ordered pairs, (i, j) and (j, i) do not represent the same element unless $i = j$. The pair $G' = (V, E')$, $E' \subseteq Q$, is called a *directed graph*; the elements of E' are called *directed edges.*

In the pictorial representation of directed graphs, arrowheads are attached to the lines to indicate direction. The directed edge (i, j) is represented by a line joining vertices i and j with an arrowhead pointing to j. Directed graphs are needed to handle asymmetric relationships between pairs of elements. For example, in a road network, a one-way street can be represented by a directed edge.

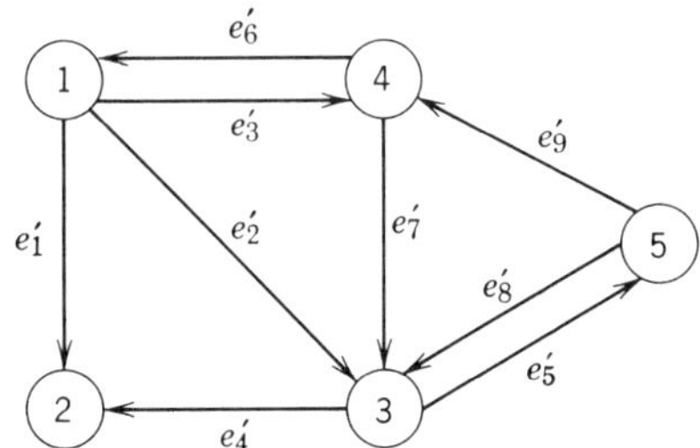

Figure 3

Figure 3 shows a directed graph with $V = \{1, 2, \ldots, 5\}$, and $E' = \{e'_1, e'_2, \ldots, e'_9\} = \{(1, 2), (1, 3), \ldots, (5, 4)\}$. Frequently, the notation $G = (V, E)$ will be used to represent both directed and undirected graphs, with the specific case either being stated or obvious from the context.

Incidence Matrices

If $e_j = (i, k)$, edge j is said to be *incident* to vertices i and k, which are, in turn, said to be incident to edge j. The incidence relationship can be represented by a matrix. Let $A = \{a_{ij}\}$, $i = 1, \ldots, |V|$, $j = 1, \ldots, |E|$ be a matrix defined on an undirected graph as

$$a_{ij} = \begin{cases} 1 & \text{if edge } j \text{ and vertex } i \text{ are incident.} \\ 0 & \text{otherwise} \end{cases}$$

The matrix A is called the *incidence matrix* of the graph $G = (V, E)$. The ith row of A corresponds to vertex i, and its ones indicate all of the edges incident to vertex i. The jth column of A corresponds to edge j and is either a unit vector or has two elements equal to one. The former case occurs if and only if $e_j = (i, i)$ for some i. The incidence matrix for the graph of Figure 2 is

$$A = \begin{bmatrix} 1 & 0 & 0 & 0 & 0 & 0 & 0 & 0 & 0 & 0 & 0 \\ 1 & 1 & 1 & 0 & 0 & 0 & 0 & 0 & 0 & 0 & 1 \\ 0 & 1 & 0 & 1 & 1 & 1 & 0 & 0 & 0 & 0 & 0 \\ 0 & 0 & 1 & 1 & 0 & 0 & 1 & 0 & 0 & 0 & 0 \\ 0 & 0 & 0 & 0 & 1 & 0 & 0 & 1 & 1 & 0 & 0 \\ 0 & 0 & 0 & 0 & 0 & 1 & 1 & 1 & 0 & 1 & 0 \\ 0 & 0 & 0 & 0 & 0 & 0 & 0 & 0 & 1 & 1 & 0 \end{bmatrix}$$

Similarly, an incidence matrix is defined for a directed graph. Let $A' = \{a'_{ij}\}$, $i = 1, \ldots, |V|$, $j = 1, \ldots, |E'|$, be the incidence matrix for a directed graph $G' = (V, E')$ defined as follows:

$$a'_{ij} = \begin{cases} -1 & \text{if } e_j = (k, i), \quad k \neq i \\ 1 & \text{if } e_j = (i, k), \quad k \neq i \\ 0 & \text{otherwise} \end{cases}$$

In the above definition it is assumed, for simplicity, that $(i, i) \notin E'$ for any $i \in V$. Thus every column of A' has two nonzero entries, one $(+1)$ and one (-1). The incidence matrix for the graph of Figure 3 is

$$A' = \begin{bmatrix} 1 & 1 & 1 & 0 & 0 & -1 & 0 & 0 & 0 \\ -1 & 0 & 0 & -1 & 0 & 0 & 0 & 0 & 0 \\ 0 & -1 & 0 & 1 & 1 & 0 & -1 & -1 & 0 \\ 0 & 0 & -1 & 0 & 0 & 1 & 1 & 0 & -1 \\ 0 & 0 & 0 & 0 & -1 & 0 & 0 & 1 & 1 \end{bmatrix}$$

Paths, Cycles, and Trees

In an undirected graph, two edges are said to be *adjacent* if they are both incident to the same vertex. (In a directed graph, e_i and e_k are *adjacent* if $e_i = (h, j)$ and $e_k = (j, p)$ for some h, j and p). A sequence of distinct edges $(e_{j1}, e_{j2},...,e_{jp})$ is called a path if the induced sequence of vertices $(i_0,...,i_p)$ are distinct; where $e_{j1} = (i_0, i_1),...,e_{jp} = (i_{p-1}, i_p)$. The edge sequence is also called a *path connecting vertices* i_0 *and* i_p. In a directed graph, the path is said to be *from vertex* i_0 *to vertex* i_p.

In Figure 2, $p_1 = (e_2, e_6, e_{10})$ and $p_2 = (e_3, e_7, e_6, e_5, e_8, e_{10})$ are two different paths connecting vertices 2 and 7. In the directed graph of Figure 3, (e'_3, e'_7, e'_5) is a path from vertex 1 to vertex 5. However, (e'_1, e'_4, e'_5) is not a path, since e'_4 is not adjacent to e'_1 or e'_5.

A path that begins and ends at the same vertex is called a *cycle*. For example, in Figure 2, $c_1 = (e_6, e_5, e_8)$ is a cycle. The path p_2 given above contains the cycle c_1. Paths without cycles are called *simple paths*; thus p_1 above is simple, but p_2 is not.

A path can also be represented by its vertex sequence. For example, the vertex representations of p_1 and p_2 are (2, 3, 6, 7) and (2, 4, 6, 3, 5, 6, 7).

An undirected graph is called *connected* if for all $i, j \in V$, $i \neq j$, there is at least one path connecting vertices i and j. The graph of Figure 2 is obviously connected.

A graph $\hat{G} = (\hat{V}, \hat{E})$ is called a *subgraph* of $G = (V, E)$ if $\hat{G}$ can be obtained from G by removing from G the subset $E - \hat{E}$ of edges and any subset $V - \hat{V}$ of vertices that are not incident to any edges of $\hat{E}$.

A connected, undirected graph is said to be a *tree* if it contains no cycles. A subgraph of $G = (V, E)$ which is a tree and whose vertex set is V, is called a *spanning tree* of G. Clearly, the graph of Figure 2 is not a tree. However, a connected graph always contains a subgraph that is a spanning tree. When $\hat{E} = \{4, 6, 8, 10, 11\}$ is deleted from the graph of Figure 2, we obtain the subgraph $\hat{G} = (V, E - \hat{E})$, which is the spanning tree shown in Figure 4. It can be shown that if $G = (V, E)$ is a tree, then $|E| = |V| - 1$ (see Exercise 4).

Given an undirected graph $G = (V, E)$ and $i \in V$, we define

1. $E^i = \{e_j |$there exists a path through vertex i containing $e_j\}$.
2. $V^i = \{t |$there exists $r \in V$ such that $(r, t) \in E^i\}$.

The subgraph $G^i = (V^i, E^i)$ is called a *component* of G. Clearly, G is connected if and only if $G^i = G$ for all $i \in V$. Note that the graph of Figure 5 is not connected and contains three components.

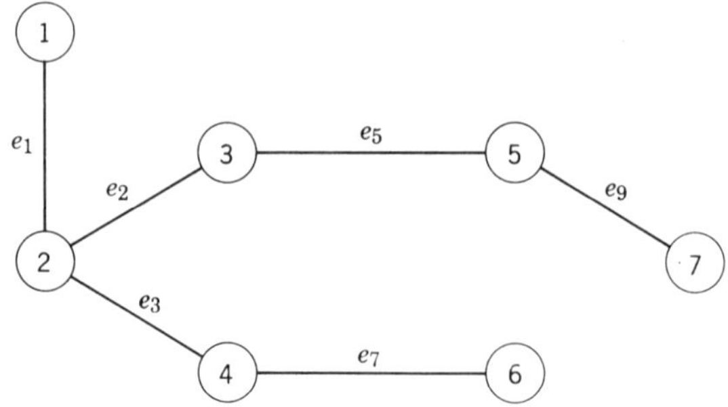

Figure 4

Some additional properties of graphs will be introduced in subsequent sections as they are required.

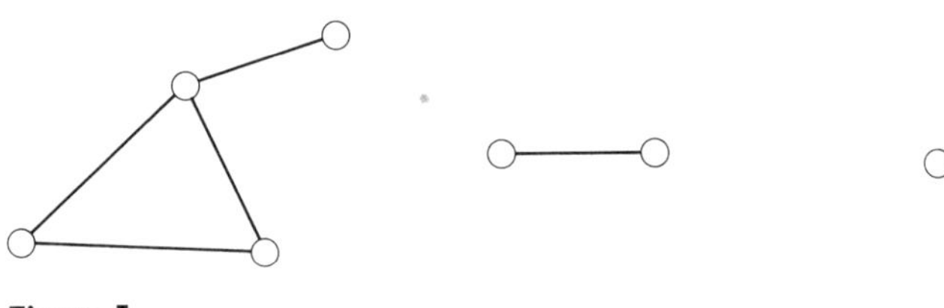

Figure 5

3.3. UNIMODULARITY

The purpose of this section is to give some necessary and sufficient conditions for an LP with integer data to have an optimal solution that is integer.

Consider an LP with constraints

$$Ax = b, \qquad x \geq 0 \tag{1}$$

where A and b are assumed to be integer matrices. As in Chapter 2, we partition A as (B, N), where B is nonsingular, and write the equations of (1) as

$$Bx_B + Nx_N = b$$

Recall that a basic solution to (1) is

$$x_B = B^{-1}b, \qquad x_N = 0$$

Thus a sufficient condition for a basic solution to be integer is for B^{-1} to be an integer matrix.

To derive conditions for B^{-1} to be an integer matrix, we introduce the notion of unimodularity. A square, integer matrix B is called *unimodular* if $D = |\det B| = 1$. An integer $m \times n$ matrix A is *totally unimodular* if every square, nonsingular submatrix of A is unimodular.

In Section 2.7, it was mentioned that if B is nonsingular, then

$$B^{-1} = \frac{B^{+}}{\det B}$$

where B^{+} is an integer matrix. Thus B unimodular implies that B^{-1} is an integer matrix. Now if A is totally unimodular, every basis matrix B is unimodular and every basic solution $(x_B, x_N) = (B^{-1}b, 0)$ is integer. We have just proved

Theorem 1: *If A is totally unimodular, then every basic solution of* (1) *is integer.*

The example below shows that total unimodularity of A is not necessary for integer basic solutions for all integer b. Consider

$$\begin{aligned} 5x_1 + 2x_2 &= b_1 \\ 2x_1 + x_2 &= b_2 \\ b_1, b_2 &\text{ integer} \end{aligned} \tag{2}$$

Although

$$A = \begin{bmatrix} 5 & 2 \\ 2 & 1 \end{bmatrix}$$

is not totally unimodular, the unique solution to (2), is integer.

For LP's with inequality constraints, Theorem 1 can be strengthened. In particular, given the constraint set $S(b) = \{x | Ax \le b, x \ge 0\}$, total unimodularity of A is necessary and sufficient for all extreme points of S to be integer for every integer vector b. This result is the crux of Theorem 2.

Theorem 2: *Let A be an integer matrix. The following statements are equivalent:*

1. *A is totally unimodular.*
2. *The extreme points (if any) of $S(b) = \{x | Ax \le b, x \ge 0\}$ are integer for arbitrary integer b.*
3. *Every square nonsingular submatrix of A has an integer inverse.*

PROOF: ($1 \Rightarrow 2$, the sufficiency of total unimodularity.) It suffices to show that A totally unimodular implies (A, I) totally unimodular. Given that (A, I) is totally unimodular, Theorem 1 can be applied to the constraints

$$\begin{aligned} Ax + Iy &= b \\ x \geq 0, y &\geq 0 \end{aligned}$$

To show that (A, I) is totally unimodular, take an arbitrary square and nonsingular submatrix P of (A, I). If P does not contain columns of I, then $|\det P| = 1$, since A is totally unimodular. If P does not contain columns of A, then clearly $|\det P| = 1$. Suppose then that

$$P = (\bar{A}, \bar{I})$$

where $\bar{A}$ is a submatrix of A and $\bar{I}$ is a submatrix of I. Permute the rows of P so that

$$\bar{I} = \begin{bmatrix} 0 \\ \hat{I} \end{bmatrix} \qquad \text{and} \qquad \bar{A} = \begin{bmatrix} \bar{A}_1 \\ \bar{A}_2 \end{bmatrix}$$

yielding

$$P' = \left[\begin{array}{c|c} \bar{A}_1 & 0 \\ \hline \bar{A}_2 & \hat{I} \end{array}\right]$$

Note that $|\det P'| = |\det P|$ and

$$|\det P'| = |\det \bar{A}_1| \cdot |\det \hat{I}| = |\det \bar{A}_1|$$

Now A totally unimodular implies

$$|\det \bar{A}_1| = 0, 1$$

and since P is assumed to be nonsingular, $|\det P'| = 1$.

The necessity of total unimodularity is, of course, equivalent to $2 \Rightarrow 1$. To prove $2 \Rightarrow 1$, we prove $2 \Rightarrow 3$ and then $3 \Rightarrow 1$.

($2 \Rightarrow 3$): Let B be a basis matrix for $Ax + Iy = b$ and $\hat{b}_i$ be the ith column of B^{-1}. Our first objective is to prove that $\hat{b}_i$ is an integer vector. Let t be any integer vector such that $t + \hat{b}_i \geq 0$ and $b(t) = Bt + e_i$, where e_i is the ith unit vector. Consider $Ax \leq b(t)$. Note that

$$B^{-1}b(t) = B^{-1}(Bt + e_i) = t + B^{-1}e_i = t + \hat{b}_i \geq 0$$

Thus $(x_B, x_N) = (t + \hat{b}_i, 0)$ is an extreme point of $S(b(t))$; it follows from statement 2 of the theorem that $t + \hat{b}_i$ is an integer vector. Since t is an integer vector, $\hat{b}_i$ is an integer vector. Thus B^{-1} is an integer matrix. We have proved that every square submatrix of A that is a basis matrix has an integer inverse.

Now consider F, an arbitrary square nonsingular submatrix of A. Using F and appropriate columns from I the basis matrix

$$B = \left[\begin{array}{c|c} F & 0 \\ \hline F_1 & I \end{array}\right]$$

can be constructed. Note that B is a basis matrix since $|\det B| = |\det F| \neq 0$, by hypothesis. Furthermore,

$$B^{-1} = \left[\begin{array}{c|c} F^{-1} & 0 \\ \hline -F_1F^{-1} & I \end{array}\right]$$

However B^{-1} is an integer matrix, which implies that F^{-1} is also.

(3 ⇒ 1): Let F be an arbitrary square, nonsingular submatrix of A. From 3, F^{-1} is an integer matrix. Therefore, $|\det F|$ and $|\det F^{-1}|$ are integers. But

$$|\det F| \cdot |\det F^{-1}| = |\det FF^{-1}| = 1$$

which implies $|\det F| = |\det F^{-1}| = 1$. ■

Theorems 1 and 2 are important in integer programming because any ILP with a totally unimodular constraint matrix can be solved as an LP. Specifically, if A is totally unimodular, the ILP

$$\begin{aligned} &\max cx \\ &\quad Ax = b \\ &\quad x \geq 0 \text{ integer} \end{aligned} \tag{3}$$

can be solved by solving the corresponding LP

$$\begin{aligned} &\max cx \\ &\quad Ax = b \\ &\quad x \geq 0 \end{aligned} \tag{4}$$

If x^* is an optimal basic solution to (4), then x^* is an optimal solution to (3). In this sense, the integrality constraints of (3) are superfluous.

An obvious necessary condition for A to be totally unimodular is $a_{ij} = 0, 1, -1$ for all i and j. However, in general, it is not easy to tell whether a matrix consisting only of $(0, +1, -1)$ is totally unimodular. Theorem 3 gives a sufficient condition that is extremely useful for ILP's associated with graphs. It will be applied in Section 3.4.

Theorem 3: *An integer matrix A with $a_{ij} = 0, 1, -1$, for all i and j, is totally unimodular if*

1. *No more than two nonzero elements appear in each column.*
2. *The rows can be partitioned into two subsets Q_1 and Q_2 such that*
   (a) *If a column contains two nonzero elements with the same sign, one element is in each of the subsets.*
   (b) *If a column contains two nonzero elements of opposite sign, both elements are in the same subset.*

PROOF: The proof is by induction. Clearly, any one element submatrix of A has a determinant equal to $(0, 1, -1)$. Assume that the theorem is true for all submatrices of A of order $k - 1$ or less. Consider C an arbitrary kth order submatrix. If C contains a null vector, $\det C = 0$. If C contains a column with only one nonzero element, we expand $\det C$ by that column and apply the induction hypothesis. Finally, consider the case in which every column of C contains two nonzero elements. Note that from 2a and 2b, for each column j

$$\sum_{i \in Q_1} a_{ij} = \sum_{i \in Q_2} a_{ij}, \qquad j = 1, \ldots, k \tag{5}$$

Let r_i be the ith row of C. From (5), $\sum_{i \in Q_1} r_i - \sum_{i \in Q_2} r_i = 0$, which implies $\det C = 0$. ■

Note that an incidence matrix of a directed graph satisfies the condition of Theorem 3. Every column has one $(+1)$ and one (-1). Thus Condition 1 is satisfied and Condition 2 is obviously fulfilled with Q_1 the index set of the rows, and $Q_2 = \varnothing$ (empty set).

In general the conditions of Theorem 3 are not necessary for total unimodularity. However, it has been proved that for the class of matrices which satisfy Condition 1, Condition 2 is necessary (see Exercise 8).

3.4. SOME TOTALLY UNIMODULAR LP'S ON GRAPHS

The assignment problem, introduced in Section 3.1, has an A matrix that satisfies the conditions of Theorem 3. Instead of proving this result specifically for the assignment problem, it will be shown that the assignment problem belongs to a class of LP's, associated with graphs, which have A matrices satisfying the conditions of Theorem 3. Problems in this class are called *minimum cost capacitated flow problems.*

Formulation of the Minimum Cost Capacitated Flow Problem

Consider the problem of sending quantities of a commodity, say trucks or oil, from a set of origins V_1, through a network, to a set of destinations V_3. For each $i \in V_1$, there is a supply of the commodity denoted by a_i and, for each $i \in V_3$, there is a demand for the commodity denoted by b_i. There is a cost per unit c_{ij} associated with the "flow" of the commodity through link (i, j), and a capacity d_{ij} (possibly infinite) that bounds the flow through the link. The problem of determining appropriate flows, which minimize total cost, is the minimum cost capacitated flow problem.

More precisely, the network is represented by a directed graph $G = (V, E)$, where $V = \{1, \ldots, m\}$. Let V be partitioned into V_1 (origins, sources), V_2 (intermediate points) and V_3 (destinations, sinks). For each $i \in V$, define

$$V(i) = \{j \mid (i, j) \in E\} \qquad \text{and} \qquad V'(i) = \{j \mid (j, i) \in E\}$$

A *flow* is defined by a set of numbers x_{ij}, $(i, j) \in E$, satisfying

$$\sum_{j \in V(i)} x_{ij} - \sum_{j \in V'(i)} x_{ji} \begin{cases} \leq a_i, & i \in V_1 \\ = 0, & i \in V_2 \\ \leq -b_i, & i \in V_3 \end{cases} \tag{6}$$

and

$$0 \leq x_{ij} \leq d_{ij}, \qquad (i, j) \in E \tag{7}$$

Note that (6) requires the conservation of flow at intermediate points, a net flow into sinks at least as great as demand, and a net flow out of sources equal to or less than the supply. In some applications, demand must be satisfied exactly and all of the supply must be used. If all of the constraints of (6) are equalities, the problem has no feasible solution unless

$$\sum_{i \in V_1} a_i = \sum_{i \in V_3} b_i$$

The cost of a flow satisfying (6) and (7) is given by

$$\sum_{i \in V} \sum_{j \in V(i)} c_{ij} x_{ij} \tag{8}$$

The minimum cost capacitated flow problem is to minimize (8), subject to (6) and (7). It is assumed that if $((i_1, i_2), \ldots, (i_k, i_1))$ is a cycle in G, then either

(i) $c_{i_1 i_2} + \cdots + c_{i_k i_1} \geq 0$, or
(ii) $\min \{d_{i_1 i_2}, \ldots, d_{i_k i_1}\} < \infty$

If neither (i) nor (ii) holds, an arbitrarily large flow over the cycle would yield an arbitrarily small cost.

Given that the a_i, b_i, and d_{ij} are integers, our objective is to prove that every basic solution to (6) and (7) is integer.

Some Special Cases

Before proving the main result regarding integer flows, some special cases of the minimum cost capacitated flow problem, which have received considerable attention, are discussed. First, there is the uncapacitated version in which $d_{ij} = \infty$ for all $(i, j) \in E$. This problem is sometimes called the *transshipment problem*.

Perhaps, the best-known case is the (capacitated) *transportation problem*. The transportation problem has $V_2 = \varnothing$, $V'(i) = \varnothing$, $i \in V_1$, and $V(i) = \varnothing$, $i \in V_3$. The capacitated transportation problem can be written as

$$\begin{aligned} \min \sum_{i \in V_1} \sum_{j \in V(i)} & c_{ij} x_{ij} \\ \sum_{j \in V(i)} x_{ij} &\le a_i, \quad i \in V_1 \\ \sum_{j \in V'(i)} x_{ji} &\ge b_i, \quad i \in V_3 \\ 0 \le x_{ij} &\le d_{ij} \quad \text{for all } (i, j) \in E \end{aligned} \tag{9}$$

Consider the special case of (9) in which $a_i = 1$ for all $i \in V_1$, $b_i = 1$ for all $i \in V_3$, $d_{ij} = \infty$ for all $(i, j) \in E$, and $|V_1| = |V_3|$. Note that $|V_1| = |V_3|$ implies that all of the constraints must be satisfied with equality. Therefore, this special case of (9) reduces to

$$\begin{aligned} \min \sum_{i \in V_1} \sum_{j \in V(i)} & c_{ij} x_{ij} \\ \sum_{j \in V(i)} x_{ij} &= 1, \quad i \in V_1 \\ \sum_{j \in V'(i)} x_{ji} &= 1, \quad i \in V_3 \\ x_{ij} &\ge 0 \quad \text{for all } (i, j) \in E \end{aligned} \tag{10}$$

The problem (10) is identical to the *assignment problem* introduced in Section 3.1, except that in the assignment problem $x_{ij} \ge 0$ is replaced by $x_{ij} = 0, 1$. However, because all basic solutions of (10) are integer, the assignment problem is solved by any basic optimal solution to the LP (10).

Still another important problem that can be obtained from (6), (7), and (8) is the *maximum flow problem*. In this problem, $V_1 = \{1\}$, $V_3 = \{m\}$, $V'(1) = \varnothing$, $V(m) = \varnothing$, $a_1 = \infty$, and $b_m = 0$. The capacity constraints are the essential factor. The problem is to maximize the total flow into vertex m. This objective can be stated by taking $c_{im} = -1$ for all $i \in V'(m)$ and $c_{ij} = 0$, otherwise. Thus the maximum flow problem is

$$\begin{aligned} \max \sum_{i \in V'(m)} x_{im} \\ \sum_{j \in V(i)} x_{ij} - \sum_{j \in V'(i)} x_{ji} = 0, \qquad i \in V_2 = \{2, \ldots, m-1\} \\ 0 \leq x_{ij} \leq d_{ij} \qquad \text{for all } (i, j) \in E \end{aligned} \tag{11}$$

Finally, consider the *shortest path problem*. Let c_{ij} be interpreted as the length of edge (i, j). Define the length of a path in G to be the sum of the edge lengths over all edges in the path. The objective is to find a path from vertex 1 to vertex m of minimum length. It is assumed that all cycles have nonnegative length. This problem is a special case of the transshipment problem in which $V_1 = \{1\}$, $V_3 = \{m\}$, $a_1 = 1$, and $b_m = 1$. Clearly, the demand at vertex m will be satisfied with equality by exactly one unit of flow from vertex 1. The capacity constraints $d_{ij} = 1$ are implicitly accounted for, since $a_1 = 1$, and no negative cycles imply that there are optimal solutions in which $x_{ij} \leq 1$ for all $(i, j) \in E$. If the minimal cost flow pattern is integer ($x_{ij} = 0, 1$), the unit of flow must travel over a simple path from vertex 1 to vertex m. It follows that this path is a shortest path from vertex 1 to vertex m.

Proof of Total Unimodularity

It has been shown that the minimum cost capacitated flow problem encompasses a wide variety of significant problems. Theorem 4 establishes that all of the basic solutions to (6) and (7) are integer.

Theorem 4: *The constraint matrix corresponding to* (6) *and* (7) *is totally unimodular.*

PROOF: Let the constraint matrix be $A' = \begin{pmatrix} A \\ I \end{pmatrix}$, where A is the matrix for (6) and I is an identity matrix for (7). In the proof of Theorem 2, it was shown that C totally unimodular implies (C, I) totally unimodular. It is also clear that the transpose of a totally unimodular matrix is totally unimodular. Thus it suffices to establish the total unimodularity of A.

Each variable x_{ij} appears in exactly two constraints of (6). Specifically, x_{ij} has a coefficient of $+1$ in the vertex i equation and a coefficient of -1 in the vertex j equation. Thus A is an incidence matrix for a directed graph. It follows that A satisfies the conditions of Theorem 3. ■

Corollary: *The constraint matrices for the transshipment, transportation, assignment, maximum flow, and shortest path problems are totally unimodular.*

As a consequence of Theorem 4 and its corollary, all of the flow problems mentioned in this section can be solved for integer solutions by a simplex method. However, special purpose, graph-oriented algorithms, considerably more efficient than simplex methods, are available to solve them. Except for a shortest path algorithm, these flow algorithms will not be discussed here. Because the shortest path problem is used in certain ILP algorithms (see Section 6.11), an algorithm for it is given in the next section.

3.5. AN ALGORITHM FOR THE SHORTEST PATH PROBLEM

The most efficient algorithms for the shortest path problem only work for the case $c_{ij} \geq 0$ for all $(i, j) \in E$. Since the shortest path problems that will be considered in later chapters have nonnegative edge lengths, the more general case is not considered.

Our objective is to find a shortest path from vertex 1 to vertex m in a directed graph. Define $\theta(i)$ to be the length of a shortest path from vertex 1 to vertex i, $i = 2, \ldots, m$ and $\theta(1) = 0$. Let $(j_1, j_2, \ldots, j_m)$ be a permutation of $(1, 2, \ldots, m)$ such that $j_1 = 1$ and

$$0 = \theta(j_1) \leq \theta(j_2) \leq \cdots \leq \theta(j_m)$$

The algorithm is constructed so that at iteration $k - 1$, j_k and $\theta(j_k)$ are determined. When $j_k = m$, the procedure terminates.

Discussion of the General Step

Let $S_k = \{j_1, j_2, \ldots, j_k\}$, $R_k = V - S_k$, and for all $i \in R_k$,

$$d_k(i) = \begin{cases} \min_{j \in S_k} (\theta(j) + c_{ji}) & \text{if there exists } (j, i) \in E \\ \infty & \text{otherwise} \end{cases} \tag{12}$$

and

$$d_k(i^*) = \min_{i \in R_k} d_k(i) \tag{13}$$

The algorithm to be given is based on

Theorem 5: *$d_k(i^*) = \theta(i^*)$ and $j_{k+1} = i^*$.*

PROOF: The proof of $d_k(i^*) = \theta(i^*)$ is by contradiction. Since $d_k(i^*)$ is the length of some path from vertex 1 to vertex i^*, $d_k(i^*) \geq \theta(i^*)$. Suppose $d_k(i^*) > \theta(i^*)$. Let

$$p = (1, l_1, l_2, \ldots, l_t, i^*)$$

be the vertex sequence of some shortest path from vertex 1 to vertex i^*. From (12), p must go through at least one vertex of R_k besides i^*. Thus there exists a smallest integer q, $1 \le q \le t$, such that $l_q \in R_k$. Consider the path

$$p_1 = (1, l_1, l_2, \ldots, l_{q-1}, l_q)$$

which is contained in the path p. From (12),

$$\text{length of } p_1 \ge d_k(l_q) \tag{14}$$

Furthermore, from $c_{ij} \ge 0$ and (14),

$$\theta(i^*) = \text{length of } p \ge d_k(l_q) \tag{15}$$

But (15) is a contradiction, since $\theta(i^*) < d_k(i^*)$ and $l_q \in R_k$, so that $d_k(i^*) \le d_k(l_q)$ from (13).

From the argument just given, it also follows that

$$\theta(i^*) \le \theta(i) \qquad \text{for all } i \in R_k$$

Thus we can set $i^* = j_{k+1}$. ■

The $d_k(i)$ of (12) can be calculated iteratively. In particular, since $S_k = S_{k-1} \cup \{j_k\}$

$$d_k(i) = \min\{d_{k-1}(i), \theta(j_k) + c_{j_k i}\} \tag{16}$$

for all $i \in R_k$. The algorithm given below is based on (16).

Algorithm

STEP 1: Initialize $k = 1$, $S_k = \{1\}$, $R_k = \{2, \ldots, m\}$, $d_k(1) = 0$, $d_k(i) = c_{1i}$ if $(1, i) \in E$, $d_k(i) = \infty$ if $(1, i) \notin E$. Let $p(i) = 1$, for all $i \ne 1$, where $p(i)$ is an indicator that is used to identify the edges of a shortest path. Go to Step 2.

STEP 2: Let $d_k(i^*) = \min_{i \in R_k} d_k(i)$. If $i^* \ne m$, let $R_{k+1} = R_k - \{i^*\}$ and go to Step 3. If $i^* = m$, let $\theta(m) = d_k(i^*)$ and go to Step 4.

STEP 3: For all $i \in R_{k+1}$, let

$$d_{k+1}(i) = \min\{d_k(i), d_k(i^*) + c_{i^*i}\}$$

If $d_{k+1}(i) = d_k(i^*) + c_{i^*i}$, put $p(i) = i^*$. Let $k = k + 1$ and return to Step 2.

STEP 4: To trace an optimal path, let $j = p(i^*)$ and record (j, i^*) as an edge of a shortest path. If $j = 1$, terminate. If $j \ne 1$, go to Step 5.

STEP 5: Put $i^* = j$ and return to Step 4.

Note that at the last execution of Step 2, $(p(m), m)$ is an edge of a shortest path from vertex 1 to vertex m. Thus Steps 4 and 5 identify the edges of a shortest path. If desired, the $p(i)$ list can be eliminated by a bookkeeping scheme that requires some additional calculation (see Exercise 26).

Example

Consider the graph shown in Figure 6. The c_{ij} are the numbers adjacent to the edges. It is desired to find a shortest path between vertices 1 and 6.

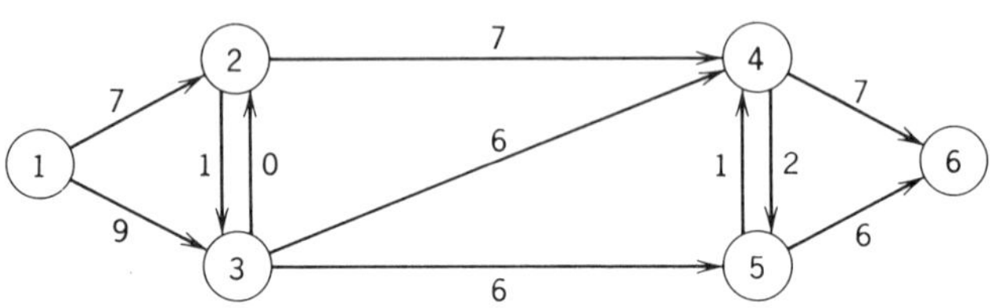

Figure 6

STEP 1: Initialize, $R_1 = \{2, 3, 4, 5, 6\}$, $d_1(1) = 0$, $d_1(2) = 7$, $d_1(3) = 9$, $d_1(4) = d_1(5) = d_1(6) = \infty$, $p(i) = 1$, $i = 2, \ldots, 6$.

STEP 2: $i^* = 2$, $\theta(2) = 7$, $R_2 = \{3, 4, 5, 6\}$.

STEP 3: $d_2(3) = 8$, $p(3) = 2$, $d_2(4) = 14$, $p(4) = 2$, $d_2(i) = d_1(i)$, otherwise.

STEP 2: $i^* = 3$, $\theta(3) = 8$, $R_3 = \{4, 5, 6\}$.

STEP 3: $d_3(5) = 14$, $p(5) = 3$, $d_3(i) = d_2(i)$, otherwise.

*STEP 2: $i^* = 4, 5$, $\theta(4) = \theta(5) = 14$, $R_4 = \{6\}$.

STEP 3: $d_4(6) = 20$, $p(6) = 5$.

STEP 2: $i^* = 6$, $\theta(6) = 20$.

STEP 4: $j = p(6) = 5$, path (5, 6).

STEP 5: $i^* = 5$.

STEP 4: $j = p(5) = 3$, path (3, 5, 6).

STEP 5: $i^* = 3$.

STEP 4: $j = p(3) = 2$, path (2, 3, 5, 6).

STEP 5: $i^* = 2$.

STEP 4: $j = p(2) = 1$, path (1, 2, 3, 5, 6).

A shortest path is (1, 2, 3, 5, 6) of length 20. Note that in the starred execution of Step 2, there are alternative solutions to (13). In this case, as shown in the example, all of the alternative solutions can be used in one iteration. Although the algorithm only finds one shortest path, it is easily modified to find all shortest paths from vertex 1 to vertex m (see Exercise 26).

3.6. MATCHING AND COVERING ON GRAPHS

The Matching Problem

Given a (undirected) graph $G = (V, E)$ and a subset $S \subseteq E$, we define the *degree* of vertex i with respect to S as

$$d_S(i) = \text{number of edges of } S \text{ incident to } i$$

A subset $M \subseteq E$ is called a *matching* of G if $d_M(i) \leq 1$ for all $i \in V$. Every graph trivially contains a matching, since $d_\varnothing(i) = 0$ for all $i \in V$. Without loss of generality, it is assumed throughout this section that G is connected; if it is not, all of the results apply separately to the components of G.

A matching M^* is a *maximum matching* if

$$|M^*| = \max \{|M|, M \text{ a matching}\}$$

More generally, let weight c_j be assigned to edge e_j and define the weight of a matching M to be

$$w(M) = \sum_{e_j \in M} c_j$$

A matching M^* is called a *maximum weighted matching* if

$$w(M^*) \geq w(M) \qquad \text{for all matchings } M$$

Another generalization is to allow $d_M(i) \leq b_i$, where b_i is a positive integer. Given the vector $b = (b_1, \ldots, b_m)$, M is called a *b-matching* if $d_M(i) \leq b_i$ for all $i \in V$.

Let c and x be $|E|$-vectors, where c_j is the weight of e_j and

$$x_j = \begin{cases} 1 & \text{if } e_j \text{ is in the } b\text{-matching} \\ 0 & \text{otherwise} \end{cases}$$

The problem of finding a *maximum weighted b-matching* is the ILP,

$$\begin{aligned} \max\ & cx \\ & Ax \leq b \\ & x \text{ binary} \end{aligned} \tag{17}$$

where A is the incidence matrix of G. When $b = (1, \ldots, 1)$, (17) specializes to the problem of maximum weighted matching and when, in addition, $c = (1, \ldots, 1)$, to maximum matching.

In general, the integrality requirements in (17) are not superfluous. Consider the maximum matching problem associated with the graph of Figure 7. Clearly, each of the three matchings $\{e_1\}$, $\{e_2\}$, and $\{e_3\}$ is optimal.

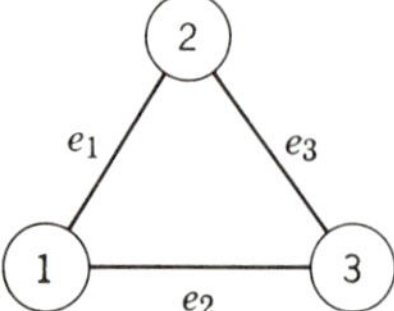

Figure 7

The ILP (17) for the graph of Figure 7 is

$$\begin{aligned}\max x_1 + x_2 + x_3 \\ x_1 + x_2 \qquad \le 1 \\ x_1 \qquad + x_3 \le 1 \\ x_2 + x_3 \le 1 \\ x_1, x_2, x_3 = 0, 1\end{aligned} \tag{18}$$

The unique optimal solution to the LP corresponding to (18), in which $x_j = 0, 1$ is replaced by $x_j \ge 0$, is $x_1 = x_2 = x_3 = \frac{1}{2}$, and the optimal basis matrix is

$$B = \begin{bmatrix} 1 & 1 & 0 \\ 1 & 0 & 1 \\ 0 & 1 & 1 \end{bmatrix}$$

Note that $|\det B| = 2$.

The example shows that the constraint matrix for a matching problem is not, in general, totally unimodular. However, as will be established later in this section, all basic feasible solutions to the LP corresponding to a matching problem are such that $x_j = 0, \frac{1}{2}, 1$ and $|\det B| = 2^k$, $k = 0, 1, \ldots$.

Matching on a Bipartite Graph

A graph $G = (V, E)$ is called *bipartite* if there exist sets V_1 and V_2 such that

$$V_1 \cup V_2 = V, \qquad V_1 \cap V_2 = \varnothing \tag{19}$$

and every edge of G is incident to one vertex of V_1 and one vertex of V_2. For bipartite graphs the matrix A of (17) is totally unimodular, since the partition of the vertex set given by (19) is equivalent to the row partition of Theorem 3.

The assignment problem can be interpreted as a weighted matching problem on a bipartite graph. Let $V_1 = \{\text{set of men}\}$, $V_2 = \{\text{set of jobs}\}$ and $|V_1| = |V_2|$. If man i can be assigned to job j, there is an edge e_k with weight c_k, joining $i \in V_1$ and $j \in V_2$. The weight c_k is equal to the value of assigning man i to job j. There is no loss of generality in assuming that $c_k > 0$, for all k, since a suitably large positive number can always be added to all of the weights, without changing the set of optimal assignments. Thus a maximum weighted matching assigns one man to each job and corresponds to an optimal solution to the assignment problem. It is also easy to establish the converse that a maximum weighted matching problem on a bipartite graph can be formulated as an assignment problem (see Exercise 15).

The Covering Problem

A subset of edges C is called a *cover* of $G = (V, E)$ if $d_C(i) \geq 1$ for all $i \in V$. A cover C^* is called a *minimum cover* if

$$|C^*| = \min \{|C|, C \text{ a cover}\}$$

Minimum weighted covers and *minimum* (*weighted*) *b-covers* are similarly defined. For the graph of Figure 7, any two edges constitute a minimum cover.

The problem of finding a minimum weighted b-cover is the ILP,

$$\begin{aligned} \min\ & cx \\ & Ax \geq b \\ & x \text{ binary} \end{aligned} \tag{20}$$

where A, b, c, and x are defined as in (17).

Relation between Covering and Matching

In the simple case, $b = (1, \ldots, 1)$ and $c = (1, \ldots, 1)$, minimum covers and maximum matchings are closely related. A graph, a minimum cover, and a maximum matching are shown in Figure 8. Note that only vertex 4 has $d_{C^*}(i) > 1$ and M^* is obtained from C^* by deleting all but one of the edges incident with vertex 4. This procedure works in general since

Theorem 6: *Let C^* be a minimum cover. For every vertex i having $d_{C^*}(i) > 1$, remove $d_{C^*}(i) - 1$ of the edges incident to it. The resulting set of edges M^* is a maximum matching.*

PROOF: Since $d_{M^*}(i) \leq 1$ for all $i \in V$, M^* is a matching. To prove that M^* is maximum, let p be the number of deleted edges so that $|M^*| =$

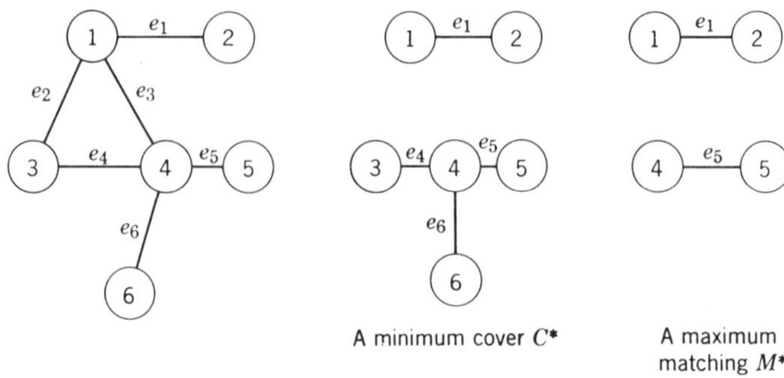

A minimum cover C^*

A maximum matching M^*

Figure 8

$|C^*| - p$. Because C^* is minimum, every deleted edge e_k of C^* is such that, for exactly one $i \in V$, say $i = i'$,

$$d_{C^*}(i') = 1 \qquad \text{and} \qquad d_{C^* - \{e_k\}}(i') = 0$$

This implies that there are exactly p vertices of V such that $d_{M^*}(i) = 0$.

Assume that M^* is not maximum; then there exists a matching M' with $|M'| = |C^*| - p + 1$. Note that $d_{M'}(i) = 0$ for exactly $p - 2$ vertices of V. Therefore $q \leq p - 2$ edges can be added to M' to obtain a cover C', with $|C'| \leq |M'| + p - 2 = |C^*| - 1$, contradicting the minimality of C^*. ■

There is an analogous theorem (Theorem 7) for converting a maximum matching into a minimum cover. The proof is asked for in Exercise 24.

Theorem 7: *Let M^* be a maximum matching. Add one edge incident to every vertex such that $d_{M^*}(i) = 0$. The resulting set of edges is a minimum cover.*

It follows from Theorems 6 and 7 that it suffices to consider either covering or matching in the case $b = (1, \ldots, 1)$ and $c = (1, \ldots, 1)$. For the remainder of this section, only matching is considered.

Augmenting Paths

Relative to a set $S \subseteq E$, an *alternating path* in G is a simple path whose edges alternate between S and $E - S$. Relative to a matching M, an *augmenting path* is an alternating path connecting vertices i and k such that $d_M(i) = d_M(k) = 0$. Note that an augmenting path must contain an odd number of edges. Figure 9 shows a matching $M = \{e_2\}$, an augmenting path $p = (e_1, e_2, e_3)$, and an improved matching $M' = \{e_1, e_3\}$. Edges in the matchings are denoted by thick lines. Certainly, if an augmenting path exists, a better matching can be obtained. But even more strongly,

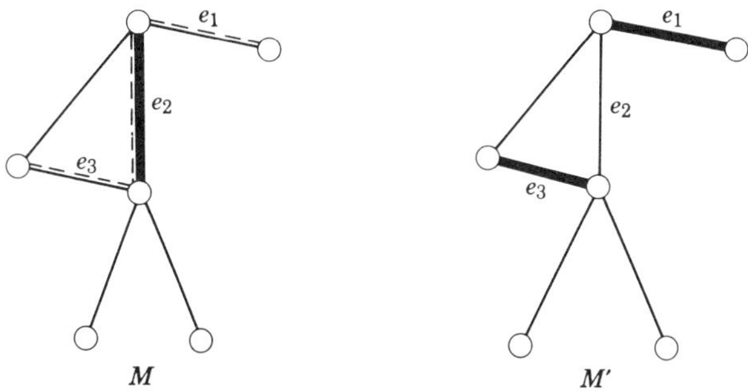

Figure 9

Theorem 8: *A matching M is maximum if and only if there is no augmenting path relative to M.*

PROOF: ($\Rightarrow$, only if) Suppose that there is an augmenting path $p = (e_1, e_2, \ldots, e_{2k+1})$. Let

$$M' = M \cup \{e_1, e_3, \ldots, e_{2k+1}\} - \{e_2, e_4, \ldots, e_{2k}\}$$

or

$$M' = (M \cup P) - (M \cap P)$$

where $P = \{e_j | e_j \text{ is in the augmenting path } p\}$. Clearly, M' is a matching and $|M'| = |M| + 1$.

($\Leftarrow$): Assume that there exists a matching M^* such that $|M^*| = |M| + 1$. Let

$$D = (M \cup M^*) - (M \cap M^*)$$

It will be shown that D contains an augmenting path relative to M.

Since $|D| = |M| + |M^*| - 2|M \cap M^*|$, it follows that $|D|$ is odd. Now D and the vertices of G incident to D form a subgraph $G^* = (V^*, D)$ which may not be connected. Let $G' = (V', D')$ be a component of G^*. We claim that G' is either a simple path or a cycle, as shown in Figure 10. Since all elements of D' are from M or M^*, no vertex of V' can be incident to more than two edges of D'. Furthermore, the edges of D' must alternate between M and M^*. Now if D' forms a cycle in G', as in case (c), it follows that $|D'|$ is even; otherwise one vertex would be incident to two edges of M or M^*. Thus $|D'|$ odd implies that D' is a simple path, and therefore an augmenting path with respect to M or M^*. Finally, because $|D|$ is odd and $|M^*| = |M| + 1$, case (a), with D' being an augmenting path with respect

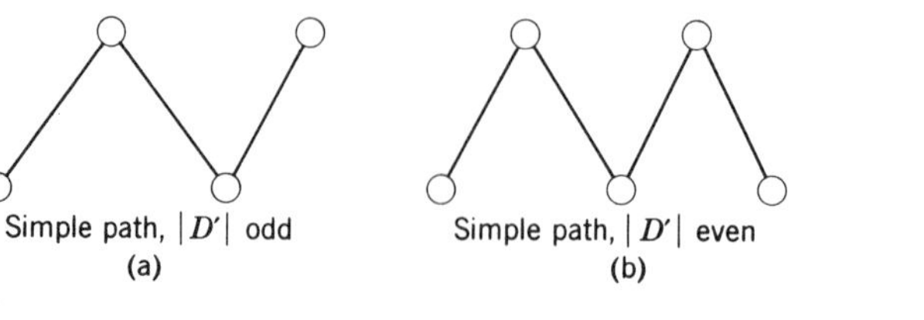

Figure 10

to M, must occur for some G' (see Exercise 17). ■

Example

A maximum matching problem on a small graph is solved in Figure 11. The search for augmenting paths is by inspection. Initially, $M_0 = \varnothing$. Clearly, e_4 is an augmenting path. Put $M_1 = M_0 \cup \{e_4\} = \{e_4\}$. An augmenting path with respect to M_1 is (e_3, e_4, e_5). Thus $M_2 = M_1 \cup \{e_3, e_5\} - \{e_4\} = \{e_3, e_5\}$. An augmenting path with respect to M_2 is e_7. Thus $M_3 = \{e_3, e_5, e_7\}$. There are no augmenting paths with respect to M_3, so M_3 is maximum.

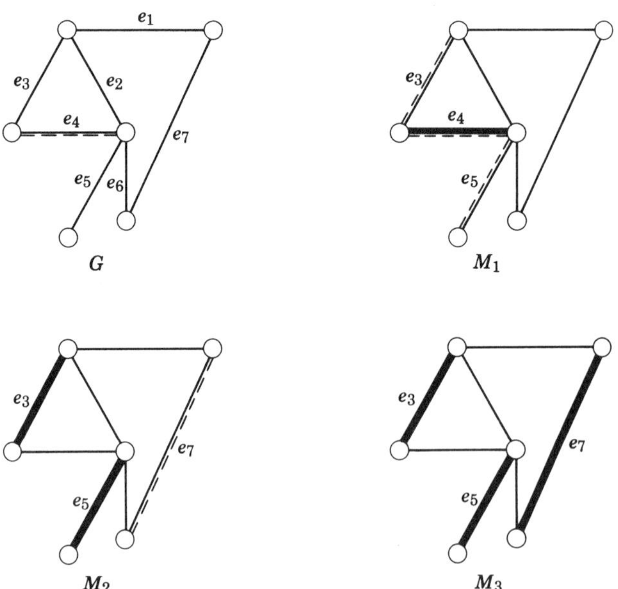

Figure 11

In general, the problem of searching for an augmenting path is not trivial. To develop an efficient procedure for finding one or showing that none exists, it is necessary to introduce some additional concepts.

Odd Cycles

A cycle in G is called *odd* if it contains an odd number of edges. Odd cycles are important in explaining the relationship between the ILP formulation of the matching problem and its corresponding LP, and also in discovering augmenting paths.

Theorem 9: *A graph G is bipartite if and only if it has no odd cycles.*

PROOF: ($\Rightarrow$) Suppose G is bipartite and there is an odd cycle with the vertex sequence $(i_1, i_2, \ldots, i_k, i_1)$. Since G is bipartite, there exist V_1, V_2 such that

$$V_1 \cup V_2 = V, \qquad V_1 \cap V_2 = \varnothing$$

and every edge of G is incident to one vertex of V_1 and another of V_2. Arbitrarily put $i_1 \in V_1$, which implies $i_j \in V_1$, j odd, and $i_j \in V_2$, j even, $j = 1, \ldots, k$. Then $i_k \in V_1$ implies $i_1 \in V_2$, which is a contradiction.

($\Leftarrow$): Assume that there are no odd cycles. For a path p connecting vertices i and j, define $d_{ij}(p)$ to be the number of edges in p. Let

$$d_{ij} = d_{ij}(p_{ij}) = \min\,(d_{ij}(p), p \text{ a path connecting } i \text{ and } j)$$

Select any vertex $i^* \in V$ and define

$$V_1 = \{i^*\} \cup \{k \mid d_{i^*k} \text{ even}\}, \qquad V_2 = V - V_1$$

If G is not bipartite, then there exist vertices i and j in either V_1 or V_2, and $e_k = (i, j)$. Without loss of generality assume $i, j \in V_1$. Consider the paths p_{ji^*} and p_{i^*i} shown in Figure 12a. If these two paths have no common vertices except i^*, $c = (e_k, p_{ji^*}, p_{i^*i})$ is an odd cycle, since it contains $d_{ji^*} + d_{i^*i} + 1$ edges and d_{ji^*} and d_{i^*i} are both even.

Suppose p_{ji^*} and p_{i^*i} intersect for the first time at i'. We consider the case of $i' \in V_2$, as shown in Figure 12b. (The same argument applies if $i' \in V_1$.) The path $c' = (e_k, p_{ji'}, p_{i'i})$ is a cycle in G. To show that c' is an odd cycle, note that $p_{ii^*} = (p_{ii'}, p_{i'i^*})$ and $p_{ji^*} = (p_{ji'}, p_{i'i^*})$. Since $i' \in V_2$, it follows that $d_{i^*i'}$ is odd, which implies that $d_{ji'} = d_{ji^*} - d_{i^*i'}$ and $d_{ii'} = d_{ii^*} - d_{i^*i'}$ are both odd numbers. Thus the assumption that e_k exists must be false, since G contains no odd cycles. ■

Theorem 9 demonstrates that the difficulty in solving a matching problem by linear programming is caused by odd cycles in G. Specifically, if

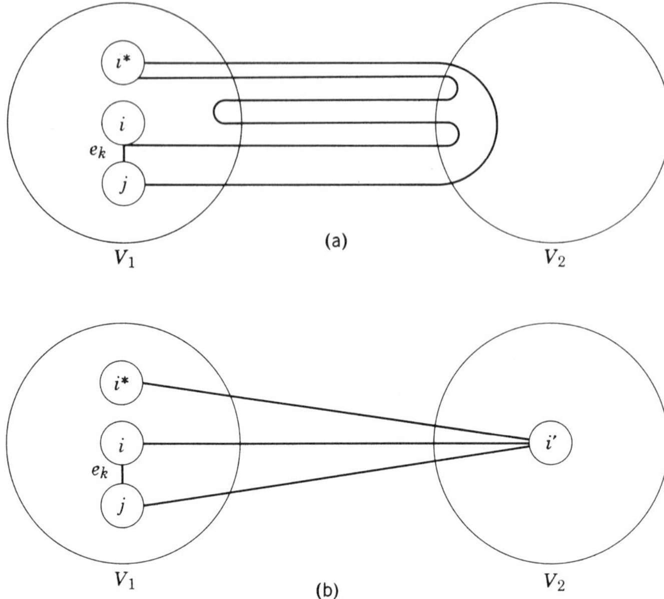

Figure 12

$(e_1, e_2, \ldots, e_{2k+1})$ is an odd cycle with vertex sequence $(i_1, i_2, \ldots, i_{2k+1}, i_1)$, the constraints in (17) corresponding to $i_j, j = 1, \ldots, 2k + 1$, are satisfied by

$$x_1^0 = x_2^0 = \cdots = x_{2k+1}^0 = \tfrac{1}{2} \tag{21}$$

Note that (21) implies

$$\sum_{j=1}^{2k+1} x_j^0 = k + \tfrac{1}{2}$$

However, in a matching, no more than k edges from an odd cycle of $2k + 1$ edges can be used. Thus every feasible integer solution to (17) with $b = b^o = (1, \ldots, 1)$ satisfies

$$\sum_{j=1}^{2k+1} x_j \leq k \tag{22}$$

where $(e_1, e_2, \ldots, e_{2k+1})$ is an odd cycle.

The LP corresponding to (17) with $b = b^o$ is

$$\begin{aligned} \max\ & cx \\ Ax &\leq b^o \\ x &\geq 0 \end{aligned} \tag{23}$$

Solutions to (23) which assign a value of one-half to each edge in an odd cycle, as in (21), are by no means unusual. The occurrence of these solutions is justified by

Theorem 10: *Let B be a basis matrix for a basic feasible solution to $Ax + s = b^{\circ}$, $x \geq 0$, $s \geq 0$, from* (23). *Each component of $x_B = B^{-1}b^{\circ}$ is* $(0, \frac{1}{2}, 1)$.

PROOF: Let $A' = (A, I)$ and $B = (C, U)$ where the columns of C are a subset of the columns of A and the columns of U are a subset of the columns of I. The columns of C represent an edge set $E_c \subseteq E$ and C is the incidence matrix of the subgraph $G_c = (V, E_c)$. Let G_c^t, $t = 1, \ldots, T$, be the components of G_c. By appropriately numbering the vertices of G, C can be written as

$$C = \begin{bmatrix} C^1 & 0 & \cdots & 0 \\ 0 & C^2 & \cdots & 0 \\ \vdots & \vdots & & \vdots \\ 0 & 0 & \cdots & C^T \end{bmatrix}$$

If we let $x_B = (x_B^1, \ldots, x_B^T)$ and

$$U = \begin{bmatrix} U^1 & 0 & \cdots & 0 \\ 0 & U^2 & \cdots & 0 \\ \vdots & \vdots & & \vdots \\ 0 & 0 & \cdots & U^T \end{bmatrix}$$

$Bx_B = b^{\circ}$ decomposes into

$$H^t x_B^t = b_0^t, \qquad t = 1, \ldots, T$$

where $H^t = (C^t, U^t)$ and $b_0^t = (1, \ldots, 1)$. Thus the proof reduces to showing that

$$x_{B_i}^t = (H^t)_i^{-1} b_0^t = 0, \tfrac{1}{2}, 1 \tag{24}$$

where C^t is the incidence matrix for a connected graph, U^t is a matrix of unit vectors and $(H^t)_i^{-1}$ is the ith row of $(H^t)^{-1}$.

We will prove that $|\det H^t| = 1, 2$, which implies that every element of $(H^t)^{-1}$ is an integer or an integer divided by 2. Then from $0 \leq x_{B_i} \leq 1$ in (23), (24) follows.

The proof of $|\det H^t| = 1, 2$ is by induction on the order of H^t. Note that B nonsingular implies that H^t is nonsingular. If H^t has only one element, clearly $\det H^t = 1$. Assume $|\det H^t| = 1, 2$ if H^t is of order $p - 1$ or less and consider H^t of order p. If H^t contains a row or column that is a unit vector, we expand det H^t by that column or row and apply the induction hypothesis.

The remaining case is the one in which no row or column contains fewer than two 1's. Thus $H^t = C^t$ and every column and row of C^t contains exactly two 1's. Since C^t is the incidence matrix for the connected graph G_c^t, by an appropriate numbering of its vertices and edges, we obtain

$$C^t = \begin{bmatrix} 1 & 0 & \cdots & 0 & 1 \\ 1 & 1 & \cdots & 0 & 0 \\ 0 & 1 & \cdots & 0 & 0 \\ \vdots & \vdots & & \vdots & \vdots \\ 0 & 0 & \cdots & 1 & 1 \end{bmatrix}$$

where $c_{11}^t = c_{1p}^t = 1$, $c_{1j}^t = 0$, otherwise

$$c_{ii}^t = c_{i,i-1}^t = 1,\ c_{ij} = 0, \text{ otherwise}, \qquad i = 2, \ldots, p$$

Thus (see Exercise 2) G_c^t is a cycle.

Let

$$Q = \{i(1), \ldots, i(p) | i(1), \ldots, i(p) \text{ is a permutation of } 1, \ldots, p\}$$

and recall that

$$|\det C^t| = \sum_Q \pm \prod_{j=1}^{p} c_{i(j),j}^t$$

Clearly

$$\prod_{j=1}^{p} c_{i(j),j}^t = \begin{cases} 1 & c_{i(j),j}^t = 1 \text{ for all } j \\ 0 & \text{otherwise} \end{cases}$$

Consider how $\prod_{j=1}^{p} c_{i(j),j}^t = 1$ can be achieved. The only two cases are:

$[i(1) = 1]$: $i(1) = 1 \Rightarrow i(p) = p \Rightarrow i(j) = j, j = p - 1, \ldots, 2$

$[i(1) = 2]$: $i(1) = 2 \Rightarrow i(j) = j + 1, j = 2, \ldots, p - 1 \Rightarrow i(p) = 1$

Since there are exactly two nonzero products and C^t is nonsingular, $|\det C^t| = 2$. ■

Theorem 10 explains why (23) may have basic feasible solutions with basic variables equal to one-half. The proof of Theorem 10 also explains why the one-halves always occur in odd cycles. A determinant of 2 must come from a cycle. The cycle must be odd, since graphs without odd cycles have totally unimodular incidence matrices (see Theorem 9).

The constraints (22) can be used to exclude solutions to (23) that assign a value of one-half to each edge in an odd cycle. However, by adding the constraints (22) to (23), new fractional extreme points can be introduced (see Exercise 19).

Let

$$E(V') = \{e_j | e_j \text{ incident to two vertices in } V',\ V' \subseteq V\}$$

and consider the constraint

$$\sum_{e_j \in E(V')} x_j \leq k \tag{25}$$

where $|V'|$ is odd and $k = (|V'| - 1)/2$. Constraints of the form (25) are clearly satisfied by every matching. Furthermore, every constraint of the form (22) is implied by a constraint (25), since $E(V')$ contains the edges of an odd cycle with vertices V'. Thus the constraints (25) can be used to exclude odd cycles, but they accomplish considerably more since

Theorem 11: *The convex hull of* $\{x | Ax \leq b^o,\ x \text{ binary}\}$, *where* A *is the incidence matrix of* G, *is given by the constraints of* (23) *and a constraint* (25) *for each* $V' \subseteq V$ *such that* $|V'| > 1$ *is odd.*

Outline of Proof: Although the details of the proof of Theorem 11 are too involved to be given here, a mode of proof is the following: Let $x_0^o = cx^o$ be the value of a maximum matching and

$$\begin{aligned} y^o = \max \quad & cx \\ & Ax \leq b^o \\ & \sum_{e_j \in E(V')} x_j \leq \frac{(|V'| - 1)}{2} \qquad \text{all } V' \text{ such that } |V'| > 1 \text{ is odd} \\ & x \geq 0 \end{aligned} \tag{26}$$

Since every matching satisfies the constraints (25), it follows that $y^o \geq x_0^o$. Now (26) is an LP; thus from the duality theorem of linear programming, $y^o \leq$ (objective value of every feasible solution to the dual of (26)). Then, to prove that (26) has an integer optimal solution, it is sufficient to show that the dual of (26) has a feasible solution with objective value equal to x_0^o. Since this can be done for arbitrary edge weights, it follows that all basic feasible solutions to (26) are integer. ■

Maximum matching and some generalizations are some of a very small number of ILP's for which the convex hull of feasible solutions has been explicitly identified. Note that Theorem 11 states that the maximum matching problem can be solved using the LP formulation (26).

An algorithm for maximum matching will be developed, using the concepts of augmenting paths and odd cycles. This algorithm is considerably more efficient than solving the LP (26) by a simplex algorithm.

3.7. AN ALGORITHM FOR MAXIMUM MATCHING

For a connected graph $G = (V, E)$ and a matching M, the algorithm to be given is based on a procedure for finding an augmenting path relative to M, or establishing that none exists. From Theorem 8 it follows that if such a procedure is finite, the algorithm will find a maximum matching in a finite number of steps. After "examining" each edge at most once, an augmenting path is either found or it is established that the current matching is maximum. Since $|M^*| \leq |E|$, the procedure is repeated at most $|E|$ times, and thus the number of edge examinations cannot exceed $|E|^2$.

It has been discussed at some length in Section 3.6 that odd cycles in G make the matching problem difficult. First, an algorithm will be presented for finding augmenting paths in bipartite graphs. In showing that this algorithm finds an augmenting path if one exists, the problem caused by odd cycles will be established. The algorithm will then be generalized to treat arbitrary connected graphs. This is done by incorporating a step that identifies odd cycles and eliminates them by considering a "reduced" graph. The matching on the reduced graph is then translated back to a matching on the original graph.

Labeling

Vertex i is said to be *exposed* relative to M if $d_M(i) = 0$ and *unexposed* otherwise. In the algorithm, there are instructions for "labeling" vertices. A label $L(i) = (L_1(i), L_2(i))$ consists of two components; $L_1(i)$ is a sign ($+$ or $-$) and $L_2(i)$ is a vertex number. An exposed vertex i_0 can only be labeled $(+, i_0)$. An unexposed vertex i^* can only be labeled if an adjacent vertex i' has already been labeled. Under specific rules to be given in the algorithm, $L(i^*) = (-L_1(i'), i')$.

Given a matching M, labels are produced so that $L(i_k) = (+, i_{k-1})$, $L(i_{k-1}) = (-, i_{k-2}), \ldots, L(i_{j+1}) = (+, i_j)$ only if $(i_j, i_{j+1}, \ldots, i_{k-1}, i_k)$ is an alternating path relative to M. It then follows that if $L(i_0) = (+, i_0)$, $L(i_1) = (-, i_0)$, $L(i_2) = (+, i_1), \ldots, L(i_{2q}) = (+, i_{2q-1})$, and i_{2q+1} is exposed, then if the edge (i_{2q}, i_{2q+1}) exists, $(i_0, i_1, \ldots, i_{2q}, i_{2q+1})$ is an augmenting path relative to M (see Figure 13, where $q = 2$).

An Algorithm for Maximum Matching on Bipartite Graphs

STEP 1: Choose an arbitrary matching M_0 (possibly, $M_0 = \varnothing$). Let $k = 0$ (k is the iteration counter) and $S_0 = \{i | d_{M_0}(i) = 0\}$. Go to Step 2.

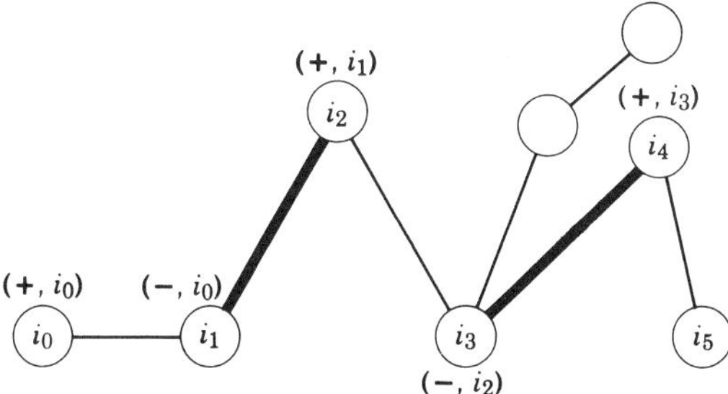

Figure 13

STEP 2: If $|S_k| \leq 1$, go to Step 8. Otherwise, arbitrarily choose $i_0 \in S_k$. Go to Step 3.

STEP 3: Let $L(i_0) = (+, i_0)$ and go to Step 4.

STEP 4: Attempt to find a labeled vertex i' and an edge $e_j = (i', i^*)$ such that e_j has not yet been considered at Step 4 of iteration k and i^* is unlabeled. If such an i' and e_j exist, go to Step 5. Otherwise go to Step 7.

STEP 5: If

(a) $L_1(i') = (+)$ and i^* is exposed; go to Step 6.
(b) $L_1(i') = (+)$ and i^* is unexposed; let $L(i^*) = (-, i')$ and go to Step 3.
(c) $L_1(i') = (-)$ and $e_j \in M_k$; let $L(i^*) = (+, i')$ and go to Step 3.
(d) $L_1(i') = (-)$ and $e_j \notin M_k$; do not label i^*. Go to Step 4.

STEP 6: An augmenting path is given by the vertex sequence $(i_0, i_1, \ldots, i_{2q}, i_{2q+1})$, where $i^* = i_{2q+1}$ and $i' = i_{2q}$; if $L(i_{2q}) = (+, i_{2q})$, then $i_{2q} = i_0$; otherwise $L(i_{2q}) = (+, i_{2q-1})$ and $L(i_{2q-1}) = (-, i_{2q-2})$. If $L(i_{2q-2}) = (+, i_{2q-2})$, then $i_{2q-2} = i_0$; otherwise $L(i_{2q-2}) = (+, i_{2q-3})$ and $L(i_{2q-3}) = (-, i_{2q-4})$, etc. Define a new matching M_{k+1} based upon this augmenting path. Restore any deleted vertices and edges to the graph. Let $k = k + 1$, $S_k = \{i | d_{M_k}(i) = 0\}$; delete all labels and return to Step 2.

STEP 7: Remove from the graph all labeled vertices and every edge incident to at least one labeled vertex. Let $S_k = S_k - \{i_0\}$. Return to Step 2.

STEP 8: Terminate; M_k is a maximum matching.

Theorem 12: *The algorithm given above finds a maximum matching for a bipartite graph.*

PROOF: First it will be shown that the vertex sequence $p = (i_0, i_1, \ldots, i_{2q}, i_{2q+1})$ identified in Step 6 must be an augmenting path. Since an unlabeled vertex can only be labeled from an adjacent vertex that has already been labeled, the set of labeled vertices and the edges used to label them form a subgraph of G that is a tree (see Exercise 25). Thus p is a simple path. Furthermore, $(i_0, i_1) \notin M_k$, by the choice of i_0 in Step 2 and $(i_{2q}, i_{2q+1}) \notin M_k$, by Step 5, rule (a). It is also clear that p must be alternating with respect to M_k because $L_1(i_t) = -L_1(i_{t-1})$, $t = 1, \ldots, 2q$. By rules (b) and (c) of Step 5, $L_1(i_t) = (+)$ implies $(i_{t-1}, i_t) \in M_k$ and $L_1(i_t) = (-)$ implies $(i_{t-1}, i_t) \in E - M_k$.

Suppose that there is an augmenting path with respect to M_k and the algorithm fails to find it. It will first be shown that the deletion of edges and vertices in Step 7 cannot be the cause of failure. In Figure 14, it is assumed that

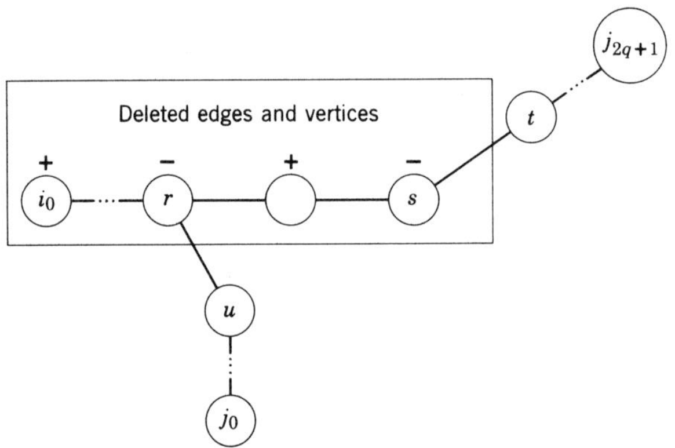

Figure 14

because there is no augmenting path containing the exposed vertex i_0, when i_0 was chosen at Step 2, Step 7 caused the deletion of the edges and vertices contained within the box. Step 7 can only be reached from Step 4 under the condition that every unlabeled vertex i^*, adjacent to a labeled vertex i', is such that $L_1(i') = (-)$ and $(i', i^*) \notin M_k$ (see Step 5d). Thus neither (r, u) nor (s, t) are in M_k. Now suppose there is an augmenting path

$$p = (j_0, \ldots, u, r, \ldots, s, t, \ldots, j_{2q+1}),$$

as shown in Figure 14, which contains some of these deleted vertices and edges. Since $L_1(r) = L_1(s) = (-)$, $(r, \ldots, s)$ contains an even number of edges. Then, because p contains an odd number of edges, either (u, r) or (s, t) must be in M_k, which is a contradiction. Thus, if the algorithm is otherwise capable of finding an augmenting path (when one exists), this capability will not be destroyed by the vertex and edge removals of Step 7.

Given that there is an augmenting path relative to M_k, there must be at least two exposed vertices so that false termination cannot occur at Step 2 with $|S_k| \leq 1$. In particular, suppose that $i_0 \in S_k$ is chosen at Step 2 and there is an augmenting path $(i_0, i_1, \ldots, i_{2q+1})$ which is not found.

Let us digress momentarily and show that this situation can occur when the graph has an odd cycle. In the graph of Figure 15, there is the augmenting path (1, 2, 3, 4, 5, 6). After determining $L(3) = (+, 2)$, we can put $L(4) = (-, 3)$ or $L(5) = (-, 3)$. If $L(5) = (-, 3)$, then vertex 6 cannot be labeled and the augmenting path remains undiscovered. Note that it is also

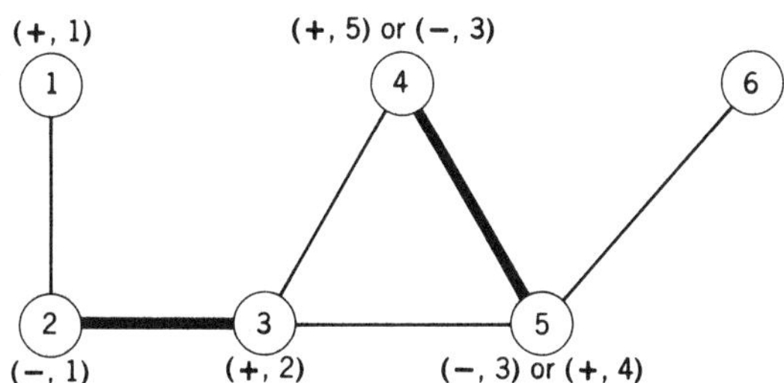

Figure 15

possible to set $L(5) = (+, 4)$, given $L(4) = (-, 3)$. In this case, the augmenting path is discovered. However, a method whose success depends on arbitrary decisions is not satisfactory.

Returning to the proof, since G is bipartite, there exists a partition $V = (V_1, V_2)$ and all edges of G join one vertex of V_1 with another of V_2. Suppose that $i_0 \in V_1$. Since the augmenting path contains an odd number of edges, it follows that $i_{2q+1} \in V_2$. Furthermore, $i \in V_1$ and labeled implies $L_1(i) = (+)$, and $i \in V_2$ and labeled implies $L_1(i) = (-)$. The status of the labels is shown in Figure 16.

Since the point has been reached where no further labeling is possible, there can be no edges from V_1^+ to V_2° [Step 5, rules (a) and (b)]. From Step 5, rule (c), there can be no edges in M_k from V_2^- to V_1°. But an augmenting path $(i_0, i_1, \ldots, i_{2q+1})$ must either contain an edge in $E - M_k$ from V_1^+ to V_2° or one in M_k from V_2^- to V_1° or both, which is a contradiction. ■

Example

Consider the graph shown in Figure 17. Clearly this graph contains no odd cycles and is bipartite with $V_1 = \{1, 3, 4, 6, 7\}$ and $V_2 = \{2, 5, 8\}$.

STEP 1: $M_0 = \{(2, 4)\}$, $S_0 = \{1, 3, 5, 6, 7, 8\}$.

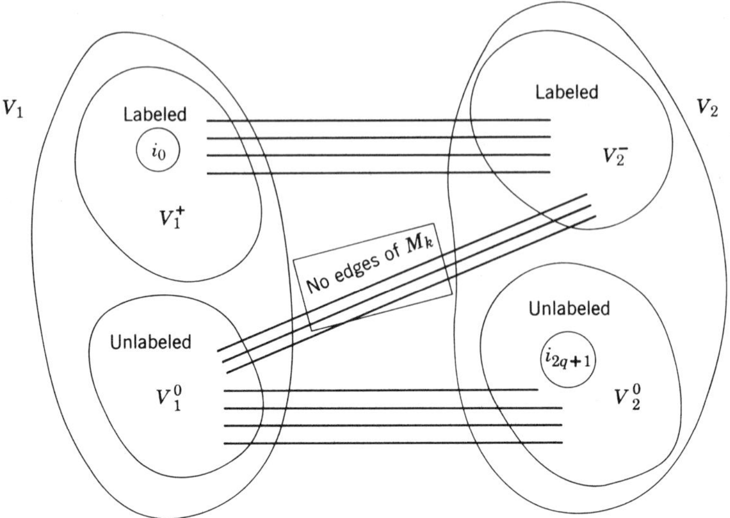

Figure 16

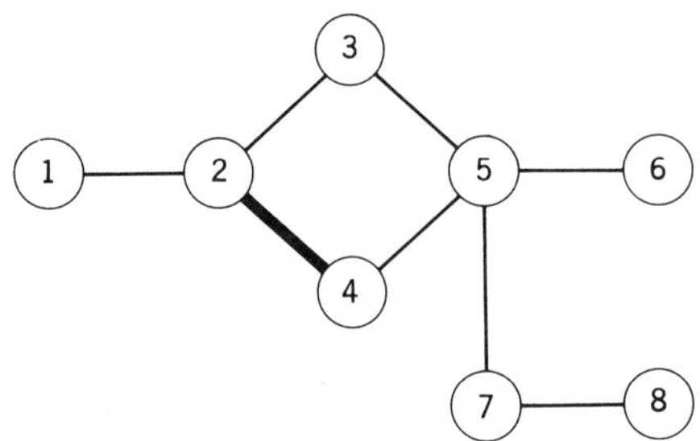

Figure 17

STEP 2: $i_0 = 1$.
STEPS 3–5: The labels are shown in Figure 18.

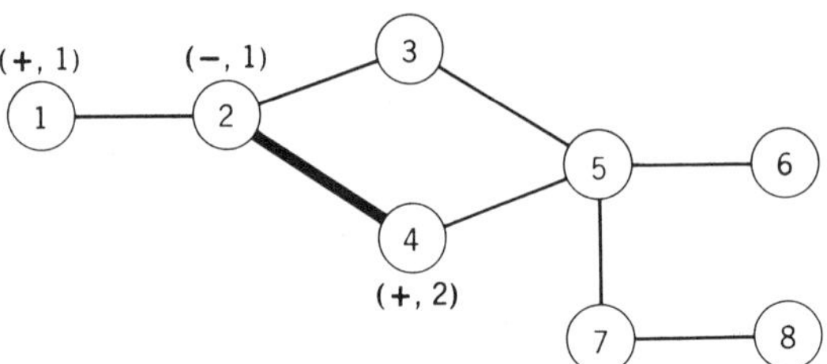

Figure 18

STEP 6: An augmenting path is given by the vertex sequence (1, 2, 4, 5). The new matching $M_1 = \{(1, 2), (4, 5)\}$ is shown in Figure 19. Also, $S_1 = \{3, 6, 7, 8\}$.

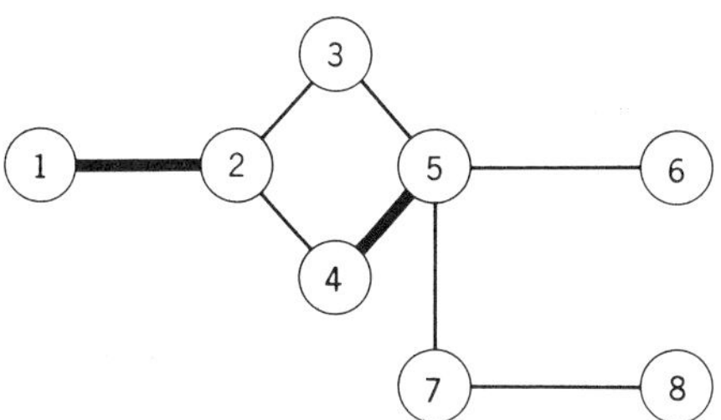

Figure 19

STEP 2: $i_0 = 3$.
STEPS 3–5: The labels are shown in Figure 20.

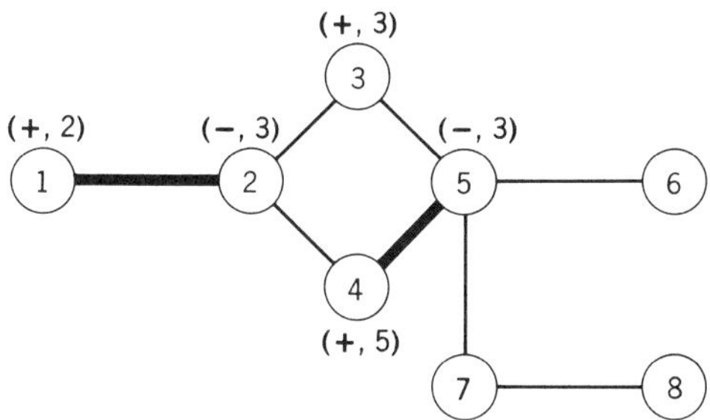

Figure 20

STEP 7: Delete vertices 1, 2, 3, 4 and 5 and all edges incident to these vertices (see Figure 21). Also, $S_1 = \{6, 7, 8\}$.

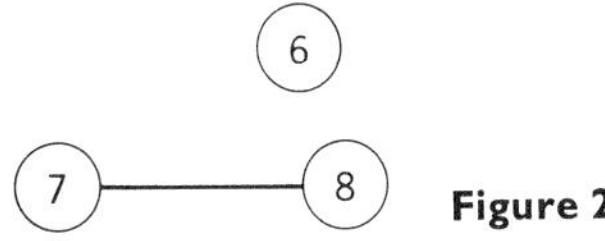

Figure 21

STEP 2: $i_0 = 7$.
STEPS 3–5: At Step 5, $L(7) = (+, 7)$ and then rule (a) applies with $i^* = 8$.

STEP 6: An augmenting path is (7, 8). The matching $M_2 = \{(1, 2), (4, 5), (7, 8)\}$ is shown in Figure 22. Also, $S_2 = \{3, 6\}$.

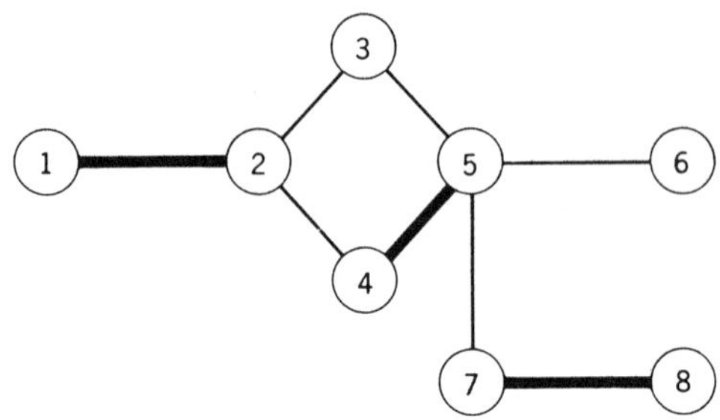

Figure 22

STEP 2: $i_0 = 3$.
STEPS 3–5: The labels are the same as in Figure 20.
STEP 7: $S_2 = \{6\}$.
STEP 2: Go to Step 8, since $|S_2| \leq 1$.
STEP 8: Terminate; M_2 is a maximum matching.

Odd Cycles

Our objective is to extend the matching algorithm to nonbipartite graphs. Suppose that the algorithm for bipartite graphs were applied to a graph with at least one odd cycle.

Lemma 1: *Let $e_j = (i', \hat{\imath})$ be an edge not yet considered at Step 4 of the algorithm. If*

(i) $L_1(i') = L_1(\hat{\imath}) = (+)$ *and* $e_j \in E - M$, *or*
(ii) $L_1(i') = L_1(\hat{\imath}) = (-)$ *and* $e_j \in M$,

then e_j is contained in an odd cycle c of labeled vertices containing $2q + 1$ edges and q of these edges are in M.

PROOF: The set of labeled vertices and the edges used to label them form a subgraph that is a tree of alternating paths. Thus if e_j has not been chosen at Step 4 and i' and $\hat{\imath}$ are labeled, a cycle is indicated. In particular, there is a labeled vertex i^o such that $p' = (i^o, \ldots, s, i')$ and $\hat{p} = (\hat{\imath}, r, \ldots, i^o)$ are alternating paths, so that $c = (\hat{\imath}, \ldots, i^o, \ldots, i', \hat{\imath}) = (\hat{p}, p', \hat{\imath})$ is a cycle. Since $L_1(i') = L_1(\hat{\imath})$, p' and $\hat{p}$ are both odd or both even. Therefore, c is an odd cycle. If $L_1(i') = L_1(\hat{\imath}) = (+)$, (s, i') and $(r, \hat{\imath})$ are in M. Then, $e_j \in E -$

M and q of the $2q + 1$ edges of c are in M (see Figure 23a). If $L_1(i') = L_1(\hat{\imath}) = (-)$, (s, i') and $(r, \hat{\imath})$ are in $E - M$. Therefore if $e_j \in M$, q edges of c are in M (see Figure 23b). ■

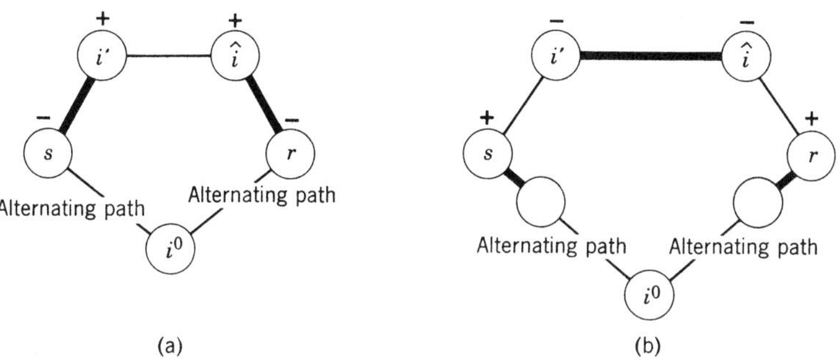

Figure 23

Lemma 2: *Let $(i_0, i_1, \ldots, i_{2j+1})$ be an augmenting path not found by the algorithm when no further labeling is possible. Then there exists $e_j = (i', \hat{\imath})$ on the augmenting path which satisfies the conditions of Lemma 1.*

PROOF: Since $L_1(i_0) = (+)$, if $L_1(i_1) = (-), \ldots, L_1(i_{2j}) = (+)$, the augmenting path is found. In this case we shall say that the labels on the augmenting path are "correct." Furthermore i_1 can be labeled so that if the augmenting path is not found, there exists i_t, $1 \leq t \leq 2j$ such that i_t is labeled and $L_1(i_t)$ is "wrong." Let

$$t^* = \min_{t \in T} t, \qquad T = \{t \,|\, L_1(i_t) \text{ wrong}\}$$

Since $L_1(i_{t^*-1})$ is correct, $L_1(i_{t^*-1}) = L_1(i_{t^*})$. If $L_1(i_{t^*-1}) = (+)$, then $e_j = (i_{t^*-1}, i_{t^*}) \in E - M$ satisfies condition (*i*) of Lemma 1. Alternatively if $L_1(i_{t^*-1}) = (-)$, then $e_j \in M$ satisfies condition (ii) of Lemma 1. ■

Lemmas 1 and 2 establish that if the algorithm fails to discover an existing augmenting path, it can be used to discover an odd cycle of $2q + 1$ edges that has edges in common with the augmenting path and q edges in the matching.

Lemma 3: *Let $c = (\hat{p}, p', \hat{\imath})$, where $\hat{p} = (\hat{\imath}, r, \ldots, i^o)$ and $p' = (i^o, \ldots, s, i')$ are as in Lemma 1, and let Q be the edge set of c. Then i^o is the unique vertex of c exposed relative to $Q \cap M$ and $L_1(i^o) = (+)$.*

PROOF: Let V' be the vertex set of c, so that $|V'| = |Q| = 2q + 1$. Since $|Q \cap M| = q$, exactly one vertex of c is exposed relative to $Q \cap M$. Since $\hat{p}$ and p' are alternating paths, and from Lemma 1, $d_{Q \cap M}(i') = d_{Q \cap M}(\hat{\imath}) = 1$, it follows that $d_{Q \cap M}(i^{\circ}) = 0$.

If condition (i) of Lemma 1 holds, it was shown that (s, i') and $(r, \hat{\imath})$ are in M. Then, since $d_{Q \cap M}(i^{\circ}) = 0$, p' and $\hat{p}$ have an even number of edges. Thus $L_1(\hat{\imath}) = L_1(i') = (+)$ implies $L_1(i^{\circ}) = (+)$. Similarly, if condition (ii) of Lemma 1 is obtained, (s, i') and $(r, \hat{\imath})$ are in $E - M$. Thus p' and $\hat{p}$ have an odd number of edges and $L_1(\hat{\imath}) = L_1(i') = (-)$ implies that $L_1(i^{\circ}) = (+)$. ■

Lemma 3 establishes that either $i^{\circ} = i_0$ or there is an edge (t, i°), $t \notin V'$, such that $(t, i^{\circ}) \in M$. It also follows from Lemma 3 that given c and i°, $Q \cap M$ is uniquely determined.

Reduced Graphs

Suppose that no further labeling is possible and an odd cycle c, with edge set Q and vertex set V', where $|V'| = |Q| = 2q + 1$ and $|Q \cap M| = q$, has been identified by the conditions of Lemma 1. A *reduced graph* G_Q is defined by shrinking c into a single vertex. To obtain G_Q from G, delete from G all

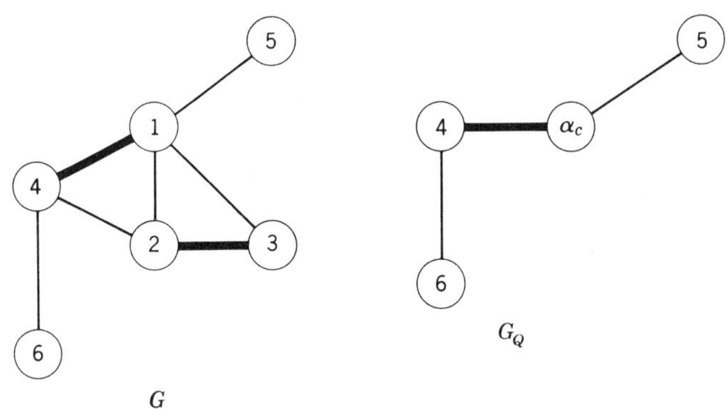

Figure 24

$i \in V'$ and all edges incident to elements of V'; then add an artificial vertex α_c and edges (j, α_c) for all $j \in V - V'$ such that for some $i \in V'$, $(j, i) \in E$. The edge $(j, \alpha_c) \in M$, if $(j, i) \in M$ for some $i \in V'$. The shrinking process is illustrated in Figure 24 for the odd cycle $c = (1, 2, 3)$.

The algorithm that follows is based on shrinking odd cycles.

The General Algorithm

Only the modifications of the bipartite algorithm are given. In the "otherwise" statement of Step 4, replace "go to Step 7" by "go to Step 9."

STEP 9: If no e_j satisfying the conditions of Lemma 1 exists, go to Step 7. If e_j satisfying the conditions of Lemma 1 exists, an odd cycle c, as in Lemma 1, is indicated. Identify c and store it together with i^o (see Lemma 3). Reduce the graph as explained above. Keep all labels associated with vertices not in V'. Let $L(\alpha_c) = L(i^o)$. Return to Step 4.

When Step 6 occurs, it is still the case that an augmenting path with *respect to G* and relative to M_k is found, even though we may be working on a reduced graph (possibly reduced several times). If an augmenting path contains artificial vertices, Step 6 must include a method for determining the appropriate edges to choose from the cycles. The modifications follow.

STEP 6′: Identify the augmenting path as in Step 6. If it does not contain artificial vertices, continue with Step 6. Otherwise, assume that the augmenting path is $p = (i_0, \ldots, i_{r-1}, i_r, i_{r+1}, \ldots, i_{2j+1})$ and i_r is an artificial vertex. Retrieve the cycle $c = (l_0, \ldots, l_s, \ldots, l_t, \ldots, l_{2q+1}, l_0)$ associated with i_r and assume that i^o of c is l_s. From Lemma 3, exactly one of the following three cases must hold:

(a) $i_r = i_0$, so that $p = (i_r, i_{r+1}, \ldots, i_{2j+1})$ and $d_{M_k}(l_s) = 0$ (see Figure 25).

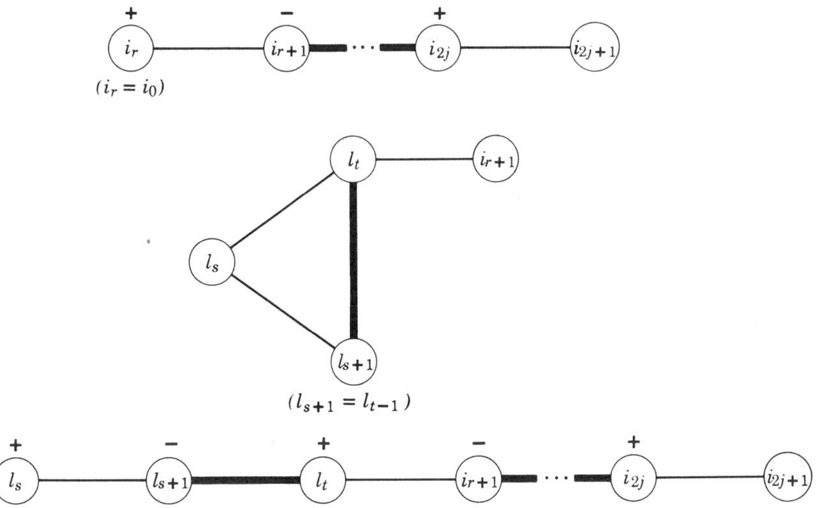

Figure 25

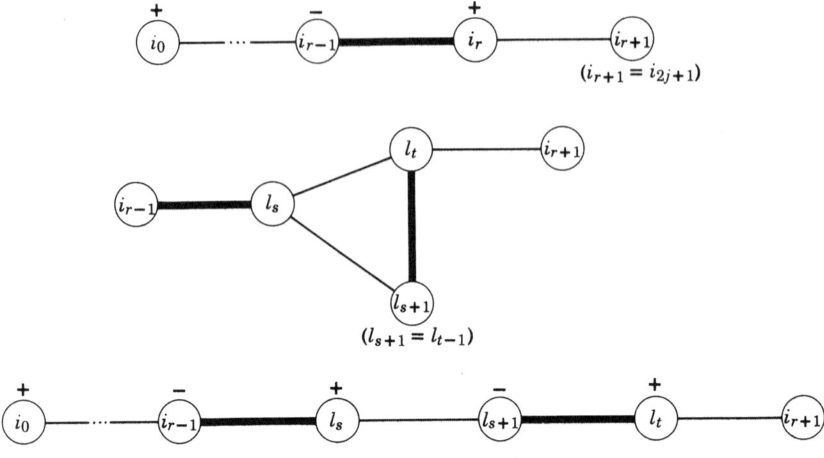

Figure 26

(b) $(i_{r-1}, l_s) \in M_k$ (see Figure 26).

(c) $(i_{r+1}, l_s) \in M_k$ (see Figure 27).

In cases (a) and (b), there must be an edge from the cycle to i_{r+1} that is not in the matching. Thus there exists l_t such that $(l_t, i_{r+1}) \in E - M_k$ and $(l_{t-1}, l_t) \in M_k$. In case (a), replace p by $p^* = (l_s, \ldots, l_t, i_{r+1}, \ldots, i_{2j+1})$ and label

$$L(l_s) = (+, l_s), \qquad L(l_u) = (-L_1(l_{u-1}), l_{u-1}), \qquad u = s + 1, \ldots, t$$

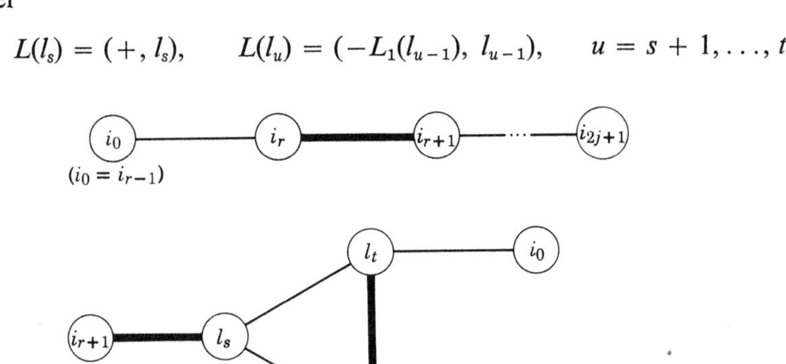

Figure 27

In case (b), replace p by $p^* = (i_0, \ldots, i_{r-1}, l_s, \ldots, l_t, i_{r+1}, \ldots, i_{2j+1})$ and label

$$L(l_s) = (-L_1(i_{r-1}), i_{r-1}), \qquad L(l_u) = (-L_1(l_{u-1}), l_{u-1}), \qquad u = s + 1, \ldots, t$$

In case (c), $(i_{r+1}, l_s) \in M_k$ implies that there exists l_t such that $(l_t, i_{r-1}) \in E - M_k$ and $(l_{t-1}, l_t) \in M_k$. Replace p by

$$p^* = (i_0, \ldots, i_{r-1}, l_t, \ldots, l_s, i_{r+1}, \ldots, i_{2j+1})$$

and label

$$L(l_t) = (-L_1(i_{r-1}), i_{r-1}), L_u = (-L_1(l_{u+1}), l_{u+1}), \qquad u = t - 1, \ldots, s$$

In each of these three cases the artificial vertex is replaced by an alternating path containing an even number of edges, since $(l_s, l_{s+1}) \in E - M$ from Lemma 3 and $(l_{t-1}, l_t) \in M$. Thus p^* contains an odd number of edges. Furthermore, the constructions associated with cases (a), (b), and (c) insure that p^* is alternating and therefore augmenting. Note that p may contain several artificial vertices and in the process of replacing an artificial vertex by an alternating path, other artificial vertices can be included in p^*. Thus the process of removing artificial vertices may have to be repeated several times. After all artificial vertices of p have been removed, define a new matching M_{k+1} based on this augmenting path. Let $k = k + 1$, $S_k = \{i | d_{M_k}(i) = 0\}$; delete all labels and return to Step 2.

It has been shown that an augmenting path on a reduced graph can be transferred back to the original graph to obtain an improved matching. The proof that the general algorithm finds an augmenting path, if one exists, is based upon the fact that if no augmenting path is found, the graph is eventually reduced to a bipartite graph to which the argument of Theorem 12 applies. The details of the proof are asked for in Exercise 23.

Example

Consider the graph shown in Figure 28.

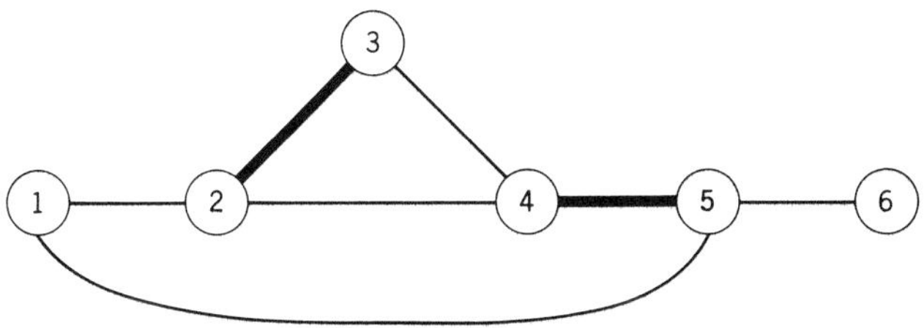

Figure 28

STEP 1: $M_0 = \{(2, 3), (4, 5)\}$, $S_0 = \{1, 6\}$.
STEP 2: $i_0 = 1$, $L(1) = (+, 1)$.
STEPS 3–5: The labels are shown in Figure 29.

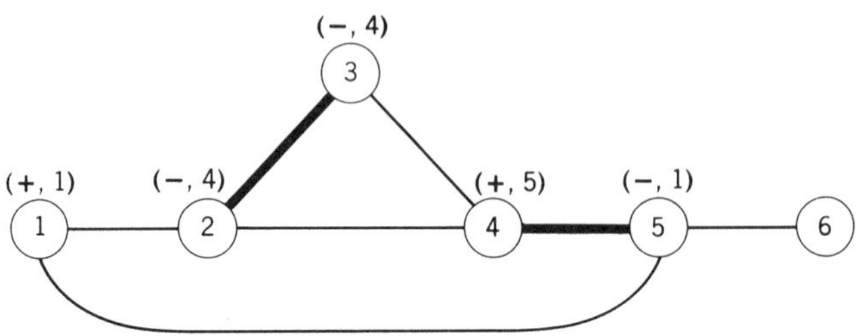

Figure 29

STEP 9: The odd cycle $c_1 = (2, 3, 4, 2)$ is identified, since $L_1(2) = L_1(3) = (-)$ and $(2, 3) \in M_0$. Also $i^o = 4$. Reduce the graph as shown in Figure 30 and let $L(\alpha_{c_1}) = (+, 5)$.

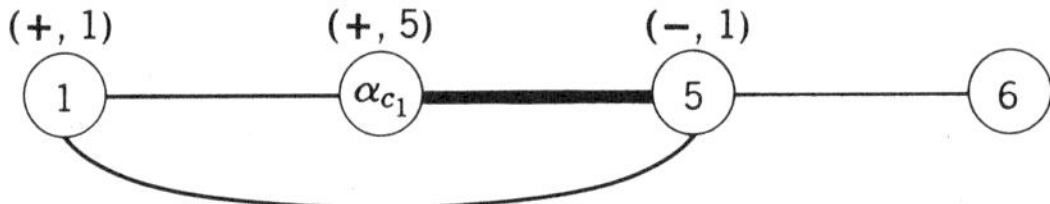

Figure 30

STEP 4: No labeling.

STEP 9: The odd cycle $c_2 = (1, \alpha_{c_1}, 5, 1)$ is identified, since $L_1(1) = L_1(\alpha_{c_1}) = (+)$ and $(1, \alpha_{c_1}) \notin M_0$. Also $i^o = 1$. Reduce the graph as shown in Figure 31 and let $L(\alpha_{c_2}) = (+, 1)$.

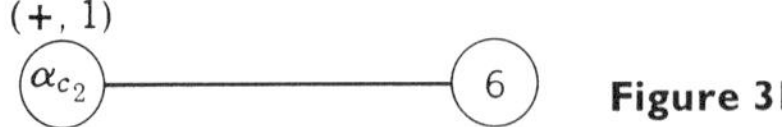

Figure 31

STEP 6: $p = (\alpha_{c_2}, 6)$ is an augmenting path.

STEP 6′: Retrieve $c_2 = (1, \alpha_{c_1}, 5, 1)$, and let $l_s = 1$ and $l_t = 5$. This yields the augmenting path $p^* = (1, \alpha_{c_1}, 5, 6)$ and labels $L(1) = (+, 1)$, $L(\alpha_{c_1}) = (-, 1)$, and $L_5 = (+, \alpha_{c_1})$, as in Figure 32. Retrieve $c_1 = (2, 3, 4, 2)$,

Figure 32

and let $l_s = 4$ and $l_t = 2$. This yields the augmenting path $p^{**} = (1, 2, 3, 4, 5, 6)$, which does not contain artificial vertices. Thus define $M_1 = M_0 \cup \{(1, 2), (3, 4), (5, 6)\} - \{(2, 3), (4, 5)\}$ yielding $S_1 = \varnothing$ as in Figure 33.

STEP 2: Go to Step 8, since $|S_1| \leq 1$.

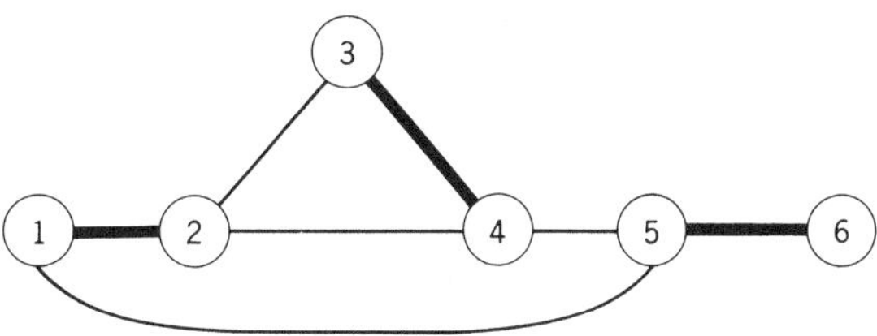

Figure 33

STEP 8: Terminate; M_1 is maximum.

3.8. EXERCISES

1. Construct a directed graph having the incidence matrix,

$$\begin{bmatrix} 0 & 1 & 0 & 1 & -1 & 0 & 0 & 0 & 0 & 0 & 0 & 0 & 0 & 0 \\ 0 & 0 & 0 & 0 & 0 & 1 & 0 & 1 & 1 & 0 & 0 & -1 & -1 & 0 \\ 0 & -1 & 1 & 0 & 0 & 0 & 0 & -1 & 0 & 0 & 0 & 0 & 0 & 0 \\ 0 & 0 & 0 & 0 & 0 & 0 & 0 & 0 & 0 & 0 & 0 & 1 & 0 & 0 \\ 0 & 0 & 0 & -1 & 0 & 0 & 0 & 0 & -1 & -1 & 0 & 0 & 0 & 1 \\ 0 & 0 & 0 & 0 & 0 & 0 & 1 & 0 & 0 & 1 & 1 & 0 & 0 & -1 \\ 0 & 0 & 0 & 0 & 0 & -1 & -1 & 0 & 0 & 0 & 0 & 0 & 1 & 0 \\ 1 & 0 & -1 & 0 & 0 & 0 & 0 & 0 & 0 & 0 & -1 & 0 & 0 & 0 \\ -1 & 0 & 0 & 0 & 1 & 0 & 0 & 0 & 0 & 0 & 0 & 0 & 0 & 0 \end{bmatrix}$$

2. Let $G = (V, E)$ be a graph that is a cycle. Show that by an appropriate numbering of vertices and edges, the incidence matrix of G can be written as the $|V| \times |V|$ matrix A with

$$a_{11} = a_{1,|V|} = 1, \; a_{1j} = 0 \qquad \text{otherwise}$$
$$a_{ii} = a_{i,i-1} = 1, \; a_{ij} = 0 \qquad \text{otherwise } i \geq 2$$

Prove that A is nonsingular if and only if $|V|$ is odd, and in this case $|\det A| = 2$.

3. Find the distinct components of the graph with the incidence matrix,

$$A = \begin{bmatrix} 1 & 0 & 0 & 1 \\ 0 & 1 & 0 & 0 \\ 1 & 0 & 1 & 0 \\ 0 & 1 & 0 & 0 \\ 0 & 0 & 1 & 1 \end{bmatrix}$$

4. Prove that if $G = (V, E)$ is a tree, then $|E| = |V| - 1$.

5. Let $G = (V, E)$ be a connected graph. Assigned to each edge e_j is a weight c_j. Consider the problem of finding a subgraph $G = (V, E^*)$ that is a spanning tree and has the property,

$$\sum_{e_j \in E^*} c_j \leq \sum_{e_j \in E'} c_j, \qquad E' \text{ an edge set for a spanning tree.}$$

Prove that the following construction solves the minimum weight spanning tree problem given above.

STEP 1: Choose an edge with smallest weight.

STEP 2: If $|V| - 1$ edges have been chosen, terminate. The chosen set of edges constitutes a minimal weight spanning tree. Otherwise go to Step 3.

STEP 3: Of the edges not yet chosen, pick one with the smallest weight that does not make a cycle with the edges already chosen. Go to Step 2.

6. Is the following matrix totally unimodular?

$$A = \begin{bmatrix} 1 & 1 & 0 & 0 & 0 & 0 & 0 \\ 1 & 0 & 1 & 0 & 1 & 0 & 0 \\ 0 & 1 & 0 & 1 & 0 & 0 & 0 \\ 0 & 0 & 1 & 1 & 0 & 1 & 0 \\ 0 & 0 & 0 & 0 & 1 & 0 & 1 \\ 0 & 0 & 0 & 0 & 0 & 1 & 1 \end{bmatrix}$$

7. Prove that the following statements are equivalent:
(a) A is totally unimodular.
(b) The transpose of A is totally unimodular.
(c) (A, A) is totally unimodular.

8. Prove that for the class of matrices that satisfy Condition 1 of Theorem 3, Condition 2 is necessary for total unimodularity.

9. Give an example which shows that for arbitrary matrices with elements of $(0, 1, -1)$, the conditions of Theorem 3 are not necessary for total unimodularity.

10. There is given a directed graph $G = (V, E)$ with nonnegative integer capacities d_{ij}, $(i,j) \in E$. Consider the problem of finding a set of edges E^* with minimum capacity such that no flow can be sent from vertex 1 to vertex m in the subgraph $G^* = (V, E - E^*)$. Formulate this problem as an ILP. How is it related to minimum cost flow problems?

11. Consider some commodity that deteriorates over time. In particular, a new item has value Δ and an item that is k periods old has value $c_k\Delta$, where $1 = c_0 > c_1 > \cdots > c_n$. A firm has in inventory a_k items of age $k = 0, 1, \ldots, n - m$. In each of the next m periods, the firm must supply b_i items of any age, $i = 1, \ldots, m$. The objective is to minimize the value of the items supplied. Formulate this problem as a minimum cost flow problem.

12. Consider a minimum cost flow problem in which more than one commodity is to be shipped through the network. For commodity k, $k = 1, \ldots, p$, there is a set of origins V_1^k and a set of destinations V_2^k. For each $i \in V_1^k$, there is a supply a_i^k, and for each $i \in V_2^k$ there is a demand b_i^k. The unit shipping cost on edge (i, j) is c_{ij}^k. Let x_{ij}^k be the quantity of flow of commodity k through edge (i, j). The capacity constraints are

$$\sum_k x_{ij}^k \leq d_{ij} \qquad \text{for all } (i, j) \in E$$

Formulate this *multicommodity, minimum cost flow problem* as an LP. Is the constraint matrix totally unimodular? Look at the special case of two-commodity flows and provide examples of constraint matrices that are, and are not, totally unimodular.

13. Find a shortest path from vertex 1 to vertex 7 in the graph of Figure 34.

14. Give an example with one edge of negative length, but no cycle of negative length, for which the algorithm of Section 3.5 fails.

15. Prove that a maximum weighted matching problem on a bipartite graph can be formulated as an assignment problem.

16. For the problem of maximum weighted matching, define an augmenting path p, relative to M, to be an alternating path or alternating cycle having no

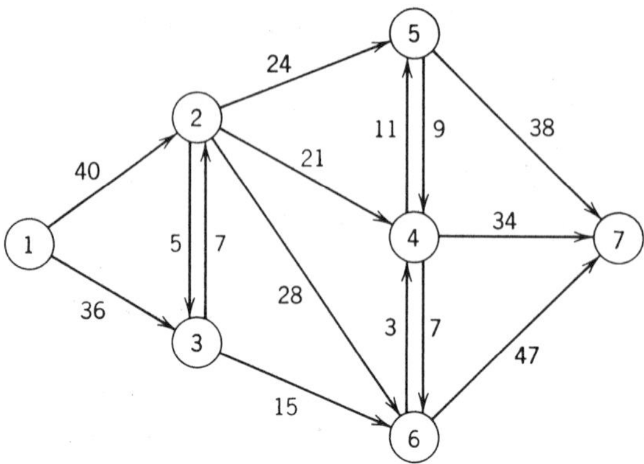

Figure 34

edge of $M - P$ incident to P, and with the property,

$$\sum_{e_j \in P - M} c_j - \sum_{e_j \in P \cap M} c_j > 0$$

where P is the set of edges contained in the path p. Prove that M is a maximum weighted matching if and only if M admits no augmenting path.

17. In the proof of Theorem 8, why must case (a) occur with D' an augmenting path with respect to M?

18. Give an appropriate generalization of Theorem 10 for the case where b is an arbitrary vector of positive integers.

19. Consider the graph G with incidence matrix

$$A = \begin{array}{c} \begin{matrix} e_1 & e_2 & e_3 & e_4 & e_5 & e_6 \end{matrix} \\ \begin{bmatrix} 1 & 0 & 0 & 0 & 1 & 0 \\ 1 & 1 & 0 & 0 & 0 & 1 \\ 0 & 1 & 1 & 0 & 0 & 0 \\ 0 & 0 & 1 & 1 & 0 & 1 \\ 0 & 0 & 0 & 1 & 1 & 0 \end{bmatrix} \end{array}$$

Note that G contains exactly two cycles $(e_1, e_2, e_3, e_4, e_5)$ and (e_2, e_3, e_6). Show that

$$\begin{aligned} Ax &\leq b^o, \; b^o = (1, 1, 1, 1, 1) \\ x_1 + x_2 + x_3 + x_4 + x_5 \qquad &\leq 2 \\ x_2 + x_3 \qquad + x_6 &\leq 1 \\ x &\geq 0 \end{aligned}$$

has a basic feasible solution with $x = (\frac{1}{3}, \frac{1}{3}, \frac{1}{3}, \frac{1}{3}, \frac{2}{3}, \frac{1}{3})$.

20. Using the algorithm of Section 3.7, solve the maximum matching problem (unweighted) for the graph of Figure 34. (Ignore the arrowheads.)

21. Use the results of Exercise 20 to solve the minimum covering problem for the graph of Figure 34.

22. Consider the problem of finding a maximum matching on $G = (V, E)$, where $i^* \in V$ is such that $d_E(i^*) = 1$. Let $e_j = (i^*, i')$. Prove that there exists a maximum matching that contains e_j.

23. Give all of the details of the proof that the general maximum matching algorithm of Section 3.7 works.

24. Prove Theorem 7 of Section 3.6.

25. In the algorithm of Section 3.7, prove that when Step 6 is reached, the set of labeled vertices and the edges used to label them form a subgraph of G that is a tree.

26. Modify the shortest path algorithm of Section 3.5 so that

(a) It will find all shortest paths from vertex 1 to vertex m.

(b) The indicators $p(i)$ can be eliminated.

27. Let A be a binary matrix that satisfies:

$a_{ij} = 1$ and $a_{kj} = 1$ for $k > i + 1$ imply that $a_{i+1,j} = \cdots = a_{k-1,j} = 1$, for all i, k, and j. Prove that A is totally unimodular.

28. A graph $G = (V, E)$ is said to have a perfect matching if there exists $M \subseteq E$ such that $d_M(i) = 1$ for all $i \in V$. Prove: Let $P(S)$, $S \subseteq V$, denote the number of components with an odd number of vertices in the graph G_S obtained from G by suppressing S and all edges incident with S. Then G has a perfect matching if and only if $P(S) \leq |S|$ for all $S \subseteq V$.

3.9. NOTES

3.2. Some texts on graph theory are Berge (1962), Busacker and Saaty (1965), and Harary (1969).

3.3. A version of Theorem 2 was first proved in Hoffman and Kruskal (1958). Our development follows the much simpler approach given in Veinott and Dantzig (1968).

Theorem 3 is proved in Heller and Tompkins (1958). In an appendix of this paper written by Gale, the necessity of Condition 2 of Theorem 3, given that Condition 1 holds, is established. Other characterizations of total unimodularity are given in Hoffman and Kruskal (1958); Hoffman and

Heller (1962); Heller (1957), (1963), and (1964); Camion (1965); and Chandrasekaran (1969).

Veinott (1968) identifies a class of problems in which A is not totally unimodular, but the extreme points of $\{x|Ax = b, x \geq 0\}$ are integer for all integer $b \geq 0$.

3.4. The basic reference on minimum cost capacitated flow problems is Ford and Fulkerson (1962). Elmaghraby (1970) gives a survey.

3.5. A critical survey of shortest path algorithms is given in Dreyfus (1969). The algorithm of this section comes from Dijkstra (1959). A generalization is given in Nemhauser (1972). An algorithm closely related to Dijkstra's, but based on (12) rather than (16), is given in Dantzig (1960b).

3.6. A more general class of matching and covering problems is considered in Chapter 8.

An introductory reference on graph matching is Edmonds (1967).

A different covering problem on a graph is the covering of edges by vertices. Given $G = (V, E)$, the objective is to find $V^\circ \subseteq V$ of minimum cardinality such that every edge is incident to at least one vertex of V°. This problem is much more difficult than the corresponding problem of covering vertices by edges. It is studied in Balinski (1970b).

Theorems 6, 7, and 8 are from Norman and Rabin (1959), and Theorem 8 is also proved in Berge (1957). Theorem 9 is proved in Harary (1969), Theorem 10 in Balinski (1970a). Other results on the nature of fractional extreme points of binary matrices are given in Rebman (1969) and Padberg (1971). Theorem 11 is proved in Edmonds (1965b) and in Balinski (1969b).

3.7. The algorithm of this section is based on Edmonds (1965a). A modified version of Edmond's algorithm is given in Witzgall and Zahn (1965). A different algorithm for maximum matching is presented in Balinski (1969a); the Balinski algorithm uses a bookkeeping procedure to identify augmenting paths that does not involve the shrinking of odd cycles.

Edmonds (1965b) has generalized his algorithm to solve maximum weighted matching problems. Edmonds and Johnson (1970) give a further generalization to maximum weighted b-matching problems and some variations.

White (1971) has developed an algorithm for finding minimum weighted edge covers of fixed cardinality.

Specialized algorithms for matching on bipartite graphs are given in Glover (1967b), Desler and Hakimi (1969), Morrison (1969), and De Werra (1970).

Further Notes

The problem of Exercise 5 is solved in Kruskal (1956). The Theorem of Exercise 28 was first proved by Tutte (1947); see also Anderson (1971). Multicommodity integer flows (see Exercise 12) are discussed by Hu (1963), Rothschild and Whinston (1966a) and (1966b), and Bozoki (1969).

4 Enumeration Methods

4.1 ENUMERATION TREES

In this chapter we are concerned with enumerative approaches to integer programming. These approaches take advantage of the fact that in a bounded ILP or MILP, the set of values of the integer variables is finite. The basic idea of enumerative methods can be explained using a tree.

Example

In how many different ways can distinct positive integers be chosen such that they sum to 7? Letting

$$x_j = \begin{cases} 1 & \text{if } j \text{ is one of the integers chosen} \\ 0 & \text{otherwise} \end{cases}$$

we require all solutions to

$$\sum_{j=1}^{7} jx_j = 7$$
$$x_j = 0, 1 \qquad \text{for all } j \tag{1}$$

The solutions are given by the unique paths from vertex 0 (v_0) to each of the vertices marked by an asterisk in Figure 1. The vertices 0, . . ., 18 (sometimes referred to as $v_0, \ldots, v_{18}$) are numbered by the order in which they were considered. Each edge imposes a constraint, and each vertex j represents the constraint set (1) in addition to the constraints given by the edges along the unique path P_j from v_0 to v_j. A line underneath a vertex indicates that no further exploration from that vertex can be profitable. Such a vertex is said to be *fathomed.*

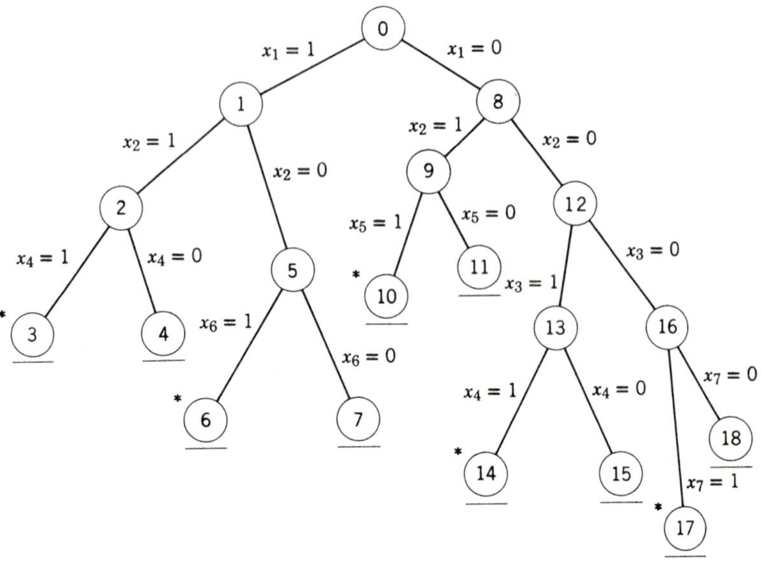

Figure 1

More generally, if the problem is to find every $x \in S$, then vertex j restricts x to S_j, where S_j is the intersection of S with the set of points satisfying the constraints given by the edges of P_j. If P_j has $k + 1$ vertices denoted by

$$v_0 = v_{j(0)}, v_{j(1)}, \ldots, v_{j(k-1)}, v_{j(k)} = v_j$$

then

$$S = S_{j(0)} \supseteq S_{j(1)} \supseteq \cdots \supseteq S_{j(k)} = S_j$$

We call $v_{j(k-1)}$ the *predecessor* of v_j, which in turn is called a *successor* of its predecessor. Note that a vertex has a unique predecessor but generally more than one successor.

Separation

The constraints of the edges from v_j to its successors determine a finite set S_j^* of subsets of S_j such that

$$\bigcup_{T \in S_j^*} T = S_j$$

The set S_j^* is called a *separation* of S_j. Each of the $|S_j^*|$ edges emanating from v_j corresponds to a constraint restricting x to one of the elements of S_j^*. Although not necessary, in many cases S_j^* is a *partition* of S_j.

An additional condition on the separation is required to guarantee the finiteness of the tree. Assume that S_0 is finite. (This assumption will be relaxed in Section 4.2.) For all j, let $H_j \supseteq S_j$ be a finite set such that if $|H_j| > 1$ and separation is required

$$S_k \in S_j^* \Rightarrow |H_k| < |H_j| \tag{2}$$

For $|H_j| \leq 1$, we assume that vertex j can be fathomed. Any separation satisfying (2) guarantees bounded enumeration, in the sense that no path can contain more than $|H_0|$ edges.

Branching

A vertex that is not fathomed and whose corresponding constraint set has not been separated is called a *live* vertex. *Branching* means choosing a live vertex to consider next for fathoming or separation. There are many possible rules for branching. A common one is branching to one of the successor vertices of the vertex currently being considered. If the current vertex j is fathomed, one simply *backtracks* along P_j until a vertex having at least one live successor is encountered. One of these successor vertices is chosen for branching. If there are no live vertices the enumeration is complete.

4.2. THE CONCEPT OF BRANCH AND BOUND

Branch and *bound* is an optimization technique that uses the basic tree enumeration described in Section 4.1. It involves calculating *upper bounds* and *lower bounds* on the objective function, in order to accelerate the fathoming process and thereby curtail the enumeration. In general, for the problem

$$\max z(x), \qquad x \in S \tag{3}$$

bounds are calculated as follows.

Upper Bounds

Suppose that the enumeration is at vertex j in the tree. The problem considered at v_j is

$$\max z(x), \qquad x \in S_j \tag{4}$$

Let

$$z_j^* = \begin{cases} z(x^*(j)) & \text{if } x^*(j) \text{ solves (4)} \\ -\infty & \text{if } S_j = \varnothing \\ \infty & \text{if (4) is unbounded} \end{cases}$$

An *upper bound* $\bar{z}_j \geq z_j^*$ may be calculated by considering the *relaxation* of (4)

$$\max z(x), \qquad x \in T_j \supseteq S_j \tag{5}$$

and letting

$$\bar{z}_j = \begin{cases} \infty & \text{if (5) is unbounded} \\ -\infty & \text{if } T_j = \varnothing \\ z_j^0 = z(x^\circ(j)) & \text{if } x^\circ(j) \text{ solves (5)} \end{cases}$$

The choice of T_j is one of the critical parts of any branch and bound algorithm. It must be chosen such that (5) is relatively easy to solve, but at the same time must yield an upper bound which is small enough to be worthwhile. Note that an upper bound at a vertex is valid for any of its successors, since, if v_k is a successor of v_j, then $T_j \supseteq S_j \supseteq S_k$.

Of particular interest for ILP's is

$$\begin{aligned} S_j &= \{x \mid A^j x = b^j,\ x \geq 0 \text{ integer}\} \\ T_j &= \{x \mid A^j x = b^j,\ x \geq 0\} \end{aligned} \tag{6}$$

so that $\bar{z}_j$ is calculated by solving the corresponding LP.

Lower Bounds

A lower bound $\underline{z}_j$ satisfies $\underline{z}_j \leq z_j^*$. One way to calculate a lower bound is to find any $x' \in S_j$ and let $\underline{z}_j = z(x')$. If v_k is the predecessor of v_j, then $\underline{z}_j \leq z_k^*$, which yields the important result that $\underline{z}_j \leq z_0^*$.

Fathoming by Bounds

Vertex j is fathomed if

(a) $\bar{z}_j = \underline{z}_j$

(b) $\bar{z}_j \leq \underline{z}_0$

In case (a) no better solution can be found to (4). When case (b) occurs, no successor of v_j can yield a solution that improves on the best known solution to (3). Note that the case $T_j = \varnothing$ is included in case (b), since $\bar{z}_j = -\infty$.

A General Branch and Bound Algorithm

A branch and bound algorithm involves enumeration by tree as described in Section 4.1, and bounding as described above.

STEP 1: (Initialization.) Begin at the live vertex 0, where $S_0 = S$, $\bar{z}_0 = \infty$, $\underline{z}_0 = -\infty$. Go to Step 2.

STEP 2: (Branching.) If no live vertices exist, go to Step 7; otherwise select a live vertex j. If (5) has been solved at v_j, go to Step 3; otherwise go to Step 4.

STEP 3: (Separation.) Choose a separation S_j^*, which determines the successor vertices of v_j. Go to Step 2.

STEP 4: (Calculating bounds.) Solve (5). If $T_j = \varnothing$, v_j is fathomed and we go to Step 2. If (5) has an optimal solution $x^0(j)$, let $\bar{z}_j = z_j^0$ and go to Step 5. If (5) is unbounded, let $\bar{z}_j = \infty$ and go to Step 2.

STEP 5: (Fathoming, case a.) If $x^0(j) \notin S_j$, go to Step 6. Otherwise let $\underline{z}_j = z_j^0$ and v_j is fathomed. Let $\underline{z}_0 = \max\{\underline{z}_0, \underline{z}_j\}$. Go to Step 6.

STEP 6: (Fathoming, case b.) Any vertex i such that $\bar{z}_i \leq \underline{z}_0$ is fathomed. Go to Step 2.

STEP 7: (Termination.) If $\underline{z}_0 = -\infty$, there is no feasible solution. If $\underline{z}_0 > -\infty$, that feasible solution which yielded $\underline{z}_0$ is optimal.

To see that the algorithm finds an optimal solution to (3) if one exists, note that at termination every vertex j whose constraint set has not been separated has $\bar{z}_j \leq \underline{z}_0$. Clearly, the sets associated with such vertices cannot contain a solution better than the best solution on hand.

When S_0 is finite, bounded enumeration is guaranteed by choosing separations that satisfy (2). In many cases it is not necessary to determine whether $|H_j| > 1$. If

$$|H_j| \leq 1 \Rightarrow T_j = S_j \tag{7}$$

v_j must be fathomed when $|H_j| \leq 1$, since $T_j = S_j = \varnothing \Rightarrow \bar{z}_j = -\infty$, and $T_j = S_j \neq \varnothing \Rightarrow \bar{z}_j = \underline{z}_j$.

When S_0 is not finite, an extension of (2) frequently can be used to establish bounded enumeration. Let $f(S_j)$ be a function whose domain is all the subsets of S and whose range is the nonnegative integers. If $f(S_j) > 1$, corresponding to (2) we require

$$S_k \in S_j^* \Rightarrow f(S_k) < f(S_j) \tag{2a}$$

For $f(S_j) \leq 1$, we assume that vertex j can be fathomed. Any separation satisfying (2a) guarantees bounded enumeration, in the sense that no path can contain more than $f(S_0)$ edges. Note that (2) is a special case of (2a) having $f(S_j) = |H_j|$.

4.3. BRANCH AND BOUND FOR THE ILP

The general branch and bound algorithm of the last section specializes very neatly for ILP's with S_j and T_j given by (6), so that (5) is an LP.

Separation

Suppose that the LP (5) is solved at v_j and $x^o(j)$ is not all-integer. In particular, some variable $x_{B_i} = [y_{io}] + f_{io}$, $0 < f_{io} < 1$. Then a separation of S_j, which is a partition, is

$$S_j^* = \{S_j \cap \{x | x_{B_i} \leq [y_{io}]\}, S_j \cap \{x | x_{B_i} \geq \langle y_{io} \rangle\}\} \tag{8}$$

where $\langle a \rangle$ denotes the smallest integer greater than or equal to a, and $[a]$ denotes the largest integer less than or equal to a.

Finiteness

Assume that an upper bound u_j is known for each variable x_j. Let

$$\begin{aligned} S_k &= \{x | Ax = b, 0 \leq \alpha_j^k \leq x_j \leq \beta_j^k \leq u_j, x_j \text{ integer}, j = 1, \ldots, n\} \\ H_k &= \{x | 0 \leq \alpha_j^k \leq x_j \leq \beta_j^k \leq u_j, x_j \text{ integer}, j = 1, \ldots, n\} \end{aligned} \tag{9}$$

where the α_j^k and β_j^k are integers determined from (8) and $\alpha_j^0 = 0$ and $\beta_j^0 = u_j$. Letting

$$N_j(k) = \beta_j^k - \alpha_j^k + 1$$

yields

$$|H_k| = \prod_{j=1}^{n} N_j(k)$$

If v_p is a successor of v_k, then from (8) it follows that for some t

$$N_t(p) < N_t(k)$$

and

$$N_j(p) = N_j(k), \qquad j \neq t$$

Thus $|H_p| < |H_k|$ and (2) is satisfied. It follows that the algorithm is finite, since no path can contain more than $|H_0| = \prod_{j=1}^{n} (1 + u_j)$ edges.

To see that the separation satisfies (7), observe that

$$|H_k| = 1 \Rightarrow \alpha_j^k = \beta_j^k, \qquad j = 1, \ldots, n$$

If $(\alpha_1^k, \ldots, \alpha_n^k) \in S_k$, then $S_k = T_k = \{(\alpha_1^k, \ldots, \alpha_n^k)\}$. If $(\alpha_1^k, \ldots, \alpha_n^k) \notin S_k$, then $S_k = T_k = \varnothing$.

The case $|H_k| = 0$ never occurs. This follows because any vertex k having $|H_k| = 1$ is fathomed and if $|H_k| > 1$, then the two successors, v_p and v_t have $1 \leq |H_p| < |H_k|$ and $1 \leq |H_t| < |H_k|$.

The LP's are particularly tractable, since the added constraints of (8) are simply lower and upper bounds on individual variables. They are solved by the dual simplex algorithm for bounded variables described in Section 2.11.

A Rudimentary Branch and Bound Algorithm for the ILP

STEP 1: (Initialization.) Begin with one live vertex 0, where S_0 is given in (9), $\bar{z}_0 = \infty, \underline{z}_0 = -\infty$. Go to Step 2.

STEP 2: (Branching.) If no live vertices exist, go to Step 7; otherwise select a live vertex j. If the LP (5) has been solved, go to Step 3; otherwise go to Step 4.

STEP 3: (Separation.) Choose a row with $f_{i0} > 0$ and partition S_j as in (8). Go to Step 2.

STEP 4: (Solving the LP.) Solve the LP (5). If (5) has no feasible solution, v_j is fathomed and we go to Step 2. If (5) has an optimal solution $x^o(j)$, let $\bar{z}_j = [z_j^o]$ and go to Step 5. (Note that z_j^o can be "rounded down," since every integer solution yields an integer value for the objective function.)

STEP 5: (Fathoming by integrality.) If $x^o(j)$ is not all-integer, go to Step 6. If $x^o(j)$ is integer, let $\underline{z}_j = z_j^o$ and v_j is fathomed. Let $\underline{z}_0 = \max\{\underline{z}_0, \underline{z}_j\}$. Go to Step 6.

STEP 6: (Fathoming by bounds.) Any vertex i such that $\bar{z}_i \leq \underline{z}_0$ is fathomed. Go to Step 2.

STEP 7: (Termination.) Terminate. If $\underline{z}_0 = -\infty$, there is no feasible solution. If $\underline{z}_0 > -\infty$, that feasible solution which yielded $\underline{z}_0$ is optimal.

Example

$$
\begin{aligned}
\max x_0 = -7x_1 - 3x_2 - 4x_3 & \\
x_1 + 2x_2 + 3x_3 - x_4 \quad & = 8 \\
3x_1 + x_2 + x_3 \quad - x_5 & = 5 \\
x_1, \ldots, x_5 & \geq 0 \text{ integer}
\end{aligned}
$$

To begin, the only live vertex is v_0 and we solve the corresponding LP. The optimal LP tableau is given in Table 1.

Table I

		$-x_3$	$-x_4$	$-x_5$
x_0	$-\frac{71}{5}$	$\frac{3}{5}$	$\frac{2}{5}$	$\frac{11}{5}$
x_1	$\frac{2}{5}$	$-\frac{1}{5}$	$\frac{1}{5}$	$-\frac{2}{5}$
x_2	$\frac{19}{5}$	$\frac{8}{5}$	$-\frac{3}{5}$	$\frac{1}{5}$

The solution is not all-integer and $\bar{z}_0 = [-\frac{71}{5}] = -15$. We must now partition S_0 based on x_1 or x_2 and arbitrarily choose x_2. The resulting tree is shown in Figure 2.

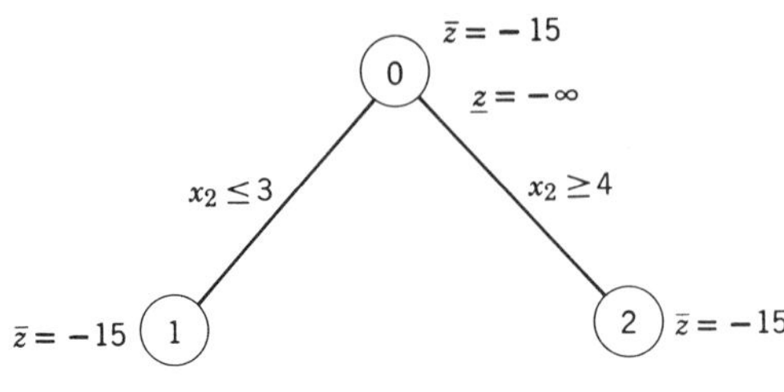

Figure 2

Taking the left branch we add the constraint $x_2 \leq 3$. The first dual simplex iteration yields Table 2.

Table 2

		$\overset{*}{-x_2}$	$-x_4$	$-x_5$	
x_0	$-\frac{29}{2}$	$-\frac{3}{8}$	$\frac{5}{8}$	$\frac{17}{8}$	$x_2 = 3$
x_1	$\frac{1}{2}$	$\frac{1}{8}$	$\frac{1}{8}$	$-\frac{3}{8}$	
x_3	$\frac{1}{2}$	$\frac{5}{8}$	$-\frac{3}{8}$	$\frac{1}{8}$	

The solution in Table 2 is primal feasible but not integer. There is no change in the bounds at v_1 or v_2, and we choose to partition S_1, using x_1 as in Figure 3.

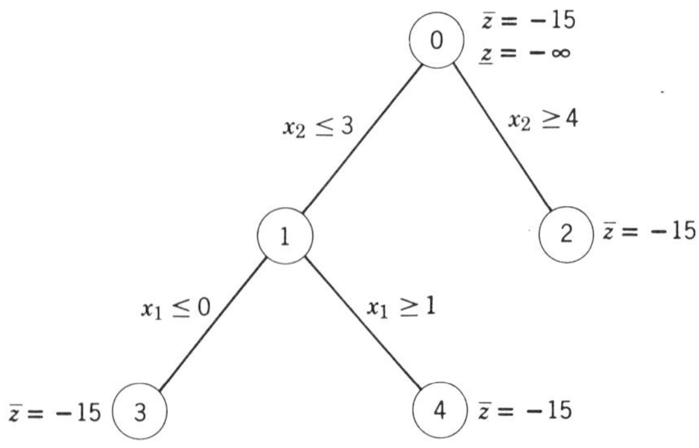

Figure 3

Taking the left branch and solving the LP, we obtain Table 3, which is primal feasible and all-integer. Thus $\underline{z}_0 = \underline{z}_3 = -17$. Also, v_3 is fathomed, but v_2 and v_4 remain live. We branch to v_4. Two dual simplex iterations yield Table 4. Thus $\bar{z}_4 = -17 < \underline{z}_0$ and v_4 is fathomed. The resulting tree is in Figure 4. The only live vertex is v_2.

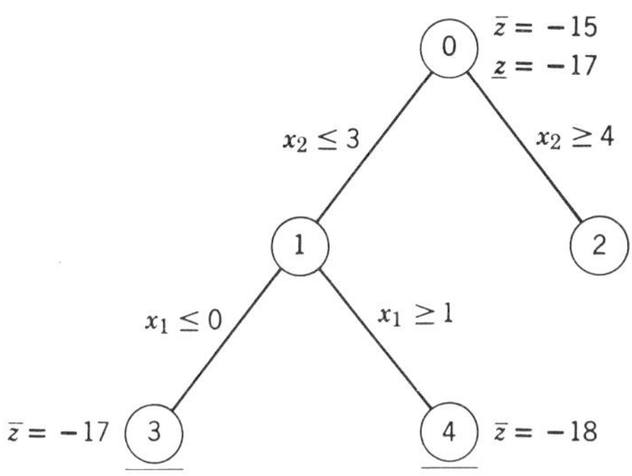

Figure 4

Table 3

		$\overset{*}{-x_1}$	$\overset{*}{-x_2}$	$-x_5$	
x_0	-17	-5	-1	4	$x_1 = 0$
x_3	2	3	1	-1	$x_2 = 3$
x_4	4	8	1	-3	

Table 4

		$-x_1$	$-x_2$	$-x_4$	
x_0	$-\frac{49}{3}$	$\frac{17}{3}$	$\frac{1}{3}$	$\frac{4}{3}$	
x_3	$\frac{7}{3}$	$\frac{1}{3}$	$\frac{2}{3}$	$-\frac{1}{3}$	$x_1 = 1$
x_5	$\frac{1}{3}$	$-\frac{8}{3}$	$-\frac{1}{3}$	$-\frac{1}{3}$	

Adding $x_2 \geq 4$ to Table 1 and doing one dual simplex iteration, we obtain Table 5. The resulting tree is in Figure 5 and we choose to branch on x_1. Taking the left branch, one dual simplex iteration yields Table 6.

Table 5

		$-x_2$	$-x_3$	$-x_5$	
x_0	$-\frac{43}{3}$	$\frac{2}{3}$	$\frac{5}{3}$	$\frac{7}{3}$	$x_2 = 4$
x_1	$\frac{1}{3}$	$\frac{1}{3}$	$\frac{1}{3}$	$-\frac{1}{3}$	
x_4	$\frac{1}{3}$	$-\frac{5}{3}$	$-\frac{8}{3}$	$-\frac{1}{3}$	

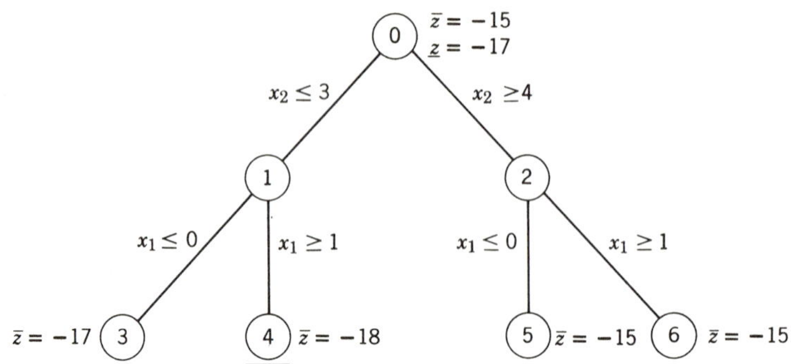

Figure 5

Table 6

		$-x_1$ *	$-x_3$	$-x_5$	
x_0	-15	-2	1	3	$x_1 = 0$
x_2	5	3	1	-1	
x_4	2	5	-1	-2	

The solution in Table 6 is primal feasible and integer. Thus $z_0 = \bar{z}_5 = -15$, and v_5 and v_6 are fathomed. An optimal solution is $x_1 = 0$, $x_2 = 5$, $x_3 = 0$, $x_4 = 2$, $x_5 = 0$, and $x_0 = -15$.

4.4. EFFICIENT BRANCHING AND BOUNDING

The rudimentary algorithm of Section 4.3 is readily modified to admit more efficient branching, partitioning, and bounding.

Branching to a Newest Vertex

At each vertex of the tree, information must be stored. This involves keeping track of the basis inverse, as well as the relevant upper and lower bounds. One can think of having a *list* of vertices, with newly generated information being put at the top of the list. Branching involves retrieving the required information about a vertex from somewhere in the list. Thus one is motivated to branch to a most recently generated vertex to facilitate searching the list. When branching away from v_0, this rule implies that the current problem remains in the computer core, and dual simplex iterations are performed. Note that when S_j is partitioned, the question of which one of v_j's successors to branch to is still open. This problem will be considered after we show how to calculate better bounds.

Calculating Better Bounds

Consider any vertex such that (5) has a noninteger optimal solution given by

$$x_{B_i} = y_{i0} - \sum_{j \in R} y_{ij} x_j, \qquad i = 0, \ldots, m$$

and $x_j = 0$ for all $j \in R$. For simplicity of exposition we assume that the LP has been solved by the ordinary dual (as opposed to the upper bounded dual) simplex algorithm. The modifications required for the upper bounded algorithm are asked for in Exercise 10.

Consider i such that $f_{i0} > 0$. If the constraint $x_{B_i} \leq [y_{i0}]$ is imposed, it is introduced into the tableau as

$$s_i = -f_{i0} + \sum_{j \in R} y_{ij} x_j$$

The pivot column will be determined by

$$\frac{y_{0k}}{y_{ik}} = \min_{j \in R} \left[\frac{y_{0j}}{y_{ij}}, \ y_{ij} > 0 \right]$$

and the corresponding decrease in y_{00} after the first dual simplex iteration is $f_{i0} y_{0k}/y_{ik}$. Thus we say that

$$D_i = \min_{j \in R} \left[\frac{f_{i0} y_{0j}}{y_{ij}}, \ y_{ij} > 0 \right]$$

is a *down penalty* for partitioning on row i. It is a lower bound on the decrease in y_{00}, since further dual simplex iterations may be required to restore primal feasibility. If $y_{ij} \leq 0$ for all $j \in R$, let $D_i = \infty$; x_{B_i} is called a *monotone increasing* variable. Monotone variables are discussed later in this section.

An *up penalty* U_i is derived similarly. Note that

$$x_{B_i} \geq \langle y_{i0} \rangle \Leftrightarrow s_i = -1 + f_{i0} - \sum_{j \in R} y_{ij} x_j$$

and thus

$$U_i = \min_{j \in R} \left[(f_{i0} - 1) \frac{y_{0j}}{y_{ij}}, \ y_{ij} < 0 \right]$$

If $y_{ij} \geq 0$ for all $j \in R$, let $U_i = \infty$ and x_{B_i} is a *monotone decreasing* variable.

Now, since one or the other of the penalties must be incurred, a lower bound on the decrease in y_{00} when the partitioning is based on x_{B_i} is

$$P_i = \min\{U_i, D_i\}$$

The penalties D_i and U_i were derived without taking into account the integrality requirement on the nonbasic variables. For instance, in order to have an integer solution, some nonbasic variable must become positive, and therefore not less than one. If x_q is such a variable, the corresponding decrease in y_{00} is at least y_{0q} so that

$$P = \min_{j \in R} y_{0j}$$

is a valid bound which does not depend on the partitioning row.

Another bound is available from the constraint of the method of integer forms (see Section 5.3):

$$s_i = -f_{i0} + \sum_{j \in R} f_{ij} x_j$$

where f_{ij} is the fractional part of y_{ij}. If this constraint were added to the tableau, the first dual simplex iteration would yield a decrease in y_{00} of

$$P_i' = \min_{j \in R}\left[\frac{f_{i0} y_{0j}}{f_{ij}}, f_{ij} \neq 0\right]$$

Finally, a lower bound on the decrease in y_{00} is

$$P^* = \max\left\{P, \max_{i, f_{i0} > 0} \max\{P_i, P_i'\}\right\}$$

Thus if P^* is obtained at vertex j, $\bar{z}_j = [z_j^0 - P^*]$ in Step 4.

Example

From Table 1 the following bounds are available:

$$D_1 = \tfrac{4}{5},\ U_1 = \min\{\tfrac{9}{5}, \tfrac{33}{10}\} = \tfrac{9}{5},\ P_1 = \tfrac{4}{5}$$
$$D_2 = \min\{\tfrac{3}{10}, \tfrac{44}{5}\} = \tfrac{3}{10},\ U_2 = \tfrac{2}{15},\ P_2 = \tfrac{2}{15}$$
$$P = \tfrac{2}{5}$$
$$P_1' = \min\{\tfrac{3}{10}, \tfrac{4}{5}, \tfrac{22}{15}\} = \tfrac{3}{10}$$
$$P_2' = \min\{\tfrac{4}{5}, \tfrac{4}{5}, \tfrac{44}{5}\} = \tfrac{4}{5}$$
$$P^* = \tfrac{4}{5}$$

Partitioning and Branching Based on the Improved Bounds

In the algorithm given above, partitioning has the effect of replacing one vertex by two and thus making the tree larger. It may happen that for large problems, the tree becomes so unwieldy as to make the branch and bound approach unattractive. However, consider the case where a *monotone variable* x_{B_i} is present at v_j. If row i is chosen as the partitioning row, v_j will have only one successor. If x_{B_i} is monotone increasing (decreasing), which means that D_i (U_i) is infinite, one need only consider $x_{B_i} \geq \langle y_{i0} \rangle$ $(x_{B_i} \leq [y_{i0}])$. For example, in Table 5 the basic variable x_4 is monotone increasing and we could have partitioned v_2 using only $x_4 \geq 1$, instead of increasing the size of the tree as we did by partitioning based on x_1.

When monotone variables are not present, a rule for partitioning and branching that has proved very effective in practice is:

1. Partition based on row k where $P_k^o = \max_i \max \{D_i, U_i\}$.
2. If $P_k^o = D_k$ (U_k), branch to $x_{B_i} \geq \langle y_{k0} \rangle$ $(x_{B_i} \leq [y_{k0}])$.

The rationale behind this rule is to take advantage of the fact that vertices which are not branched to have small upper bounds associated with them. Hopefully, many of these will never have to be considered, if good lower bounds are found on the vertices that were branched to.

Returning to Table 1, we would find $P_1^o = U_1 = \frac{9}{5}$ and branch to $x_1 \leq 0$. The vertex corresponding to $x_1 \geq 1$ would have an upper bound of $-\frac{71}{5} - \frac{9}{5} = -16$ associated with it and would never have been considered.

Partitioning for Simple Constraints

For some common constraints a more effective partitioning can be developed. Consider

$$\sum_{k \in Q} x_k = 1 \tag{10}$$

Assume that in an optimal solution to (5), $0 < x_t^o(j) < 1$, for some $t \in Q$. The partitioning given by (8) on x_t is

$$\begin{aligned} S_j^* &= \{S_j \cap \{x | x_t = 0\}, S_j \cap \{x | x_t = 1\}\} \\ &= \{S_{j_1}, S_{j_2}\} \end{aligned}$$

We have

$$S_{j1} = S_j \cap \{x | \sum_{k \in Q - \{t\}} x_k = 1\}$$

Therefore, it is likely that $|S_{j_1}|$ is much larger than $|S_{j_2}|$ and v_{j_1} may not be much easier to fathom than v_j.

For this reason it is advisable, if possible, to use a partition of S_j, $S'_j = \{S_{j_3}, S_{j_4}\}$, in which $|S_{j_3}|$ and $|S_{j_4}|$ are nearly equal. Note that $0 < x_t^o(j) < 1$ in (10) implies that there exists $t' \in Q$ such that $0 < x_{t'}^o(j) < 1$. Consider any partition $\{Q_1, Q_2\}$ of Q, where $t \in Q_1$ and $t' \in Q_2$. Let

$$S_{j_3} = S_j \cap \{x | \sum_{k \in Q_1} x_k = 1\} \qquad \text{and} \qquad S_{j_4} = S_j \cap \{x | \sum_{k \in Q_2} x_k = 1\}$$

Note that $x^o(j) \notin T_{j_2} \cup T_{j_4}$. If $Q_1 = \{t\}$, then $S_{j_3} = S_{j_2}$ and $S_{j_4} = S_{j_1}$. But, by the reasoning given above, it would be better to choose Q_1 and Q_2 of approximately equal cardinality.

Other simple constraints, including (24) of Section 1.4, can be handled using this kind of partitioning (see Exercise 2).

4.5. IMPLICIT ENUMERATION

Implicit enumeration is the name of a class of branch and bound algorithms designed specifically for the case in which x is required to be a binary vector. In theory this is no restriction, as long as an upper bound u_j is known for each x_j [see (22) of Section 1.4]. For large u_j, binary representation is likely to yield too many variables to be practical. There are many problems, however, where the variables are naturally binary (see Sections 1.4–1.5). For these problems implicit enumeration is especially appropriate.

Consider the ILP

$$\begin{aligned} \max z = cx& \\ Ax \leq b& \\ x \text{ binary}& \end{aligned} \tag{11}$$

We also assume with no loss of generality that $c \leq 0$, since any x_j with $c_j > 0$ can be replaced by $x'_j = 1 - x_j$, yielding a binary problem with nonpositive c.

Separation

Given

$$S_0 = \{x | Ax \leq b, x \text{ binary}\}$$

the separation at v_k is determined by choosing a particular variable x_j not chosen previously along the path P_k from v_0 to v_k, and letting

$$S_k^* = \{S_k \cap \{x | x_j = 0\}, S_k \cap \{x | x_j = 1\}\} \tag{12}$$

The path P_k corresponds to an assignment of binary values to a subset of the variables. Such an assignment is called a *partial solution*. Denote the index set of the assigned variables by W_k and let

$$S_k^+ = \{j \mid j \in W_k \text{ and } x_j = 1\}$$
$$S_k^- = \{j \mid j \in W_k \text{ and } x_j = 0\}$$
$$F_k = \{j \mid j \notin W_k\}$$

A *completion* of W_k is an assignment of binary values to the *free* variables specified by the index set F_k.

The partitioning (12) satisfies (2), where

$$H_0 = \{x \mid x \text{ binary}\} \text{ and } H_k = \{x \mid x_j = 0, 1, j \in F_k\}.$$

Finiteness of an enumeration algorithm based on (12) is thus guaranteed. For the ILP (11), *total* enumeration would terminate in 2^n steps. The name *implicit* enumeration means that many of the 2^n possible combinations will hopefully be discarded by various tests without requiring explicit enumeration.

Bounding

The problem considered at v_k is

$$\begin{aligned} \max z_k &= \sum_{j \in F_k} c_j x_j + \sum_{j \in S_k^+} c_j \\ &\sum_{j \in F_k} a_{ij} x_j \le b_i - \sum_{j \in S_k^+} a_{ij} = s_i, \qquad i = 1, \ldots, m \qquad (13) \\ &x_j = 0, 1, \qquad j \in F_k \end{aligned}$$

Let $T_k = H_k$. Since $c_j \le 0$, $x^o(k)$ is obtained by setting $x_j = 0, j \in F_k$. Thus $\bar{z}_k = z_k^o = \sum_{j \in S_k^+} c_j$. If, in addition $s = (s_1, \ldots, s_m) \ge 0$, then $x^o(k)$ is feasible to (13) and $\underline{z}_k = z_k^o$.

Fathoming

Again, the fathoming cases are

(a) $\bar{z}_k = \underline{z}_k$
(b) $\bar{z}_k \le \underline{z}_0$

using the bounds derived above. Note that (a) occurs when $x^o(k)$ is feasible to (13). A simple sufficient condition for (b) is also available. Suppose that for some i

$$t_i = \sum_{j \in F_k} \min\{0, a_{ij}\} > s_i$$

In this case, no completion of W_k can satisfy constraint i, $\bar{z}_k = -\infty$, and v_k is fathomed.

For example, consider the constraint

$$\sum_{j \in F_k} a_{ij}x_j = 3x_1 - 4x_2 + 3x_3 - 5x_4 \leq -10 = s_i$$

Vertex k is fathomed, since $t_i = -9 > s_i = -10$.

Choosing the Partitioning Variable and Branching

Let $Q_k = \{i | s_i < 0\}$. If $Q_k = \varnothing$, then v_k is fathomed since $x^0(k)$ is feasible. If $Q_k \neq \varnothing$, let

$$R_k = \{j | j \in F_k \text{ and } a_{ij} < 0 \text{ for some } i \in Q_k\}$$

At least one variable whose index is an element of R_k must equal one in any feasible completion of W_k. We partition on some x_j, $j \in R_k$, and then branch to the successor vertex corresponding to $x_j = 1$. The following rule selects a $j \in R_k$ in an attempt to drive toward feasibility. Define

$$I_k = \sum_{i=1}^{m} \max\{0, -s_i\} = -\sum_{i \in Q_k} s_i$$

to be the *infeasibility* of (13). By choosing x_j, the infeasibility at the successor vertex is

$$I_k(j) = \sum_{i=1}^{m} \max\{0, -s_i + a_{ij}\}$$

and x_p is selected such that

$$I_k(p) = \min_{j \in R_k} I_k(j)$$

For example, if the constraints at v_k are

$$\begin{aligned} -7x_1 + 2x_2 + 4x_3 &\leq 3 \\ 2x_1 - 3x_2 - 5x_3 &\leq -2 \\ 8x_1 - x_2 - 4x_3 &\leq -3 \end{aligned}$$

then $R_k = \{2, 3\}$, $I_k(2) = 2$, $I_k(3) = 1$, and x_3 is chosen as the partitioning variable.

As usual we adopt the rule of branching to the most recently generated vertex. A consequence of this rule and of branching to $x_k = 1$ is that the path P_k uniquely determines the remaining enumeration required. For instance, if $n = 4$ in Figure 6, then at v_9 it is known that all solutions containing $x_3 = 1$ or $x_3 = 0$, $x_2 = 1$, and $x_4 = 1$ have been enumerated. There

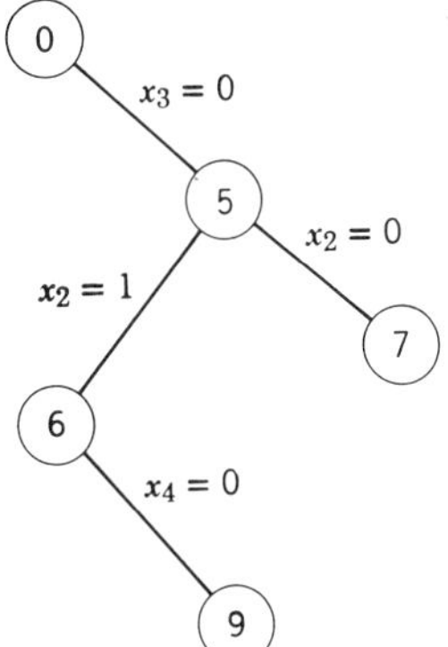

Figure 6

remain six possible solutions, $(x_1, x_2, x_3, x_4) = (1, 1, 0, 0)$, $(0, 1, 0, 0)$, $(0, 0, 0, 0)$, $(0, 0, 0, 1)$, $(1, 0, 0, 0)$, and $(1, 0, 0, 1)$.

It is possible to represent P_k concisely in vector form. In Figure 6, P_9 is represented by the vector $(\underline{3}, 2, \underline{4})$, where the order of the components represents the "level" in the tree. Indices appear in the vector representation of P_k if they are in W_k. They appear underlined if they are in S_k^-. The vector P_k is quite simple to use. When branching to $x_p = 1$ from v_k, simply change P_k to (P_k, p). In Figure 6, branching to $x_1 = 1$ from v_9 yields $(\underline{3}, 2, \underline{4}, 1)$. In backtracking, underline the rightmost nonunderlined entry and erase all entries to the right of it. In Figure 6, to backtrack to v_7 from v_9, the vector is transformed from $(\underline{3}, 2, \underline{4})$ to $(\underline{3}, \underline{2})$. If all entries of P_k are underlined and vertex k is fathomed, every element of W_k has been evaluated at zero and one, and the enumeration is complete.

A Rudimentary Implicit Enumeration Algorithm

STEP 1: (Initialization.) At v_0, $F_0 = \{1, \ldots, n\}$, $\bar{z}_0 = \infty$, $\underline{z}_0 = -\infty$. Go to Step 2.

STEP 2: (Calculating bounds.) At v_k, let $\bar{z}_k = \sum_{j \in S_k^+} c_j$. If $s \geq 0$, let $\underline{z}_k = z_k^0 = \bar{z}_k$ and let $\underline{z}_0 = \max\{\underline{z}_0, \underline{z}_k\}$. Go to Step 3.

STEP 3: (Fathoming.) If $t_i > s_i$ for any i, or if $\bar{z}_k = \underline{z}_k$, or if $\bar{z}_k \leq \underline{z}_0$, v_k is fathomed and we go to Step 4. If v_k is live, go to Step 5.

STEP 4: (Backtracking.) If no live vertex exists, go to Step 6. Otherwise branch to the newest live vertex and go to Step 2.

STEP 5: (Partitioning and branching.) Partition on x_p as in (12), where $I_k(p) = \min_{j \in R_k} I_k(j)$. Branch to the $x_p = 1$ vertex. Go to Step 2.

STEP 6: (Termination.) If $\underline{z}_0 = -\infty$, there is no feasible solution. If $\underline{z}_0 > -\infty$, that feasible solution which yielded $\underline{z}_0$ is optimal.

Example

$$\begin{aligned} \max z = -5x_1 - 7x_2 - 10x_3 - 3x_4 - x_5 & \\ -x_1 + 3x_2 - 5x_3 - x_4 + 4x_5 &\le -2 \\ 2x_1 - 6x_2 + 3x_3 + 2x_4 - 2x_5 &\le 0 \\ x_2 - 2x_3 + x_4 + x_5 &\le -1 \\ x_1, \ldots, x_5 = 0, 1 & \end{aligned}$$

Initially, $F = \{1, \ldots, 5\}$, $S_0^+ = S_0^- = \varnothing$, $\bar{z}_0 = \infty$, $\underline{z}_0 = -\infty$, $s = (-2, 0, -1)$, $I_0 = 3$, $R_0 = \{1, 3, 4\}$. For partitioning, compare $I_0(1) = 4$, $I_0(3) = 3$, $I_0(4) = 5$, and choose x_3. The tree is shown in Figure 7.

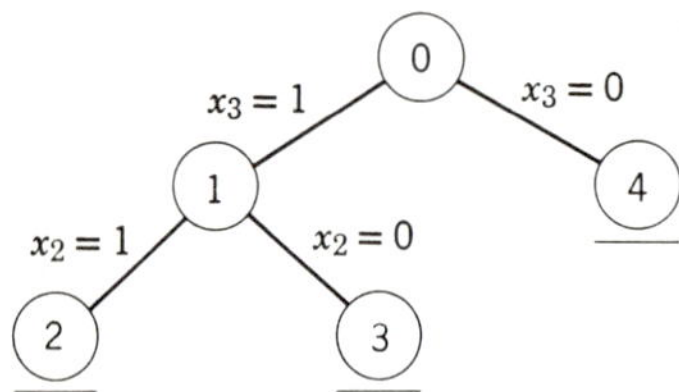

Figure 7

At v_1, $P_1 = (3)$, $\bar{z}_1 = -10$, $s = (3, -3, 1)$, $R_1 = \{2, 5\}$, $I_1(2) = 0$, $I_1(5) = 2$. Choose x_2.

At v_2, $P_2 = (3, 2)$, $\bar{z}_2 = -17 = \underline{z}_2 = \underline{z}_0$, since $s = (0, 3, 0)$; v_2 is fathomed, and we branch to v_3.

At v_3, $P_3 = (3, \underline{2})$, $s = (3, -3, 1)$, $R_3 = \{5\}$. But $t_2 = -2 > s_2 = -3$ and v_3 is fathomed. Branch to v_4.

At v_4, $P_4 = (\underline{3})$, $s = (-2, 0, -1)$, $R_4 = \{1, 4\}$. But $t_3 = 0 > s_3 = -1$ and v_4 is fathomed. There are no live vertices, and the problem is solved. The optimal solution is $x_1 = 0$, $x_2 = 1$, $x_3 = 1$, $x_4 = 0$, $x_5 = 0$; $z = -17$.

Generalized Branching

It is quite easy to modify the specific branching rules in the algorithm given above. For instance, the requirement of branching to $x_p = 1$ in Step 5 can be relaxed. The only difficulty encountered is in the vector representation of P_k. If branching to $x_p = 0$ prior to consideration of the $x_p = 1$ vertex is permitted, some revised bookkeeping is required. If $j \in W_k$, let it appear in P_k as

$$\begin{cases} j & \text{if } j \in S_k^+ \text{ and } x_j = 0 \text{ has not been considered} \\ \underline{j} & \text{if } j \in S_k^+ \text{ and } x_j = 0 \text{ has been considered} \\ -j & \text{if } j \in S_k^- \text{ and } x_j = 1 \text{ has not been considered} \\ -\underline{j} & \text{if } j \in S_k^- \text{ and } x_j = 1 \text{ has been considered} \end{cases}$$

The vector P_k is updated as before except that in backtracking, after erasing the underlined entries on the right, the rightmost remaining entry is underlined *and changes sign*. For example, if the order of vertices considered in Figure 7 had been v_1, v_3, v_2, v_4, the sequence of vectors would have been $(3), (3, -2), (3, \underline{2}), (-\underline{3})$.

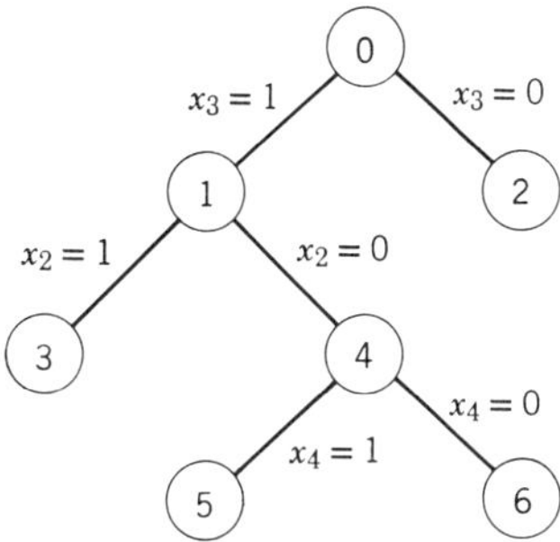

Figure 8

Branching can be further generalized by choosing to commit more than one index at a time to W_k. Instead of branching to a successor of v_k, one can branch to any vertex on the path out of v_k. As long as backtracking is done as before, there is no danger of overlooking the optimal solution. For example, in the tree of Figure 8 one could branch directly from v_1 to v_5. The vector sequence would be $(3), (3, -2, 4)$.

This option is especially valuable for starting the enumeration. It allows us to begin at a vertex other than v_0. In particular, if a good feasible solution x^* is known at the outset, one can start at the vertex corresponding to that solution. Let $\{j \mid x_j^* = 1\} = \{j_1, \ldots, j_k\}$; then the vector representation of the corresponding vertex is $(j_1, \ldots, j_k)$. Since x^* is feasible, the vertex is fathomed, $\underline{z} = cx^*$ and backtracking occurs. In Figure 8, to start at v_3, set $P_3 = (3, 2)$. A starting solution could also be chosen that is not feasible, but which is believed to be near an optimal solution. In this case, backtracking does not occur automatically.

4.6. ADDITIONAL TESTS FOR IMPLICIT ENUMERATION

Many additional tests have been proposed that can reduce the enumeration at the expense of added calculation. Some of these are presented here; others are given as exercises. In general, they are straightforward and proofs are omitted.

Augment (13) by the constraint,

$$\sum_{j \in F_k} c_j x_j + z_k^o \geq \underline{z}_0$$

which must be satisfied by any optimal solution. The resulting ILP is

$$\begin{aligned} \max z_k &= \sum_{j \in F_k} - a_{0j} x_j + z_k^o \\ &\sum_{j \in F_k} a_{ij} x_j \leq s_i, \quad i = 0, \ldots, m \\ &x_j = 0, 1, \quad j \in F_k \end{aligned} \tag{14}$$

where $a_{0j} = -c_j$ and $s_0 = z_k^o - \underline{z}_0$. It is assumed that v_k is live so that

(a) $s_0 > 0$
(b) $Q_k = \{i | s_i < 0\} \neq \varnothing$
(c) $t_i = \sum_{j \in F_k} \min \{0, a_{ij}\} \leq s_i, \quad i = 0, \ldots, m$

Test 1: Suppose for some i and $p \in F_k$,

$$\sum_{j \in F_k} \min \{0, a_{ij}\} + |a_{ip}| > s_i$$

Then $a_{ip} > 0$ $(a_{ip} < 0)$ implies that $x_p = 0$ $(x_p = 1)$ in any feasible completion of W_k.

For example, the constraint

$$-3x_1 + 2x_2 - 5x_3 + 6x_4 \leq -4$$

implies that $x_3 = 1$ and $x_4 = 0$.

Tests 2–4 are performed by first computing numbers L and U such that the constraint

$$L \leq \sum_{j \in F_k} x_j \leq U$$

is valid for any optimal completion of W_k.

For each constraint $i = 0, \ldots, m$, let the variables be reindexed so that the coefficients a_{ij} are in nondecreasing order. The reindexed coefficients for row i are $(a_{i,1(i)}, a_{i,2(i)}, \ldots, a_{i,l(i)})$ where $l = |F_k|$ and $t < p \Rightarrow a_{i,t(i)} \leq a_{i,p(i)}$. Define $T_i(0) = 0$ and let

$$T_i(p) = \sum_{j=1}^{p} a_{i,j(i)}$$

be the sum of the p smallest coefficients of row i.

For each $i \in Q_k$ determine L_i, where

$$T_i(L_i) \leq s_i < T_i(L_i - 1)$$

Then $L_i \leq L$ and we take

$$L = \max_{i, s_i < 0} L_i.$$

To determine U, for $i = 0, \ldots, m$ calculate U_i given by

$$\begin{aligned} T_i(U_i) \leq s_i < T_i(U_i + 1) \qquad & \text{if } T_i(l) > s_i \\ U_i = \infty \qquad & \text{if } T_i(l) \leq s_i \end{aligned}$$

Then $U_i \geq U$ and $U = \min_i U_i$.

Example 1

$$\begin{aligned} 3x_1 + x_2 + 4x_3 + x_4 &\leq 11 \\ -x_1 - x_2 + 2x_3 - 3x_4 &\leq -4 \\ 3x_1 - 4x_2 - 2x_3 + 6x_4 &\leq -1 \end{aligned}$$

In Example 1, $L_1 = 2$, $L_2 = 1$, $L = 2$ and $U_0 = \infty$, $U_1 = 3$, $U_2 = 3$; $U = 3$.

Test 2: If $L > U$, v_k is fathomed, since (14) cannot be satisfied. Also if $L = |F_k|$, then each of the free variables can be set to one. In other words, if $F_k = \{j(1), \ldots, j(l)\}$, then P_k is updated to $(P_k, \underline{j(1)}, \ldots, \underline{j(l)})$.

Test 3: Let

$$T_i(L, U) = \min_{L \leq p \leq U} T_i(p), \qquad i = 0, \ldots, m$$

be the smallest sum that can be obtained using at least L, but not more than U, free variables. Assume that vertex k is not fathomed by Test 2, and a_{ip} is not one of the numbers which determines $T_i(L, U)$. If $T_i(L - 1, U - 1) > s_i - a_{ip}$ then $x_p = 0$.

In Example 1, $T_2(1, 2) = -6 > s_2 - a_{24} = -7$. Therefore, $x_4 = 0$, and the new vector is $(P_k, -\underline{4})$.

Test 4: Suppose that vertex k is not fathomed by Test 2, and a_{ip} is one of the numbers that determines $T_i(L, U)$. If $T_i(L + 1, U + 1) > s_i + a_{ip}$ then $x_p = 1$.

In Example 1, $T_1(3, 4) = -5 > s_1 + a_{14} = -7$. Therefore, $x_4 = 1$, and the new vector is $(P_k, \underline{4})$. Tests 3 and 4 together showed that there is no feasible solution to Example 1.

Tests 5 and 6 seek to determine if there is a constraint that can be satisfied only at the expense of violating the zero row.

Test 5: For any constraint i and $p \in F_k$ such that $a_{ip} > s_i$, calculate

$$a_0^* = \min \{a_{0j} | j \in F_k - \{p\}, a_{ij} < 0\}.$$

If $a_{0p} + a_0^* > s_0$, then $x_p = 0$.

Example 2

$$\begin{aligned}
3x_1 + 4x_2 + 7x_3 + x_4 + 8x_5 &\leq 9 \\
-x_1 - 2x_2 - x_3 + x_4 - 3x_5 &\leq -2 \\
2x_1 - 3x_2 - 2x_3 + 3x_4 - x_5 &\leq -4 \\
x_1 - x_2 - 3x_3 + 5x_4 - 2x_5 &\leq -4
\end{aligned}$$

In Example 2, let $i = 2$ and $p = 2$ yielding $a_0^* = 7$, and $a_{02} + a_0^* = 11 > 9$ so that $x_2 = 0$.

Test 6: For $i \in Q_k$, calculate

$$r_i = \min_{j, a_{ij} < 0} \frac{s_i a_{0j}}{a_{ij}}$$

If $r_i > s_0$, vertex k is fathomed.

In Example 2, $r_3 = \frac{28}{3} > 9$ and there is no optimal completion. Note that this is the only proposed test which involves multiplications and divisions, which are much slower than additions and subtractions. Substantial parts of the tree would have to be eliminated in order to make Test 6 worthwhile.

4.7. SURROGATE CONSTRAINTS

In the binary ILP (13), consider replacing the set of constraints,

$$\sum_{j \in F_k} a_{ij} x_j \leq s_i, \qquad i = 1, \ldots, m \tag{15}$$

with the single constraint,

$$\sum_{i=1}^{m} \sum_{j \in F_k} u_i a_{ij} x_j \leq \sum_{i=1}^{m} u_i s_i \tag{16}$$

where $u = (u_1, \ldots, u_m) \geq 0$. Now, (16) is weaker than (15) because

$$S_k(u) \supseteq S_k$$

where S_k and $S_k(u)$ are the set of binary solutions to (15) and (16), respectively. Constraint (16) has the advantage, however, of containing information different from that of any individual constraint of (15), unless u is a unit vector. Fathoming tests can be performed initially on the *surrogate constraint* (16), since

$$S_k(u) = \varnothing \Rightarrow S_k = \varnothing$$

Only if vertex k is still live after testing (16) need constraints (15) be investigated.

Example

Consider the constraint set

$$\begin{aligned} x_1 - x_2 &\le -1 \\ -x_1 + 2x_2 &\le -1 \end{aligned}$$

The vertex could not be fathomed by considering either constraint individually. However, letting $u_1 = u_2 = 1$ yields

$$x_2 \le -2$$

which has no binary solution.

Defining Best Surrogate Constraints

Any $u \ge 0$ generates a valid surrogate constraint. A good surrogate should have the property of eliminating as many nonoptimal completions as possible. Let

$$g(u) = \max \sum_{j \in F_k} c_j x_j, \qquad x \in S_k(u)$$

Since $S_k(u) \supseteq S_k$, $g(u) \ge z_k - z_k^0$, for all $x \in S_k$ and $u \ge 0$.

One way to define a strongest surrogate constraint is to say that u generates a *stronger* constraint than u' if $g(u) < g(u')$. In other words, if the best solution in $S_k(u)$ is worse than the best in $S_k(u')$, it is likely that there are more nonoptimal solutions in $S_k(u')$ than in $S_k(u)$. Using the above definition, the *best* surrogate constraint would be generated by u^*, where

$$g(u^*) = \min_{u \ge 0} g(u) \tag{17}$$

Solving (17) is not easy. A reasonable estimate of u^* can be calculated, however, by dropping the integrality restrictions in $S_k(u)$. In other words, let

$$\begin{aligned} g'(u) = \max &\sum_{j \in F_k} c_j x_j \\ &\sum_{j \in F_k} \sum_{i=1}^{m} u_i a_{ij} x_j \le \sum_{i=1}^{m} u_i s_i \\ &0 \le x_j \le 1, \qquad j \in F_k \end{aligned} \tag{18}$$

and say that the best surrogate constraint is generated by u^o, where

$$g'(u^o) = \min_{u \geq 0} g'(u) \tag{19}$$

Calculating Best Surrogate Constraints

The LP corresponding to the ILP (14) at vertex k is

$$\begin{aligned} \max \gamma_k &= \sum_{j \in F_k} c_j x_j \\ &\sum_{j \in F_k} a_{ij} x_j \leq s_i, \qquad i = 1, \ldots, m \\ &0 \leq x_j \leq 1, \qquad j \in F_k \end{aligned} \tag{20}$$

where $c_j = -a_{0j}$ and the constant z_k^o has been dropped from the objective function.

The dual of (20) is

$$\begin{aligned} \min q_k &= \sum_{i=1}^{m} s_i v_i + \sum_{j \in F_k} w_j \\ &\sum_{i=1}^{m} a_{ij} v_i + w_j \geq c_j, \qquad j \in F_k \\ &v_i, w_j \geq 0 \text{ for all } i \text{ and } j. \end{aligned} \tag{21}$$

If (20) has no feasible solution, then vertex k is fathomed. If (20) has a feasible solution, it must have an optimal solution (γ_k^o, x^o), since its constraint set is bounded. Assuming that (20) is feasible, (21) has an optimal solution (q_k^o, v^o, w^o) and $\gamma_k^o = q_k^o$. We are now ready to show how to calculate the best surrogate constraint by the definition (19).

Theorem 1: *The best surrogate constraint is generated by* $u^o = v^o$.

PROOF: Clearly,

$$q_k^o = \gamma_k^o = \sum_{j \in F_k} c_j x_j^o \leq g'(u) \qquad \text{for any } u \geq 0$$

since any feasible solution to (20) is also feasible to (18). Letting $u = v^o$ and x' be any feasible solution to (18) yields

$$\sum_{j \in F_k} \sum_{i=1}^{m} v_i^o a_{ij} x_j' \leq \sum_{i=1}^{m} v_i^o s_i \tag{22}$$

Also, $x_j' \leq 1$, $w^o \geq 0$ imply that

$$\sum_{j \in F_k} x_j' w_j^o \leq \sum_{j \in F_k} w_j^o \tag{23}$$

Adding (22) and (23) yields

$$\sum_{i=1}^{m} \sum_{j \in F_k} a_{ij} v_i^o x_j' + \sum_{j \in F_k} w_j^o x_j' \leq \sum_{i=1}^{m} s_i v_i^o + \sum_{j \in F_k} w_j^o = q_k^o \tag{24}$$

Now, (v^o, w^o) is feasible to (21) and the constraints of (21) weighted by $x' \geq 0$ yield

$$\sum_{j \in F_k} \sum_{i=1}^{m} a_{ij} v_i^o x_j' + \sum_{j \in F_k} w_j^o x_j' \geq \sum_{j \in F_k} c_j x_j' \tag{25}$$

From (24) and (25), we verify that $q_k^o \geq g'(v^o)$. Finally, $g'(v^o) \leq q_k^o \leq g'(u)$ for any $u \geq 0$ implies that $g'(v^o) = \min_{u \geq 0} g'(u)$. ■

Theorem 1 says that the best surrogate constraint is generated by the dual variables corresponding to the constraints of (15). Solving the LP (20) yields not only the best surrogate constraint but also makes applicable all of the results of Sections 4.3 and 4.4. In particular, the bounds developed in these sections would now be available. Any vertex can be fathomed if the LP terminates integer, and the efficient branching rule developed in Section 4.4 may be substituted for the rule of Section 4.5.

Example

The example of Section 4.5 is

$$\begin{aligned} \max z = -5x_1 - 7x_2 - 10x_3 - 3x_4 - x_5 & \\ -x_1 + 3x_2 - 5x_3 - x_4 + 4x_5 &\leq -2 \\ 2x_1 - 6x_2 + 3x_3 + 2x_4 - 2x_5 &\leq 0 \\ x_2 - 2x_3 + x_4 + x_5 &\leq -1 \\ x_1, \ldots, x_5 &= 0, 1 \end{aligned}$$

Let $\underline{z}_0 = -26 = \sum_j c_j$, since $x_j \leq 1$.

Solving the corresponding LP yields $x_1 = 0$, $x_2 = \frac{1}{3}$, $x_3 = \frac{2}{3}$, $x_4 = x_5 = 0$; $z = -9 = \bar{z}_0$. The associated dual variables are $v_1 = 0$, $v_2 = \frac{8}{3}$, and $v_3 = 9$, and the resulting surrogate constraint is

$$\tfrac{16}{3}x_1 - 7x_2 - 10x_3 + \tfrac{43}{3}x_4 + \tfrac{11}{3}x_5 \leq -9$$

Apply Test 3, Section 4.6, to this surrogate constraint called row r. Then $L_r = 1$, $U_r = 3$, $T_r(0, 2) = -17 > -\frac{70}{3}$ and $x_4 = 0$. Branch to $P_1 = (-\underline{4})$. Now $I_0 = 3$, $R_1 = \{1, 3\}$, $I_0(1) = 4$, $I_0(3) = 3$, and we branch to $P_2 = (-\underline{4}, 3)$.

The resulting LP is

$$\begin{aligned} \max z_2 = -5x_1 - 7x_2 - x_5 - 10 \\ -x_1 + 3x_2 + 4x_5 \le 3 \\ 2x_1 - 6x_2 - 2x_5 \le -3 \\ x_2 + x_5 \le 1 \\ 0 \le x_1, x_2, x_5 \le 1 \end{aligned}$$

The optimal solution is $x_1 = 0$, $x_2 = \frac{1}{3}$, $x_5 = \frac{1}{2}$, $z_2 = -\frac{77}{6}$, yielding the bound $\bar{z}_2 = -13$. The corresponding dual variables are $v_1 = \frac{4}{9}$, $v_2 = \frac{25}{18}$, and $v_3 = 0$; the resulting surrogate constraint is

$$\tfrac{21}{9}x_1 - 7x_2 - x_5 \le -\tfrac{17}{6}$$

Test 1 of Section 4.6 applied to this surrogate constraint determines that $x_2 = 1$. The rest of the example will not be carried out, since the remaining vertices are fathomed by the test of Section 4.5.

Options

It may be more efficient to calculate new surrogate constraints periodically, rather than at every vertex. A constraint calculated at v_k is valid at every successor vertex, and one has the option of using the older constraint. The advantage of always having the best constraint at every vertex, and of having recalculated bounds, may be offset by the disadvantage of having to solve a great many LP's. However, it is generally the case that successive LP's are closely related and the solution of one LP can be used as a starting point for the next. Furthermore, one is not limited to using combinations of the constraints of (15) to form the surrogate. Any other valid constraint, including those derived from tests of Section 4.6 may also be used.

4.8. EXTENSIONS TO THE MILP

The MILP

$$\begin{aligned} x_0^* = \max x_0 = c_1x + c_2v \\ A_1x + A_2v \le b \\ x \ge 0 \text{ integer} \\ v \ge 0 \end{aligned} \tag{26}$$

can also be solved by enumeration techniques.

Branch and Bound

The most straightforward extension is that of the basic branch and bound algorithm of Section 4.3. Enumeration is restricted to the components of x, and the separation is given by (8). Although neither S_0, the constraint set of (26), nor S_k is finite, bounded enumeration is assured if an upper bound d_j is known for each x_j. Define H_k as in (9), and let $f(S_k) = |H_k|$. By applying the argument given in Section 4.3, it is easy to show that (2a) holds and fathoming occurs when $|H_k| \leq 1$. The results of Section 4.4 can also be extended rather easily to MILP's (see Exercise 23).

Extending implicit enumeration to the MILP is a more interesting problem. In Section 4.10, an algorithm is given that is based on the *partitioning* result of Section 4.9.

4.9. SOLVING THE MILP BY PARTITIONING

Consider the MILP (26) and let $x = x'$, where x' is any nonnegative integer vector. The resulting problem,

$$\begin{aligned} x_0^*(x') = c_1x' + \max\ & c_2v \\ & A_2v \leq b - A_1x' \\ & v \geq 0 \end{aligned} \tag{27}$$

is an LP. The vector x' is said to be *admissible* if (27) has an *optimal* solution. The dual of (27) is

$$\begin{aligned} u_0^*(x') = c_1x' + \min\ & u(b - A_1x') \\ & uA_2 \geq c_2 \\ & u \geq 0 \end{aligned} \tag{28}$$

Consider the possible relationships between (26), (27), and (28).

1. [The dual LP (28) has no feasible solution.] Since the dual constraint set is independent of x', there is no admissible x'. In other words, for every x', (27) is either infeasible or unbounded, which implies that (26) is either infeasible or unbounded. The MILP

$$\begin{aligned} \max x_0 = \quad & x_1 - 3v_1 + 4v_2 \\ & -2x_1 - v_1 + v_2 \leq -3 \\ & -3x_1 + v_1 - v_2 \leq 2 \\ & v_1, v_2 \geq 0,\ x_1 \geq 0 \text{ integer} \end{aligned}$$

is an example for which (26) is unbounded.

2. [The dual LP (28) is unbounded for all nonnegative integer x'.] In this case (27) is infeasible for every x'; therefore the MILP (26) has no feasible solution. An example is

$$\begin{aligned} \max x_0 &= 5x_1 + 3v_1 \\ x_1 &+ v_1 \leq -3 \\ v_1 &\geq 0,\ x_1 \geq 0 \text{ integer} \end{aligned}$$

3. [The dual LP (28) has an optimal solution $u^o(x')$ for some nonnegative integer x'.] If the corresponding optimal primal variables are $v^o(x')$, then $(x', v^o(x'))$ is a feasible solution to the MILP (26). For any such feasible x', it must be the case that

$$x_0^*(x') = u_0^*(x') \leq x_0^* \tag{29}$$

Since (28) is feasible and its constraints are independent of x', (27) cannot be unbounded for any x'. Therefore (26), if bounded, has an optimal solution.

Now since (26) can be written as

$$x_0^* = \max x_0^*(x), \qquad x \geq 0 \text{ integer and admissible}$$

it can also be written as

$$x_0^* = \max u_0^*(x), \qquad x \geq 0 \text{ integer and admissible}$$

or equivalently

$$x_0^* = \max_{\substack{x \geq 0 \text{ integer} \\ \text{and admissible}}} \begin{bmatrix} c_1x + \min u(b - A_1x) \\ uA_2 \geq c_2 \\ u \geq 0 \end{bmatrix} \tag{30}$$

For any admissible x', $u^o(x')$ can be found at an extreme point of the dual constraint set. Let

$$T = \{u^t | u^t \text{ is an extreme point of the constraint set of (28)}\}$$

be the finite set of such points. Note that T is independent of x'. Then (28) can be written as

$$u_0^*(x') = c_1x' + \min_{u^t \in T} u^t(b - A_1x') \tag{31}$$

Given that (28) has a feasible solution, then we may add constraints on the integer variables that will eliminate inadmissible x or equivalently ensure that (28) is bounded. For any inadmissible x' (28) is unbounded, and it must be the case that $u(b - A_1x')$ decreases along some extreme ray. In other words, there exists an extreme point u^t and a direction y such that every point on the extreme ray $u^t + \theta y$, $\theta \geq 0$, is feasible to (28), and such that

$$(u^t + \theta y)(b - A_1x')$$

decreases with θ or

$$\theta y(b - A_1x')$$

decreases with θ or

$$y(b - A_1x') < 0$$

Let

$$Q = \{y^q | u^t + \theta y^q, \theta \geq 0 \text{ is an extreme ray for some } u^t \in T\}$$

and require x' to satisfy

$$y^q(b - A_1x') \geq 0 \qquad \text{every } y^q \in Q$$

Now, from (30) and (31), the MILP (26) can be written as

$$x_0^* = \max_{x \geq 0 \text{ integer}} c_1x + \begin{bmatrix} \min_{u^t \in T} u^t(b - A_1x) \\ y^q(b - A_1x) \geq 0 \qquad \text{every } y^q \in Q \end{bmatrix} \tag{32}$$

Finally, (32) can be written as an MILP by introducing the objective function as a variable u_0 such that

$$u_0 \leq c_1x + u^t(b - A_1x) \qquad \text{every } u^t \in T$$

Then (32) is equivalent to

$$\begin{aligned} x_0^* = \max\ & u_0 \\ u_0 &\leq c_1x + u^t(b - A_1x) \qquad \text{every } u^t \in T \\ 0 &\leq y^q(b - A_1x) \qquad \text{every } y^q \in Q \\ & x \geq 0 \text{ integer} \end{aligned} \tag{33}$$

Now, (33) is an MILP having one continuous variable u_0. It will be convenient to refer to (33) as an ILP, keeping in mind that any algorithm for ILP's would have to be modified slightly to solve it. Essentially, the MILP (26) has been partitioned into an LP (28) and an ILP (33). In order to determine the constraints of (33) however, all of the extreme points and extreme rays of (28) must be known. The ILP (33) is really of theoretical interest only, since it will generally have an enormous number of constraints. However, (33) can be solved iteratively by generating constraints only when they are needed.

An Iterative Solution Technique

At iteration k, let the set of *known* extreme points of (28) be $T(k)$; and the set of directions of *known* extreme rays $Q(k)$. Assume that

$$\{u | uA_2 \geq c_2, u \geq 0\} \neq \varnothing;$$

otherwise the MILP has no optimal solution.

STEP 1: (Initialization.) Let $T(1)$ and $Q(1)$ be any known subsets of T and Q, respectively. Let $k = 1$, where k counts the iterations. If $T(1) = Q(1) = \varnothing$, let u_0^1 be arbitrarily large, x^1 be any nonnegative integer vector, and go to Step 3. Otherwise go to Step 2.

STEP 2: (Solving the ILP.) Solve the ILP

$$\begin{aligned} &\max u_0 \\ &\quad u_0 \leq c_1 x + u^t(b - A_1 x) \qquad \text{every } u^t \in T(k) \\ &\quad 0 \leq \qquad\qquad y^q(b - A_1 x) \qquad \text{every } y^q \in Q(k) \\ &\qquad\qquad x \geq 0 \text{ integer} \end{aligned} \tag{34}$$

(a) If no feasible solution to (34) exists, then there is no feasible solution to (33), since the constraints of (34) are included in those of (33). Therefore, (26) has no feasible solution.
(b) If (34) has an optimal solution, call it (u_0^k, x^k). Go to Step 3.
(c) If (34) is unbounded, let u_0^k be arbitrarily large and x^k be any feasible solution to (34) with $u_0 = u_0^k$. Go to Step 3.

In cases (b) and (c), $x_0^* \leq u_0^k \leq u_0^{k-1}$, since (34) at iteration $k - 1$ is a relaxation of (34) at iteration k, and since (34) at any iteration is a relaxation of (33).

STEP 3: (Solving the dual LP.) Solve the LP

$$\begin{aligned} u_0^*(x^k) = c_1 x^k + \min\ & u(b - A_1 x^k) \\ & uA_2 \geq c_2 \\ & u \geq 0 \end{aligned} \tag{35}$$

(a) If $u_0^*(x^k) \to -\infty$ along an extreme ray $u^k + \theta y^k$, let $T(k + 1) = T(k) \cup \{u^k\}$ and $Q(k + 1) = Q(k) \cup \{y^k\}$. Let $k = k + 1$ and return to Step 2.
(b) If (35) has an optimal solution, call it u^k, and let the associated solution to (27) be v^k. Using (29), we obtain $u_0^*(x^k) \leq x_0^* \leq u_0^k$. It follows that if $u_0^*(x^k) = u_0^k$, an optimal solution to (26) is given by (x^k, v^k). If $u_0^*(x^k) < u_0^k$, let $T(k + 1) = T(k) \cup \{u^k\}$, set $k = k + 1$, and go to Step 2.

The number of iterations in this algorithm is bounded from above by $|T| + |Q|$, since an additional extreme point or extreme ray is generated at each iteration. To see this, consider the possible results of Step 3 at iteration k:

(1) $u_0^*(x^k) \to -\infty$, which implies that $y^k(b - A_1x^k) < 0$. It follows that $y^k \notin Q(k)$, since x^k is feasible to (34). Thus $|Q(k + 1)| = |Q(k)| + 1$.

(2) $-\infty < u_0^*(x^k) < u_0^k$. In this case it follows that $u^k \notin T(k)$, since

$$c_1x^k + u^k(b - A_1x^k) = u_0^*(x^k) < u_0^k \leq c_1x^k + u^t(b - A_1x^k),$$

for every $u^t \in T(k)$. Thus $|T(k + 1)| = |T(k)| + 1$.

(3) $-\infty < u_0^*(x^k) = u_0^k$ implies termination.

Example

$$\begin{aligned} \max x_0 &= -x_1 - 4x_2 - 2v_1 - 3v_2 \\ &x_1 - 3x_2 + v_1 - 2v_2 \leq -2 \\ &-x_1 - 3x_2 - v_1 - v_2 \leq -3 \\ &x_1, x_2 \geq 0 \text{ integer} \\ &v_1, v_2 \geq 0 \end{aligned}$$

Initially, $T(1) = Q(1) = \varnothing$, $k = 1$ and u_0^1 arbitrarily large.

ITERATION 1:
Since $T(1) = Q(1) = \varnothing$, arbitrarily choose $x_1^1 = x_2^1 = 0$. Solve the LP,

$$\begin{aligned} u_0^*(x^1) = \min &-2u_1 - 3u_2 \\ &u_1 - u_2 \geq -2 \\ &-2u_1 - u_2 \geq -3 \\ &u_1, u_2 \geq 0 \end{aligned}$$

The dual constraint set is shown in Figure 9. The solution is $u^1 = (\frac{1}{3}, \frac{7}{3})$, $u_0^*(x^1) = -\frac{23}{3}$. Since $u_0^*(x^1) < u_0^1$, let $T(2) = \{(\frac{1}{3}, \frac{7}{3})\}$ and $k = 2$.

ITERATION 2:
Consider the ILP

$$\begin{aligned} &\max u_0 \\ &u_0 \leq -\tfrac{23}{3} + x_1 + 4x_2 \\ &x_1, x_2 \geq 0 \text{ integer} \end{aligned}$$

which is unbounded. Since u_0 increases with x_1 and x_2, choose $x_1^2 = x_2^2 = d$ (d arbitrarily large) so that u_0^2 is also arbitrarily large.

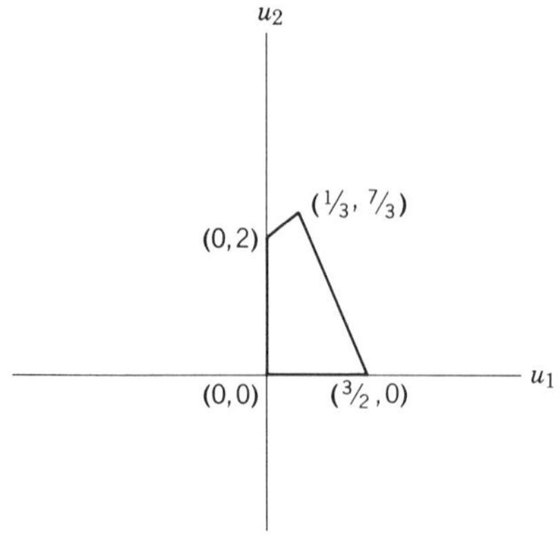

Figure 9

Solve the LP

$$u_0^*(x^2) = -5d + \min\,(2d-2)u_1 + (4d-3)u_2$$
$$\begin{aligned} u_1 \qquad - u_2 &\geq -2 \\ -2u_1 \qquad - u_2 &\geq -3 \\ u_1, u_2 &\geq 0 \end{aligned}$$

The optimal solution is $u_1^2 = u_2^2 = 0$, $u_0^*(x^2) = -5d$. Since $u_0^*(x^2) < u_0^2$, let $T(3) = \{(\frac{1}{3}, \frac{7}{3}), (0, 0)\}$ and $k = 3$.

ITERATION 3:
Solve the ILP

$$\max u_0$$
$$\begin{aligned} u_0 &\leq -\tfrac{23}{3} + x_1 + 4x_2 \\ u_0 &\leq \qquad - x_1 - 4x_2 \\ x_1, x_2 &\geq 0 \text{ integer} \end{aligned}$$

which has alternative optimal solutions $x_1^3 = 0$, $x_2^3 = 1$, $u_0^3 = -4$, and $x_1^3 = 4$, $x_2^3 = 0$, $u_0^3 = -4$.

Using $x_1^3 = 0$, $x_2^3 = 1$, solve the LP

$$u_0^*(x^3) = -4 + \min u_1$$
$$\begin{aligned} u_1 - u_2 &\geq -2 \\ -2u_1 - u_2 &\geq -3 \\ u_1, u_2 &\geq 0 \end{aligned}$$

An optimal solution is $u_1^3 = u_2^3 = 0$, $u_0^*(x^3) = -4 = u_0^3$. Terminate; an optimal solution to the MILP is $x_1 = 0$, $x_2 = 1$, $v_1 = v_2 = 0$, where the v's are the dual variables from the LP at iteration 3.

To illustrate the case in which the dual space is unbounded, consider the following example.

Example

$$\begin{aligned}\max x_0 = -2x_1 - 3x_2 - 2v_1 - 6v_2 \\ -3x_1 + x_2 + v_1 - 2v_2 \leq -5 \\ -2x_1 - 2x_2 - v_1 + 3v_2 \leq -4 \\ x_1, x_2 \geq 0 \text{ integer} \\ v_1, v_2 \geq 0\end{aligned}$$

Initially, $T(1) = Q(1) = \varnothing$, $k = 1$ and u_0^1 arbitrarily large.

ITERATION 1:

Since $T(1) = Q(1) = \varnothing$, we arbitrarily choose $x_1^1 = x_2^1 = 0$. Solve the LP

$$\begin{aligned}u_0^*(x^1) = \min -5u_1 - 4u_2 \\ u_1 - u_2 \geq -2 \\ -2u_1 + 3u_2 \geq -6 \\ u_1, u_2 \geq 0\end{aligned}$$

The dual constraint set has three extreme points and two extreme rays (see Figure 10). A solution is $u_0^*(x^1) \to -\infty$ along the extreme ray $(3, 0) + \theta(3, 2)$ (Ray 1 in Figure 10). Let $T(2) = \{(3, 0)\}$, $Q(2) = \{(3, 2)\}$, and $k = 2$.

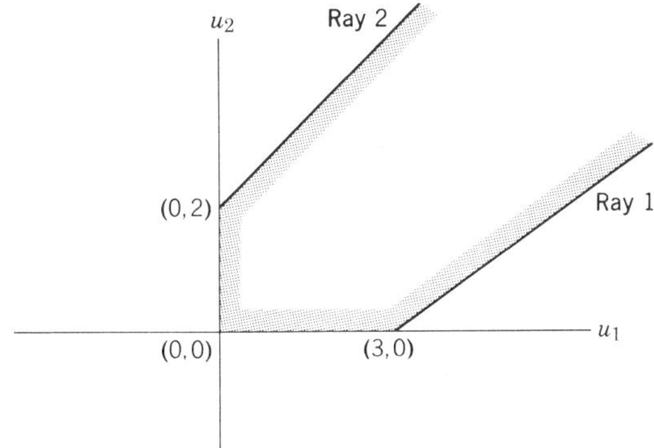

Figure 10

ITERATION 2:
Consider the ILP

$$\begin{aligned} &\max u_0 \\ &u_0 \le -15 + 7x_1 - 6x_2 \\ &0 \le -23 + 13x_1 + x_2 \\ &x_1, x_2 \ge 0 \text{ integer} \end{aligned}$$

which is unbounded. Since u_0 increases with x_1, choose $x_1^2 = d$, $x_2^2 = 0$ (d arbitrarily large) so that u_0^2 is also arbitrarily large.

Solve the LP

$$\begin{aligned} u_0^*(x^2) = -2d + \min\ (3d - 5)\, u_1 + (2d - 4)\, u_2 & \\ u_1 \qquad - u_2 &\ge -2 \\ -2u_1 \qquad + 3u_2 &\ge -6 \\ u_1, u_2 &\ge 0 \end{aligned}$$

The solution is $u_1^2 = 0$, $u_2^2 = 0$, $u_0^*(x^2) = -2d$. Since $u_0^*(x^2) < u_0^2$, let $T(3) = \{(3, 0), (0, 0)\}$, $Q(3) = Q(2)$, $k = 3$, and go to Step 2.

ITERATION 3:
Solve the ILP

$$\begin{aligned} &\max u_0 \\ &u_0 \le -15 + 7x_1 + 6x_2 \\ &u_0 \le \qquad - 2x_1 - 3x_2 \\ &0 \le -23 + 13x_1 + x_2 \\ &x_1, x_2 \ge 0 \text{ integer} \end{aligned}$$

which yields $u_0^3 = -4$, $x_1^3 = 2$, $x_2^3 = 0$.

Solve the LP

$$\begin{aligned} u_0^*(x^3) = -4 + \min u_1 & \\ u_1 - u_2 &\ge -2 \\ -2u_1 + 3u_2 &\ge -6 \\ u_1, u_2 &\ge 0 \end{aligned}$$

A solution is $u_1^3 = u_2^3 = 0$, $u_0^*(x^3) = -4 = u_0^3$. Terminate; the solution to the MILP is $x_1 = 2$, $x_2 = 0$, $v_1 = 0$, $v_2 = 0$.

An advantage of the algorithm of this section is that feasible solutions to the MILP are achieved at Step 3 when the LP has an optimal solution. Also, since Step 2 yields upper bounds on the objective function, the value of the optimal solution is bounded from above and below. Thus premature termination is very likely to yield a feasible solution and an estimate of how far it is from being optimal.

One obvious drawback of the algorithm is that Step 2 requires the solution of an ILP which may have a large number of constraints. Unless the number of extreme points and extreme rays that need to be enumerated is a tiny fraction of $|T| + |Q|$, the ILP is likely to be too large to solve. An approach that seeks to overcome the necessity of solving the ILP directly is given in Section 4.10.

4.10. PARTITIONING AND IMPLICIT ENUMERATION

The partitioning concept of Section 4.9 can be embedded very neatly in an implicit enumeration algorithm when the integer variables in (26) are binary. The problem of having to solve a number of ILP's will be avoided; also, many of the results of Sections 4.5–4.7 become applicable.

The basic idea is that implicit enumeration is performed on the set of binary variables. At each partial solution, a subset of the binary variables have been assigned binary values. The free variables are then tentatively fixed (for convenience at zero) to reduce the MILP to an LP as in (27). The corresponding dual LP is then solved; it is either found to have an optimal solution or to be unbounded. In the former case, a new feasible solution to (26) is found, as well as a dual extreme point; while in the latter, a dual extreme ray and extreme point are identified. Thus new constraints on the integer variables, and possibly new lower bounds on x_0^*, are found, and the enumeration continues.

The details of the algorithm are given below. It is assumed that x is binary and that $\{u|uA_2 \geq c_2, u \geq 0\} \neq \varnothing$.

Let any subsets (possibly empty) of extreme points and directions of extreme rays of $\{u|uA_2 \geq c_2, u \geq 0\}$ be T' and Q'. Let z (possibly $-\infty$) be a lower bound on x_0^*.

The lower bound z yields the constraint on x

$$z \leq c_1x + u^t(b - A_1x) \qquad \text{every } u^t \in T' \tag{36}$$

since $x_0^* \geq z$ and

$$x_0^* \leq c_1x + u^t(b - A_1x) \qquad \text{every } u^t \in T'$$

To restrict the enumeration to admissible x, we also require

$$y^q(b - A_1x) \geq 0 \qquad \text{every } y^q \in Q' \tag{37}$$

Putting (36) and (37) in the standard form for enumeration constraints, we obtain

$$
\begin{aligned}
(-c_1 + u^t A_1)x &\leq -\underline{z} + u^t b \qquad \text{every } u^t \in T' \\
y^q A_1 x &\leq y^q b \qquad \text{every } y^q \in Q' \\
x &\text{ binary}
\end{aligned} \tag{38}
$$

Algorithm

Initially, given $T' = T_0$ and $Q' = Q_0$, we begin implicit enumeration with the intention of finding a feasible solution to (38). At vertex k, let T_k and Q_k be the sets of known dual extreme points and directions of extreme rays. Also, let the binary vector x^k be such that

$$
x_j^k = \begin{cases} 1 & \text{if } j \in S_k^+ \\ 0 & \text{if } j \in S_k^- \cup F_k \end{cases}
$$

Thus x is fixed and (26) reduces to the LP

$$
\begin{aligned}
x_0^*(x^k) = c_1 x^k + \max\ & c_2 v \\
& A_2 v \leq b - A_1 x^k \\
& v \geq 0
\end{aligned}
$$

whose dual is

$$
\begin{aligned}
u_0^*(x^k) = c_1 x^k + \min\ & u(b - A_1 x^k) \\
& u A_2 \geq c_2 \\
& u \geq 0
\end{aligned} \tag{39}
$$

At vertex k, the LP (39) is solved. There are three possible outcomes.

1. An optimal solution u^k is found such that $u_0^*(x^k) > \underline{z}$. Then since (x^k, v^k) is feasible to the MILP, let $\underline{z} = u_0^*(x^k)$. Note that every constraint generated by an extreme point in (38) has its right-hand side reduced by using the new value of $\underline{z}$. Let $T_k = T_k \cup \{u^k\}$ and continue the enumeration with vertex k still live.
2. The same as (1) except that $u_0^*(x^k) \leq \underline{z}$. In this case, $\underline{z}$ remains unchanged and we proceed as in 1.
3. There exists an extreme ray $u^k + \theta y^k$ along which $u_0^*(x^k) \to -\infty$. In this case, let $T_k = T_k \cup \{u^k\}$, $Q_k = Q_k \cup \{y^k\}$ and continue the enumeration.

The fathoming of a vertex occurs only when it is discovered that no completion can satisfy (38). Separation and branching are performed exactly as in the basic implicit enumeration algorithm. When branching from vertex

k to vertex j, we let $T_j = T_k$ and $Q_j = Q_k$, since the constraints (38) apply at every vertex. Furthermore, all of the tests and techniques of Sections 4.6 and 4.7 are applicable. Clearly, the algorithm is finite, since there are only 2^n binary values for x, where n is the number of integer variables.

The two problems of Section 4.9 are now solved by this algorithm.

Example

$$\begin{aligned} \max x_0 = -x_1 - 4x_2 - 2v_1 - 3v_2 & \\ x_1 - 3x_2 + v_1 - 2v_2 &\le -2 \\ -x_1 - 3x_2 - v_1 - v_2 &\le -3 \\ x_1, x_2 \ge 0 & \text{ integer} \\ v_1, v_2 &\ge 0 \end{aligned}$$

To make the problem binary in x, let upper bounds on x_1 and x_2 be 3 and 1 respectively. Make the transformation $x_1 = x_3 + 2x_4$. The problem is

$$\begin{aligned} \max x_0 = -x_3 - 2x_4 - 4x_2 - 2v_1 - 3v_2 & \\ x_3 + 2x_4 - 3x_2 + v_1 - 2v_2 &\le -2 \\ -x_3 - 2x_4 - 3x_2 - v_1 - v_2 &\le -3 \\ x \text{ binary}, v &\ge 0 \end{aligned}$$

Initially $F = \{2, 3, 4\}$, $\underline{z} = -\infty$, $x^0 = (0, 0, 0)$, $T_0 = Q_0 = \varnothing$. At vertex 0 solve the LP

$$\begin{aligned} u_0^*(x^0) = \min u_0(x^0) = -2u_1 - 3u_2 & \\ u_1 - u_2 &\ge -2 \\ -2u_1 - u_2 &\ge -3 \\ u_1, u_2 &\ge 0 \end{aligned}$$

with solution $u^0 = (\frac{1}{3}, \frac{7}{3})$, $u_0^*(x^0) = -\frac{23}{3}$. Let $\underline{z} = -\frac{23}{3}$ and $T_0 = \{(\frac{1}{3}, \frac{7}{3})\}$. The new derived constraint is

$$x_3 + 2x_4 + 4x_2 - \tfrac{1}{3}(-2 - x_3 - 2x_4 + 3x_2) - \tfrac{7}{3}(-3 + x_3 + 2x_4 + 3x_2) \le \tfrac{23}{3}$$

or

$$-x_3 - 2x_4 - 4x_2 \le 0 \tag{40}$$

which does not yield any new information.

Arbitrarily choose $x_3 = 1$ so that $P_1 = (3)$. At vertex 1 solve the LP

$$u_0^*(x^1) = -1 + \min -3u_1 - 2u_2$$
$$u_1 - u_2 \geq -2$$
$$-2u_1 - u_2 \geq -3$$
$$u_1, u_2 \geq 0$$

Again $u^1 = (\frac{1}{3}, \frac{7}{3})$ so that $T_1 = T_0$. However, $u_0^*(x^1) = -\frac{20}{3}$ and $z = -\frac{20}{3}$, and (40) is modified to

$$-x_3 - 2x_4 - 4x_2 \leq -1 \tag{41}$$

Arbitrarily partitioning on x_4, we let $P_2 = (3, 4)$. At vertex 2 the LP is

$$u_0^*(x^2) = -3 + \min - 5u_1$$
$$u_1 - u_2 \geq -2$$
$$-2u_1 - u_2 \geq -3$$
$$u_1, u_2 \geq 0$$

The solution is $u^2 = (\frac{3}{2}, 0)$, $u_0^*(x^2) = -\frac{21}{2}$. Here z is unchanged but $T_2 = \{(\frac{1}{3}, \frac{7}{3}), (\frac{3}{2}, 0)\}$. The new constraint set is given by (41) and

$$\frac{5x_3}{2} + 5x_4 - \frac{x_2}{2} \leq \frac{11}{3} \tag{42}$$

The constraint (42) indicates $x_4 = 0$ by Test 1 of Section 4.6. Therefore, we can put $x_4 = 0$ and then eliminate it from the problem. The constraints (41) and (42) are modified to

$$\begin{aligned} -x_3 - 4x_2 &\leq -1 \\ \frac{5x_3}{2} - \frac{x_2}{2} &\leq \frac{11}{3} \end{aligned} \tag{43}$$

and wc lct $P_3 = (3, 2)$.

At vertex 3, the LP is

$$u_0^*(x^3) = -5 + \min \quad u_2$$
$$u_1 - u_2 \geq -2$$
$$-2u_1 - u_2 \geq -3$$
$$u_1, u_2 \geq 0$$

which yields $u^3 = (\frac{3}{2}, 0)$, $u_0^*(x^3) = -5$. Let $z = -5$ and modify (43) to

$$\begin{aligned} -x_3 - 4x_2 &\leq -\tfrac{8}{3} \\ \frac{5x_3}{2} - \frac{x_2}{2} &\leq 2 \end{aligned} \tag{44}$$

The first constraint of (44) indicates $x_2 = 1$. Therefore we fix $x_2 = 1$, eliminate it from the problem and let $P_4 = (-\underline{3})$.

The LP at vertex 4 is

$$\begin{aligned} u_0^*(x^4) = -4 + \min \quad & u_1 \\ & u_1 - u_2 \geq -2 \\ & -2u_1 - u_2 \geq -3 \\ & u_1, u_2 \geq 0 \end{aligned}$$

which yields $u^4 = (0, 0)$ and $u_0^*(x^4) = -4 > \underline{z}$. Since the enumeration is complete, this solution gives the optimal solution to the MILP, $x_1 = 0$, $x_2 = 1$, $v_1 = 0$, $v_2 = 0$.

Example

$$\begin{aligned} \max x_0 = \; & -2x_1 - 3x_2 - 2v_1 - 6v_2 \\ & -3x_1 + x_2 + v_1 - 2v_2 \leq -5 \\ & -2x_1 - 2x_2 - v_1 + 3v_2 \leq -4 \\ & x_1, x_2 \geq 0 \text{ integer} \\ & v_1, v_2 \geq 0 \end{aligned}$$

Let x_1 and x_2 have upper bounds of 3 so that

$$\begin{aligned} x_1 &= x_3 + 2x_4 \\ x_2 &= x_5 + 2x_6 \\ x_3, \ldots, x_6 &= 0, 1 \end{aligned}$$

The transformed MILP is

$$\begin{aligned} \max x_0 = \; & -2x_3 - 4x_4 - 3x_5 - 6x_6 - 2v_1 - 6v_2 \\ & -3x_3 - 6x_4 + x_5 + 2x_6 + v_1 - 2v_2 \leq -5 \\ & -2x_3 - 4x_4 - 2x_5 - 4x_6 - v_1 + 3v_2 \leq -4 \\ & x \text{ binary}, v \geq 0 \end{aligned}$$

At vertex 0, letting $\underline{z} = -\infty$, $T_0 = Q_0 = \varnothing$, $x^0 = (0, 0, 0, 0)$, the LP is

$$\begin{aligned} u_0^*(x^0) = \min \; & -5u_1 - 4u_2 \\ & u_1 - u_2 \geq -2 \\ & -2u_1 + 3u_2 \geq -6 \\ & u_1, u_2 \geq 0 \end{aligned}$$

which yields $u_0^*(x^0) \to -\infty$ along the ray $(3, 0) + \theta(3, 2)$.

Thus we have $T_0 = \{(3, 0)\}$, $Q_0 = \{(3, 2)\}$, and the constraints

$$-7x_3 - 14x_4 + 6x_5 + 12x_6 \le -15 + d \tag{45}$$

where d is arbitrarily large, and

$$-13x_3 - 26x_4 - x_5 - 2x_6 \le -23 \tag{46}$$

The constraint (46) indicates $x_4 = 1$, leaving

$$\begin{aligned} -7x_3 + 6x_5 + 12x_6 &\le -1 + d \\ -13x_3 - x_5 - 2x_6 &\le 3 \end{aligned} \tag{47}$$

By eliminating x_4 from the problem, the new LP at vertex 0 is

$$\begin{aligned} u_0^*(x^0) = -4 + \min \quad & u_1 \\ & u_1 - u_2 \ge -2 \\ & -2u_1 + 3u_2 \ge -6 \\ & u_1, u_2 \ge 0 \end{aligned}$$

which yields $u^0 = (0, 0)$ and $u_0^*(x^0) = -4$. Therefore, $\underline{z} = -4$ and $T_0 = \{(3, 0), (0, 0)\}$. From (47) and the new extreme point, we obtain

$$\begin{aligned} -7x_3 + 6x_5 + 12x_6 &\le 3 \\ -13x_3 - x_5 - 2x_6 &\le 3 \\ 2x_3 + 3x_5 + 6x_6 &\le 0 \end{aligned} \tag{48}$$

The last constraint of (48) indicates $x_3 = x_5 = x_6 = 0$, and enumeration is complete. The optimal solution is $v_1 = v_2 = x_2 = 0$, $x_1 = 2 = 2x_4$.

4.11. EXERCISES

1. Consider the following class of problems: It is desired to place k queens on a chessboard so that no queen can take another.

(a) Model the problem in 0, 1 variables.

(b) Find a feasible solution for $k = 4, \ldots, 8$, using enumeration by tree.

2. Suppose that we have the constraint $\sum_{k=1}^{n} \lambda_k = 1$ and the additional condition that no more than two of the λ_k can be positive and these must be of the form $(\lambda_k, \lambda_{k+1})$. Prove that a valid separation is determined by the constraints

$$\sum_{k=1}^{p-1} \lambda_k = 0 \qquad \text{or} \qquad \sum_{k=p+1}^{n} \lambda_k = 0 \qquad \text{for some } p$$

Develop a branch and bound algorithm for piecewise-linear separable programming in which the piecewise linear functions are not concave or convex. Refer to (23) of Section 1.4.

3. Assume that all aspects of a branch and bound algorithm have been determined, except for the branching rule. One is interested in devising a branching rule that minimizes the number of vertices considered. If the original problem is to find all optimal solutions to

$$\max z(x), \qquad x \in S$$

show that the appropriate rule is "branch to a vertex having the largest upper bound."

4. For the problem of Section 5.1,

$$\begin{aligned} \max x_0 = 2x_1 + x_2 & \\ x_1 + x_2 &\le 5 \\ -x_1 + x_2 &\le 0 \\ 6x_1 + 2x_2 &\le 21 \\ x_1, x_2 &\ge 0 \text{ integer} \end{aligned}$$

(a) Solve by the branch and bound algorithm of Section 4.3.
(b) Let upper bounds on x_1 and x_2 be 3. Solve the problem by the implicit enumeration algorithm of Section 4.5.

5. Solve the problem of Exercise 2, Chapter 5,
(a) By branch and bound.
(b) By implicit enumeration, using easily derived upper bounds to convert to binary variables.

6. Solve the problem

$$\begin{aligned} \max x_0 = 2x_1 + x_2 + 4x_3 + 5x_4 & \\ x_1 + 3x_2 + 2x_3 + 5x_4 &\le 10 \\ 2x_1 + 16x_2 + x_3 + x_4 &\ge 4 \\ 3x_1 - x_2 - 5x_3 + 10x_4 &\le -4 \\ x_1, \ldots, x_4 &\ge 0 \text{ integer} \end{aligned}$$

by the branch and bound algorithm of Section 4.3.

7. Consider the following modification of the branch and bound algorithm of Section 4.3. If $x_{B_i} = [y_{i0}] + f_{i0}$, $0 < f_{i0} < 1$, and u_{B_i} is an upper bound on x_{B_i}, replace (8) by

$$S_j^* = \{S_j \cap \{x | x_{B_i} = 0\}, \ldots, S_j \cap \{x | x_{B_i} = u_{B_i}\}\}$$

In branching, always go to the successor corresponding to $x_{B_i} = [y_{i0}]$ or $x_{B_i} = \langle y_{i0} \rangle$ before considering any other.

(a) Let $z_k^* = \max cx,\ x \in T_j \cap \{x | x_{B_i} = k\}$. Show that

$$z^*_{[y_{i0}]} = \max_{k=0,\ldots,[y_{i0}]} z_k^*$$

and

$$z^*_{\langle y_{i0}\rangle} = \max_{k=\langle y_{i0}\rangle,\ldots,u_{B_i}} z_k^*$$

(b) How does this algorithm compare computationally to the algorithm of Section 4.3?

(c) Do Exercises 4 and 5 by this algorithm.

8. Derive the improved bounds of Section 4.4 at every iteration for Exercises 4, 5, and 6.

9. In Section 4.4, the advantage of branching on monotone variables was pointed out.

(a) Since x_0 is monotone in an optimal LP tableau, is there any reason why it should not be considered first for branching?

(b) Show that the cut of Section 5.2 implies the constraint derived from branching on a monotone decreasing variable.

10. Derive the appropriate expressions for the penalties P, P_i, and P_i' of Section 4.4 when the LP has been solved by the dual simplex algorithm for bounded variables. Note that many additional cases occur (e.g., a penalty for decreasing a nonbasic variable, etc.)

11. Solve the following problem by the algorithm of Section 4.5.

$$\begin{aligned} \max z = \; & -7x_1 + 6x_2 + 4x_3 - 5x_4 - 6x_5 \\ & -3x_1 - x_2 + 2x_3 + 3x_4 - 3x_5 \le 1 \\ & x_2 - x_3 - 4x_4 - 2x_5 \le -1 \\ & x_1 + 4x_3 + 3x_4 \le 4 \\ & x_1, \ldots, x_5 = 0, 1 \end{aligned}$$

12. Show that the condition

$$\sum_{j \in R_k} x_j \ge 1$$

is satisfied by any feasible completion of W_k in the implicit enumeration algorithm.

13. For the algorithm of Section 4.5, derive alternative rules, which take into account the objective row, for selecting the partitioning variable.

14. Prove that Tests 1–6 of Section 4.6 are valid.

15. Solve Problem 11 using a surrogate constraint at every vertex having $u_1 = u_2 = u_3 = 1$.

16. Solve Problem 11 using the surrogate constraint from Theorem 1 at every vertex.

17. Solve the example of Section 4.7 using a surrogate constraint at every vertex having $u_1 = u_2 = u_3 = 1$.

18. Consider the extension of the surrogate constraint (16) that takes into account the objective row:

$$\sum_{i=0}^{m} \sum_{j \in F_k} u_i a_{ij} x_j \leq \sum_{i=0}^{m} u_i s_i$$
$$u_i \geq 0, \qquad i = 0, \ldots, m$$

What assumptions are necessary to make Theorem 1 hold for this constraint?

19. Consider the surrogate constraint of Exercise 18, and specify $u_0 = 1$. Then the constraint is

$$\hat{g}(u, x) = \sum_{i=1}^{m} u_i \left(\sum_{j \in F_k} a_{ij} x_j - s_i \right) - \sum_{j \in F_k} c_j x_j - s_0 \leq 0$$

Let the strength of a surrogate constraint be defined as

$$\min_{x \text{ binary}} \hat{g}(u, x) > \min_{x \text{ binary}} \hat{g}(u', x)$$

implies that u generates a stronger constraint than u'. For this definition, prove that Theorem 1 holds.

20. Prove that if the constraints $x_j \leq 1$ of (20) are all nonbinding in an optimal solution, then the optimal surrogate constraint is of the form

$$\sum_{j \in F_k} (c_j + y_{0j}) x_j \leq \lambda_k^0$$

21. Give a reasonable definition of the strength of a surrogate constraint for the MILP case. How can such a constraint be calculated?

22. How can the basic implicit enumeration algorithm be modified, when more than one set T_k and more than one suitable partitioning are available? Consider the surrogate constraint algorithm in this framework.

23. Derive penalties analogous to P, P_i, and P_i' of Section 4.4 for the MILP.

24. Solve the example of Exercise 22, Chapter 5, by the algorithm of Section 4.9. Convert the integer variables to binary variables using obvious upper bounds, and then solve the same problem by the algorithm of Section 4.10.

25. Show that, for the implicit enumeration algorithm of Section 4.5, a vertex may be fathomed if $T_i(L, U) > s_i$ for any i, where $T_i(L, U)$ is as defined in Section 4.6. In addition, show that this test is redundant if Test 2 is carried out.

26. For the algorithm of Section 4.10, suppose that $c_1 \geq 0$. Would it then be valid to fathom a vertex k for which x^k is admissible?

27. Why is it possible for dual extreme points to be rediscovered in the algorithm of Section 4.10 but not in the algorithm of Section 4.9?

4.12. NOTES

4.1. The use of a tree formalizes the common-sense approach widely used for enumeration. Polya (1957) is an excellent reference on common-sense problem solving. Basic references on solving combinatorial problems by enumeration are Golomb and Baumert (1965) and Walker (1960).

4.2. The expression "branch and bound" was first used in Little et al., (1963).

General treatments of the principles of branch and bound are given by Bertier and Roy (1964), Lawler and Wood (1966), Agin (1966), Balas (1968), and Mitten (1970).

4.3. This section is based on Dakin (1965). Earlier work was done by Land and Doig (1960), whose approach is similar to Dakin's but seems less practical (see Exercise 7). Zionts (1969) gives a relationship between the constraints used in branch and bound and cutting plane algorithms.

4.4. The basic reference for this section is Tomlin (1970), which extends the work of Beale and Small (1965). Other recent developments on penalties and branching rules are discussed in Roy, Benayoun, and Tergny (1970); Davis, Kendrick and Weitzman (1971); Tomlin (1971); and Benichou et al., (1971). The use of penalties was proposed by Driebeek (1966). The different partitioning rules for simple constraints are due to Beale and Tomlin (1969).

4.5. Basic references for implicit enumeration are Balas (1965) and Glover (1965c). The example of this section is from Balas' paper. The vector representation of the tree was introduced by Geoffrion (1967). Tuan (1971) presents a different bookkeeping scheme and a corresponding procedure for searching the tree.

Trotter and Shetty (1971) present a direct extension of implicit enumeration to general ILP's in bounded variables. An extension of binary enumeration in which problems in general integer variables are solved by generating successively tighter bounds is due to Krolak (1969).

4.6. Test 1 of this section was proposed by Geoffrion (1967) and Fleischmann (1967). Tests 2–4 and others are due to Glover (1965c). Tests 5 and 6 are found in Glover and Zionts (1965). A number of other tests, including some that are only applicable if $a_{ij} \geq 0$ for all i and j, have been presented in Petersen (1967). Fleischmann (1967) gives other tests as well.

4.7. A basic reference on surrogate constraints is Glover (1968b). The initial work was done by Glover (1965c), whose definition of the best constraint is given in (17). He proposed a technique for calculating this constraint only for $m = 2$.

The definition of (18) and (19) and the derivation of the optimal multipliers is due to Balas (1967). He suggested calculating the optimal multipliers only for the original LP and using the same surrogate constraint at every partial solution. The proposal for calculating new constraints at various partial solutions is due to Geoffrion (1969). He adds the constraint $z_k \geq \underline{z}_k$ to (16). Then, using another definition of the strength of a surrogate constraint, he has also shown that the best multipliers are the optimal dual variables (see Exercise 19).

4.8. All of the references of Sections 4.3 and 4.4 apply here as well.

4.9. The reference for this section is Benders (1962). Special methods for solving the (one continuous variable) MILP (33) are given in Balas (1970c) and Zoutendijk (1970).

4.10. The embedding of the partitioning concept in an enumeration algorithm is due to Lemke and Spielberg (1967). Similar algorithms have been developed by Balas (1970c) and Unger (1970).

Further Notes

Many other enumerative algorithms have been developed for solving ILP's and MILP's, including Thompson (1964), Harris (1964), Driebeek (1966), Shareshian (1966), Shareshian and Spielberg (1966), Zionts (1968), Rebelein (1968), Cabot and Hurter (1968), Glover (1968c), Hillier (1969b), and Dragan (1969) and (1970).

5 Cutting Plane Methods

5.1. INTRODUCTION

In Chapter 3, we delineated a class of ILP's for which the integer constraints are superfluous. This class is characterized by the property that all extreme points of the corresponding LP are integer. However, whenever the corresponding LP has nonintegral extreme points, the integer constraints imply additional linear constraints. By the use of these implied constraints, an ILP can be solved as an LP.

To demonstrate the general principle, consider the ILP

$$\max cx, \qquad x \in S = \{x | Ax = b,\ x \geq 0 \text{ integer}\} \tag{1}$$

Suppose that $\bar{A}$ and $\bar{b}$ exist such that $T = \{x | Ax = b,\ \bar{A}x = \bar{b},\ x \geq 0\}$, $S \subseteq T$ and the relaxation of (1),

$$\max cx, \qquad x \in T$$

has an optimal integer solution x^{o}. Then x^{o} is an optimal solution to (1).

The crux of the problem is to generate efficiently a suitable set T. The problem is not existential, since the particular set of constraints that describe the convex hull of feasible solutions to (1) certainly suffices.

Example

Consider the ILP (Figure 1),

$$\begin{aligned} \max x_0 = 2x_1 + \ \ &x_2 \\ x_1 + \ \ &x_2 \leq 5 \\ -x_1 + \ \ &x_2 \leq 0 \\ 6x_1 + 2&x_2 \leq 21 \\ x_1, &x_2 \geq 0 \text{ integer} \end{aligned}$$

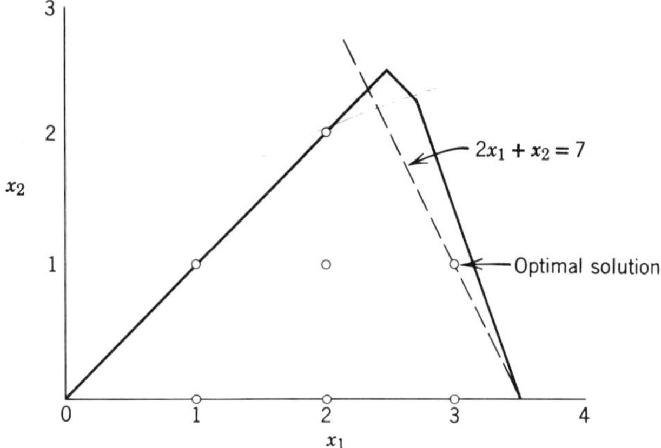

Figure 1

and the LP (Figure 2),

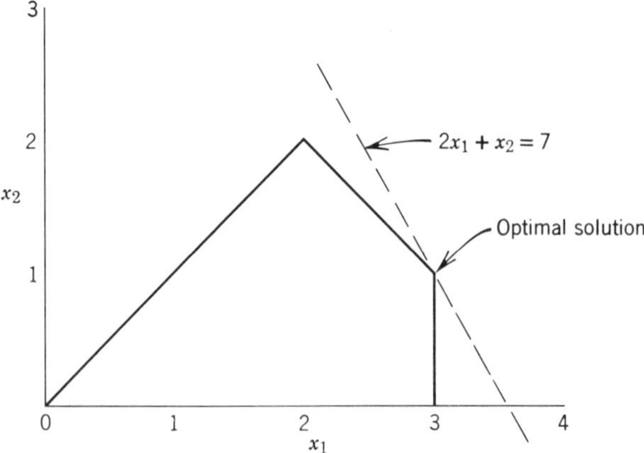

Figure 2

$$
\begin{aligned}
\max x_0 = 2x_1 &+ \ x_2 \\
x_1 &+ \ x_2 \le 5 \\
-x_1 &+ \ x_2 \le 0 \\
6x_1 &+ 2x_2 \le 21 \\
&\left.\begin{cases} x_1 + x_2 \le 4 \\ x_1 \qquad \le 3 \end{cases}\right\} \text{ new constraints} \\
x_1&, x_2 \ge 0
\end{aligned}
$$

Both problems have the same optimal solution $(x_1, x_2) = (3, 1)$, since the constraints of the LP generate the convex hull of feasible points to the ILP.

Although it is sufficient to generate the convex hull associated with the feasible integer points, it is not necessary. In fact, generating the convex hull is usually far too laborious. It is sufficient to make the optimal integer solution an extreme point and remove all noninteger extreme points with a better objective function value.

5.2. A FUNDAMENTAL CUT

Consider the ILP (1). Assume that all the data is integer and that S is nonempty and bounded. The questions of infeasibility and unboundedness are raised in Exercise 3.

Recall from Chapter 1, that the LP corresponding to (1) is defined to be

$$\max cx, \qquad x \in Q = \{x | Ax = b, x \ge 0\} \tag{2}$$

Suppose that we have a representation of (2) given by

$$x_{B_i} = y_{i0} - \sum_{j \in R} y_{ij} x_j, \qquad i = 0,1,\ldots,m \tag{3}$$

so that the basic solution determined by (3) is $x_{B_i} = y_{i0}$, $i = 0, \ldots, m$, $x_j = 0$, $j \in R$. Multiplying (3) by $h \neq 0$ yields

$$hx_{B_i} + \sum_{j \in R} hy_{ij} x_j = hy_{i0}$$

and then the requirement $x \ge 0$ implies that

$$[h]x_{B_i} + \sum_{j \in R} [hy_{ij}] x_j \le hy_{i0} \tag{4}$$

Since x is required to be integer, the left-hand side of (4) must be an integer.

Thus the left-hand side of (4) cannot exceed the integer part of the right-hand side, which implies that

$$[h]x_{B_i} + \sum_{j \in R} [hy_{ij}]x_j \leq [hy_{i0}] \tag{5}$$

Multiplying (3) by $[h]$ and then subtracting (5), yields

$$\sum_{j \in R} ([h]y_{ij} - [hy_{ij}])x_j \geq [h]y_{i0} - [hy_{i0}] \tag{6}$$

which is the fundamental constraint or *cut*. By using (6) in a variety of ways, a number of different cuts and cutting plane algorithms can be developed.

5.3. CUTS IN THE METHOD OF INTEGER FORMS

When h takes on integer values in (6), we obtain the cuts of the method of integer forms. In this section we only consider $h = 1$. Other integer values of h are considered in Section 5.15. Letting $h = 1$ in (6) yields

$$\sum_{j \in R} (y_{ij} - [y_{ij}])x_j \geq y_{i0} - [y_{i0}]$$

Letting $y_{ij} = [y_{ij}] + f_{ij}$, we obtain

$$\sum_{j \in R} f_{ij}x_j \geq f_{i0} \tag{7}$$

or

$$s = -f_{i0} + \sum_{j \in R} f_{ij}x_j, \qquad s \geq 0 \tag{8}$$

Furthermore s must be integer, since

$$x_{B_i} = -\left(-f_{i0} + \sum_{j \in R} f_{ij}x_j\right) + \left([y_{i0}] - \sum_{j \in R} [y_{ij}]x_j\right)$$

and $[y_{i0}] - \sum_{j \in R} [y_{ij}]x_j$ is integer.

A very important property of (8) is that if $f_{i0} > 0$, the basic solution determined from (3) violates (8). This is true because $x_j = 0$, for all $j \in R$, implies that $s = -f_{i0} < 0$. Thus (8) can be used to exclude a basic solution having $f_{i0} > 0$. Furthermore, (8) does not exclude any feasible integer solutions, since, as shown above, it is implied by integrality.

Example

Consider the example of Section 5.1. Rewrite the problem in equality form as

$$\begin{aligned} \max x_0 = 2x_1 &+ x_2 \\ x_1 + x_2 + x_3 &= 5 \\ -x_1 + x_2 + x_4 &= 0 \\ 6x_1 + 2x_2 + x_5 &= 21 \\ x_1, \ldots, x_5 &\geq 0 \text{ integer} \end{aligned}$$

and solve the corresponding LP. This yields the tableau of Table 1.

Table 1

		$-x_3$	$-x_5$
x_0	$\frac{31}{4}$	$\frac{1}{2}$	$\frac{1}{4}$
x_1	$\frac{11}{4}$	$-\frac{1}{2}$	$\frac{1}{4}$
x_2	$\frac{9}{4}$	$\frac{3}{2}$	$-\frac{1}{4}$
x_4	$\frac{1}{2}$	-2	$\frac{1}{2}$

Using (8) to generate cuts yields

$$\frac{x_3}{2} + \frac{x_5}{4} \geq \frac{3}{4} \quad (x_0 \text{ row})$$

$$\frac{x_3}{2} + \frac{x_5}{4} \geq \frac{3}{4} \quad (x_1 \text{ row})$$

$$\frac{x_3}{2} + \frac{3x_5}{4} \geq \frac{1}{4} \quad (x_2 \text{ row})$$

$$\frac{x_5}{2} \geq \frac{1}{2} \quad (x_4 \text{ row})$$

Each row could serve as the source of a cut or simply as a *source row*. Note, however, that the cuts generated from the first two rows happen to be identical and the cut generated from the third row is weaker than the cut from the first, since any nonnegative (x_3, x_5) that satisfies the first satisfies the third, but not conversely.

The x_4 row cut is equivalent to

$$6x_1 + 2x_2 \leq 20$$

Similarly,

$$\frac{x_3}{2} + \frac{x_5}{4} \geq \frac{3}{4}$$

is equivalent to

$$2x_1 + x_2 \leq 7$$

These two cuts together with $Ax = b, x \geq 0$ are shown in Figure 3. If both of these cuts were added to the LP, we would obtain the optimal integer solution $(x_1, x_2) = (3, 1)$.

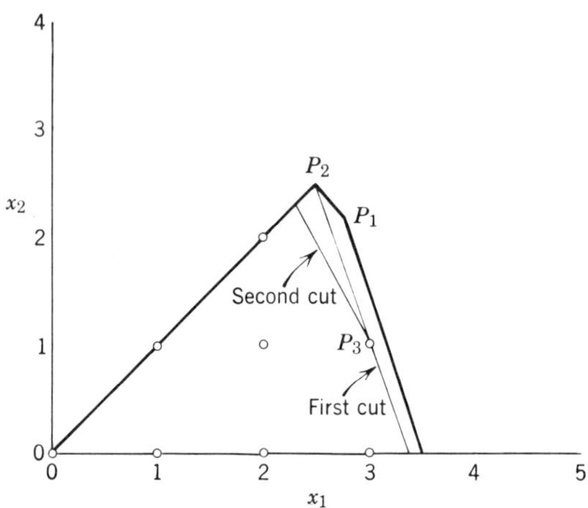

Figure 3

When adding constraints that exclude the present basic solution to an optimal (primal and dual feasible) tableau, dual feasibility is retained. Thus it is convenient to reoptimize using the dual simplex algorithm. Usually, only one constraint is added at a time.

Suppose we append the x_4 row cut, or

$$-\frac{x_5}{2} + s_1 = -\frac{1}{2}, \qquad s_1 \geq 0$$

to the tableau of Table 1. This yields Table 2, with $s_1 = -\frac{1}{2}$ as a new basic variable in the dual feasible solution.

Table 2

			$-x_3$	↓ $-x_5$
	x_0	$\frac{31}{4}$	$\frac{1}{2}$	$\frac{1}{4}$
	x_1	$\frac{11}{4}$	$-\frac{1}{2}$	$\frac{1}{4}$
	x_2	$\frac{9}{4}$	$\frac{3}{2}$	$-\frac{1}{4}$
Source	x_4	$\frac{1}{2}$	-2	$\frac{1}{2}$
← Cut	s_1	$-\frac{1}{2}$	0	$-\frac{1}{2}$

Following the dual simplex algorithm, s_1 leaves and x_5 enters the basis. After pivoting, we obtain Table 3 (excluding the bottom row).

Table 3

			↓ $-x_3$	$-s_1$
	x_0	$\frac{30}{4}$	$\frac{1}{2}$	$\frac{1}{2}$
Source	x_1	$\frac{10}{4}$	$-\frac{1}{2}$	$\frac{1}{2}$
	x_2	$\frac{10}{4}$	$\frac{3}{2}$	$-\frac{1}{2}$
	x_4	0	-2	1
	x_5	1	0	-2
← Cut	s_2	$-\frac{1}{2}$	$-\frac{1}{2}$	$-\frac{1}{2}$

The solution in Table 3 (excluding the bottom row) is primal and dual feasible, but not all-integer. Another constraint must be added. We can choose the x_0, x_1, or x_2 row as a source row. Arbitrarily selecting the x_1 row, which all yield the cut

$$\frac{x_3}{2} + \frac{s_1}{2} \geq \frac{1}{2}$$

Thus we add the constraint

$$-\frac{x_3}{2} - \frac{s_1}{2} + s_2 = -\frac{1}{2}$$

to Table 3 and reoptimize. The departing variable is s_2. Either x_3 or s_1 can enter, and we choose x_3. After pivoting, we obtain Table 4.

Table 4

		$-s_2$	$-s_1$
x_0	7	1	0
x_1	3	-1	1
x_2	1	3	-2
x_4	2	-4	3
x_5	1	0	-2
x_3	1	-2	1

The solution in Table 4 is primal and dual feasible, and integer; therefore it is optimal to the ILP. The steps from Table 2 to Table 4 are equivalent to going from P_1 to P_2 to P_3 in Figure 3, since the constraint

$$\frac{x_3}{2} + \frac{s_1}{2} \geq \frac{1}{2}$$

is equivalent to

$$2x_1 + x_2 \leq 7$$

5.4. A RUDIMENTARY ALGORITHM FOR THE METHOD OF INTEGER FORMS

The steps carried out in the example just given can be summarized as:

Algorithm

STEP 1: (Initialization.) Solve the LP (2). Go to Step 2.

STEP 2: (Optimality test.) Is the solution integer? If so, it is optimal to (1). If not, go to Step 3.

STEP 3: (Cutting and pivoting.) Choose a row r with $f_{r0} > 0$ and add, to the bottom of the tableau, the constraint (8) with $i = r$. Reoptimize using the dual simplex technique. (This may take more than one iteration.) Go to Step 2.

Step 3 is flexible because, generally, many rows may serve as the source of the cut. In the next section a source row selection rule will be given, which assures that the algorithm is finite.

5.5. A FINITE ALGORITHM FOR THE METHOD OF INTEGER FORMS

It will be shown that the algorithm given in Section 5.4 is finite if, in Step 3, we

(a) Maintain $y_j \overset{L}{>} 0$ for all $j \in R$.

(b) Delete the row corresponding to a slack variable from a cut if such a variable becomes basic during the restoration of primal feasibility.

(c) Choose the source row to be the topmost row with $f_{i0} > 0$.

At the completion of Step 1, the solution is dual feasible. However, if there is dual degeneracy in the optimal LP solution, there may be one or more columns having $y_j \overset{L}{<} 0, j \in R$. It may be the case that a simple rearrangement of the rows will yield $y_j \overset{L}{>} 0$, for all $j \in R$. If not, this can be achieved by the addition of a superfluous constraint. Since we have assumed that the constraint set of (1) is bounded, there exists a suitably large integer p such that the constraint

$$\sum_{j \in R} x_j \leq p \qquad \text{or} \qquad s = p - \sum_{j \in R} x_j, \qquad s \geq 0 \text{ integer}$$

is implied by the constraint set of (1). Add the equation

$$s = p - \sum_{j \in R} x_j$$

as the second row (just beneath the x_0 row) of the tableau. Since the coefficients in this row are $+1$ for all $j \in R$, we have $y_j \overset{L}{>} 0$ for all $j \in R$ (i.e., if $y_{0j} = 0$, $y_{1j} = 1$).

If pivot rows were never deleted, $y_j \overset{L}{>} 0$ for all $j \in R$ would be maintained by choosing the pivot column k such that

$$\alpha_k = \operatorname*{lexmax}_{j \in R_r} \alpha_j \tag{9}$$

where $r \neq 0$ is the pivot row, $\alpha_j = y_j/y_{rj}$, and $R_r = \{j \mid j \in R, y_{rj} < 0\}$. Since cut rows are put at the bottom of the tableau, they do not disturb the lexicographic property.

It is necessary to show that dropping the pivot row, after pivoting on a column corresponding to a slack variable from a cut, does not destroy $y_j \overset{L}{>} 0$ for any $j \in R$. Three cases must be considered.

1. ($j = B_r$, the column corresponding to the variable that has just left the basis.) We have $\hat{y}_{B_r} = (-y_{0k}/y_{rk}, \ldots, 1/y_{rk}, \ldots,) \overset{L}{>} 0$ and $1/y_{rk} < 0$. The deleted element $1/y_{rk}$ is negative. But the vector that remains after deleting a nonpositive element from a lexicographically positive vector is still lexicographically positive.

2. ($y_{rj} \geq 0$.) Then $\hat{y}_{rj} = y_{rj}/y_{rk} \leq 0$. The deleted element is nonpositive and the argument given above applies.

3. ($j \neq B_r$ and $y_{rj} < 0$.) We have $\hat{y}_j = (\hat{y}_{0j}, \ldots, \hat{y}_{rj}, \ldots) \overset{L}{>} 0$ and $\hat{y}_{rj} > 0$, which implies that $(\hat{y}_{0j}, \ldots, \hat{y}_{r-1,j}) \overset{L}{\geq} 0$. If $(\hat{y}_{0j}, \ldots, \hat{y}_{r-1,j}) \overset{L}{>} 0$, then $(\hat{y}_{0j}, \ldots, \hat{y}_{r-1,j}, \hat{y}_{r+1,j}, \ldots) \overset{L}{>} 0$. The difficult case is $(\hat{y}_{0j}, \ldots, \hat{y}_{r-1,j}) = 0$ or

$$\frac{y_{ij}}{y_{rj}} = \frac{y_{ik}}{y_{rk}}, \qquad i = 0, \ldots, r - 1$$

We also have $y_{rj}/y_{rj} = y_{rk}/y_{rk}$. Suppose that

$$\frac{y_{ij}}{y_{rj}} = \frac{y_{ik}}{y_{rk}}, \qquad i = 0, \ldots, p - 1 \geq r \tag{10}$$

Since a slack variable from a cut is the entering variable, k determined by (9) is unique (see Exercise 31). Thus (10) cannot hold for all i and we can assume that

$$\frac{y_{pj}}{y_{rj}} \neq \frac{y_{pk}}{y_{rk}}$$

Then (9) implies

$$\frac{y_{pj}}{y_{rj}} < \frac{y_{pk}}{y_{rk}}$$

and $\hat{y}_{pj} = y_{pj} - (y_{rj}y_{pk})/y_{rk} > 0$. Consequently, $\hat{y}_{ij} = 0, i = 0, \ldots, p - 1$, $i \neq r$, $\hat{y}_{pj} > 0$ imply that $(\hat{y}_{0j}, \ldots, \hat{y}_{r-1,j}, \hat{y}_{r+1,j}, \ldots) \overset{L}{>} 0$.

It was also shown in Section 2.10 that the solution column is lexicographically decreasing ($\hat{y}_0 \overset{L}{<} y_0$). Here, because $\hat{y}_0$ may have one fewer component than y_0, $\hat{y}_0 - y_0 \overset{L}{<} 0$ does not make sense. However, $y_{r0} < 0$ and $y_{rk} < 0$ imply $\hat{y}_{r0} > 0$ and thus $\hat{y}_0 - y_0 \overset{L}{<} 0$ before deletion of row r implies $(\hat{y}_{00}, \ldots, \hat{y}_{r-1,0}) \overset{L}{<} (y_{00}, \ldots, y_{r-1,0})$. By $\hat{y}_0 \overset{L}{<} y_0$, we will mean

$$(\hat{y}_{00}, \ldots, \hat{y}_{r-1,0}, \hat{y}_{r+1,0}, \ldots) \overset{L}{<} (y_{00}, \ldots, y_{r-1,0}, y_{r+1,0}, \ldots)$$

when a row has been deleted.

The revised Step 3 is:

STEP 3*: Choose the first row in the tableau with $f_{i0} > 0$ as the source row, say row r. Add to the bottom of the tableau the constraint (8) derived from row r. Reoptimize using the dual simplex algorithm and the column selection rule (9). If a slack variable from a cut becomes basic, delete its row. Go to Step 2.

Theorem 1: *The algorithm consisting of Steps* 1, 2, *and* 3* *yields an optimal solution to* (1) *in a finite number of simplex iterations.*

PROOF: Assume an infinite number of iterations. Let t be the iteration index. Since the sequence $\{y_0^t\}$ is lexicographically decreasing and bounded from below, the sequence $\{y_{00}^t\}$ is nonincreasing, bounded from below and has a limit point, l. Let $l = [l] + f_l$ and suppose $f_l > 0$. After some finite number of iterations, say q, there is a primal and dual optimal tableau with

$$y_{00}^q = [l] + f, \qquad f_l \leq f < 1$$

Thus the top row is the source row for the next cut

$$s - \sum_{j \in R} f_{0j} x_j = -f_{00} = -f$$

After pivoting, with s as the departing variable and x_p as the entering variable,

$$y_{00}^{q+1} = y_{00}^q - f\frac{y_{0p}}{f_{0p}}$$

But $y_{0p} \geq f_{0p} > 0$, so that $y_{00}^{q+1} \leq y_{00}^{q} - f = [l]$. This contradicts the assumption that $f_l > 0$. Therefore, l must be an integer and $y_{00}^{q+1} = l$. Now the lexicographic property guarantees that $\{y_{10}^t\}$ is nonincreasing for $t > q$. Since $y_{10}^t \geq 0$ in all primal feasible solutions, the same argument can be applied to establish that $\{y_{10}^t\}$ reaches its integer limit point in a finite number of iterations. The same argument is then applied to $\{y_{20}^t\}$, $\{y_{30}^t\}, \ldots$, and so forth. Since slack variables from cuts are deleted when they become basic, the number of rows is never more than the number of original variables. Thus an integer solution is achieved in a finite number of iterations. ■

It should be apparent from the proof that the topmost noninteger row does not have to be the source row all of the time. If the topmost noninteger row is periodically chosen as the source, the algorithm will still be finite.

In linear programming an upper bound on the number of iterations, as a function of problem size (n variables, m constraints), is given by $\binom{n}{m}$, which is an upper bound on the number of bases. Unfortunately, no such bound exists for ILP's. In fact, a class of two-variable, one-constraint ILP's has been identified in which, given any positive integer N, there is a problem in the class requiring more than N iterations.

The source row selection rule given above is sufficient for finite termination and quite flexible, since it can be applied periodically. Nevertheless, computer codes frequently use heuristic source row selection rules that ignore the termination problem. It is not known whether the source row selection rule given above is necessary to guarantee finite termination.

The idea of the heuristic rules is to have the cut remove as much of a region containing no integer points as possible. Since the cut is deeper in the x_j-direction as f_{ij} decreases and f_{i0} increases, it is desirable to have f_{i0} large and f_{ij}, $j \in R$, small. Thus, for example, we could choose the source row r from

(1) $f_{r0} = \max_i f_{i0}$.
(2) $f_{r0}/\sum_{j\in R} f_{rj} = \max_i f_{i0}/\sum_{j\in R} f_{ij}$.
(3) $f_{r0}/f_{rk} = \max_i f_{i0}/f_{ik}$ for a specified $k \in R$.

The quality of these rules must be judged empirically.

5.6. CUTS WITH UNIT COEFFICIENTS

In this section a class of cuts simpler than (8) is given. However, only a subclass of these cuts can yield a finite algorithm.

If for any $i, f_{i0} > 0$ in a basic solution determined from (3), then a necessary condition for an integer solution is that at least one of the nonbasic variables be positive. But, since the nonbasic variables are also constrained to be integers, at least one of the nonbasic variables must be equal to or greater than one. Thus

$$\sum_{j \in R} x_j \geq 1 \tag{11}$$

The cut (11) excludes the current basic solution but no feasible integer solutions. A nice property of (11) is that it does not depend on the y_{ij}. One simply checks if any basic variables are noninteger and, if so, adds the constraint (11).

However, cuts from (11) cannot, in general, yield a finite algorithm. It can be shown that if x^o is an optimal solution to (1), a necessary (but not sufficient) condition for an algorithm based on (11) to converge to x^o is that x^o be on an edge (a line joining two adjacent extreme points) of the constraint set of (2).

Improved versions of (11) are obtained by noting that $f_{i0} > 0$ and x_{B_i} integer imply that

$$\sum_{j \in R_i} x_j \geq 1, \; R_i = \{j \mid j \in R, y_{ij} \neq 0\} \tag{12}$$

and more strongly

$$\sum_{j \in R_i^o} x_j \geq 1, \; R_i^o = \{j \mid j \in R, f_{ij} > 0\} \tag{13}$$

For the cuts (12) and (13), there is a finite algorithm. The details of the algorithm are unimportant, since it is unlikely that an algorithm based on (13) could be better than an algorithm based on (7). An essential property used to prove the finiteness of the algorithm based on (7) is that if the cut is taken from the objective row, one dual simplex iteration reduces the objective function by at least its fractional part. The use of (12) or (13) from the objective row only guarantees a reduction of $1/D$ in one dual simplex iteration ($D = |\det B|$). This reduction, however, is sufficient to yield a convergent algorithm. In contrast, when (11) is used and the objective function is not an integer, one dual simplex iteration may not reduce the objective function.

The cuts (7) and (13), when taken from the same row, have positive coefficients for the same subset of nonbasic variables. But there is a much closer relationship. The cut (7) was derived from (6) with $h = 1$. By setting $h = -1$ in (6), we obtain the cut,

$$\sum_{j \in R_i^o} (1 - f_{ij}) x_j \geq 1 - f_{i0} \tag{14}$$

Adding (7) and (14) yields (13).

5.7. ROUNDOFF PROBLEMS AND INTEGRALITY TESTS

Computer arithmetic, like manual calculations done with decimals, is subject to roundoff errors. Nonterminating decimals, like the decimal representation of $\frac{1}{3}$, can only be kept accurate to a given number of places. Once a number is rounded, its error may be propagated in subsequent iterations. To avoid cumulation of these errors, in many linear programming codes, B is reinverted periodically and $x_B = B^{-1}b$ and $y_j = B^{-1}a_j$ are recomputed. Nevertheless, it is impossible to eliminate roundoff errors completely unless all calculations are done with integers.

Roundoff errors can be particularly serious in the method of integer forms, where it is essential to identify integers. On a computer, a real number r is called integer, if $\min \{1 - f_r, f_r\} < \epsilon$, where ϵ is specified by the program and f_r is the fractional part of r. Failing to recognize an integer could cause unnecessary iterations, invalid cuts, and even loss of an optimal solution. Conversely, the improper identification of an integer could cause false termination.

If it is desired to avoid the roundoff and integrality test problems completely, all calculations must be done with integers. The dual all-integer algorithm has this property.

5.8. DUAL ALL-INTEGER CUTS

Consider a dual feasible basic solution determined from (3) and an all-integer tableau. If this solution is also primal feasible, it is optimal to (1), so we assume that it is not.

We can drive toward primal feasibility using the dual simplex algorithm. If the magnitude of the pivot element for each dual simplex iteration equals 1, the tableau will remain all-integer. Then, if primal feasibility is achieved, the solution will be optimal. If, fortuitously, there is a row with $y_{i0} < 0$ and a dual simplex pivot element of -1, an ordinary dual simplex iteration can be executed. Generally, this will not be the case.

However, by using (6) as a cut, there is always an appropriate value of h, $0 < h < 1$, so that the current basic solution is eliminated and the pivot element is -1. Select any primal infeasible row, say the rth, as the source row. Rewriting (6) as an equality with $0 < h < 1$, we obtain

$$s + \sum_{j \in R} [hy_{rj}]x_j = [hy_{r0}], \qquad s \geq 0 \text{ integer} \tag{15}$$

Since $y_{r0} < 0$ and $h > 0$, $[hy_{r0}] < 0$ in (15). We must choose h so that the pivot element equals -1. Suppose that k is the pivot column. Then $h > 0$ and $[hy_{rk}] = -1$ imply that $k \in R_r$, where

$$R_r = \{j \mid y_{rj} < 0, \ j \in R\}$$

To preserve dual feasibility requires

$$y_{0j} - \frac{[hy_{rj}]y_{0k}}{[hy_{rk}]} = y_{0j} + [hy_{rj}]y_{0k} \geq 0 \tag{16}$$

for all $j \in R_r - \{k\}$. Since $[hy_{rj}] \leq -1$ for all $j \in R_r$, a necessary condition for dual feasibility is $y_{0k} \leq y_{0j}$ for all $j \in R_r$. Therefore, the pivot column must satisfy

$$y_{0k} = \min_{j \in R_r} y_{0j} \tag{17}$$

Now h must be chosen so that (16) holds and $[hy_{rk}] = -1$. Since y_{rk} is a negative integer,

$$h \leq -\frac{1}{y_{rk}} \leq 1 \tag{18}$$

If $y_{0k} > 0$, define

$$M_j = \min \{\alpha \mid y_{0j} + \alpha y_{0k} \geq 0, \alpha \text{ integer}\}, \qquad j \in R_r$$

If $y_{0k} = 0$, define $M_j = -\infty$ for all $j \in R_r - \{k\}$ and $M_k = -1$. Then if $y_{0k} > 0$,

$$M_j = \min \{\alpha \mid \alpha \geq -\frac{y_{0j}}{y_{0k}}, \alpha \text{ integer}\}$$

so that

$$M_j = -\left[\frac{y_{0j}}{y_{0k}}\right] \tag{19}$$

Now (16), (19), and $M_j = -\infty$ when $y_{0k} = 0$ imply that

$$h \leq \frac{M_j}{y_{rj}} \qquad \text{for all } j \in R_r - \{k\} \tag{20}$$

Since $M_k = -1$, (18) is equivalent to $h \leq M_k/y_{rk}$. Thus (18) and (20) can be represented as

$$h \leq \min_{j \in R_r} \frac{M_j}{y_{rj}} \equiv h^* \tag{21}$$

Any $h < 1$ that satisfies (21) suffices, but there is some justification for choosing $h = h^*$ when $h^* < 1$. After adding the cut (15) and executing a dual simplex iteration with s departing and x_k entering, we have

$$\hat{y}_{00} - y_{00} = -\frac{[hy_{r0}]y_{0k}}{[hy_{rk}]} = [hy_{r0}]y_{0k}$$

The justification for choosing $h = h^*$ is that it yields the largest decrease in the objective function.

When $h^* = 1$, pivoting on row r and column k yields a dual feasible all-integer tableau, so that no cut is necessary. This follows since $h^* = 1$ implies that $y_{rk} = -1$ and $M_j/y_{rj} \geq 1$ so that $[y_{0j}/y_{0k}] \geq -y_{rj}$ for all $j \in R_r$. Therefore,

$$-\frac{y_{0j}}{y_{rj}} \geq y_{0k} \tag{22}$$

Dividing both sides of (22) by $y_{rk} = -1$ validates the condition for dual feasibility

$$\frac{y_{0k}}{y_{rk}} \geq \frac{y_{0j}}{y_{rj}} \qquad \text{for all } j \in R_r$$

Example

Consider the tableau given in Table 5. If the x_4 row is chosen as the source, $R_4 = \{1, 3\}$ and x_1 enters. This yields $M_1 = -1$, $M_3 = -[\frac{4}{3}] = -1$, and $h^* = \min\{1, \frac{1}{4}\} = \frac{1}{4}$. With $h = h^*$, the cut (15) is

$$s - x_1 - x_3 = -2$$

If the x_5 row is chosen as the source, $R_5 = \{1, 2\}$ and x_2 is the entering variable. We have $M_1 = -[\frac{3}{1}] = -3$, $M_2 = -1$, and $h^* = \min(\frac{3}{9}, \frac{1}{6}) = \frac{1}{6}$. With $h = \frac{1}{6}$, (15) is

$$s - 2x_1 - x_2 + x_3 = -1$$

Table 5

		$-x_1$	$-x_2$	$-x_3$
x_0	8	3	1	4
x_4	-5	-1	0	-4
x_5	-2	-9	-6	8

After adding a cut and pivoting, either $y_{i0} \geq 0$, $i = 1, \ldots, m$ and the solution is optimal or another cut is added. Before working through a complete example, the question of finiteness is resolved.

5.9. A FINITE DUAL ALL-INTEGER ALGORITHM

We will show that sufficient conditions for the finiteness of an algorithm based on the cut (15) are

(i) $y_j \overset{L}{>} 0$ for all $j \in R$ and for all tableaus.
(ii) Periodically, the source row is the topmost row (excluding the x_0 row) with $y_{i0} < 0$.
(iii) The row corresponding to a slack variable from a cut is deleted if such a variable becomes basic after a dual simplex iteration.

To maintain (i), given that we start with lexicographically positive columns, we require a tighter pivot column selection rule than (17) and a somewhat more stringent condition on h than (21).

Suppose that k is the index of the pivot column. Transforming column $j \in R_r - \{k\}$ with $[hy_{rk}] = -1$ yields

$$\hat{y}_j = y_j + [hy_{rj}]y_k$$

except for the pivot row which is the cut row and is put at the bottom of the tableau. Since $[hy_{rj}] \leq -1$ for $j \in R_r - \{k\}$, a necessary condition for $\hat{y}_j \overset{L}{>} 0$ is $y_j \overset{L}{>} y_k$ for all $j \in R_r - \{k\}$. Thus

$$y_k = \operatorname*{lexmin}_{j \in R_r} y_j \tag{23}$$

Clearly, (23) implies (17) and also $\hat{y}_j \overset{L}{>} 0$ for all $j \in R - R_r$.

We must now choose h to satisfy $y_j + [hy_{rj}]y_k \overset{L}{>} 0$ for $j \in R_r - \{k\}$. Let

$$\overline{M}_j = \min\{\alpha | y_j + \alpha y_k \overset{L}{>} 0, \alpha \text{ integer}\}, j \in R_r - \{k\}. \tag{24}$$

When $y_{0k} = 0$, $\overline{M}_j$ in (24) may not exist. In this case define $\overline{M}_j = -\infty$ for all $j \in R_r - \{k\}$. Note $y_{0k} > 0$ implies $\overline{M}_j \geq M_j$ and $\overline{M}_j = M_j$, except (possibly) when $y_{0j} + M_j y_{0k} = 0$. As in (20), we require $h \leq \overline{M}_j/y_{rj}$ for all $j \in R_r - \{k\}$. With $\overline{M}_k = -1$ we obtain

$$h \leq \min_{j \in R_r} \frac{\overline{M}_j}{y_{rj}} \equiv \bar{h} \tag{25}$$

An argument similar to the one given in Section 5.8 for the case $h^* = 1$ shows that if $\bar{h} = 1$, then $y_{rk} = -1$ and $y_j + y_{rj}y_k \overset{L}{>} 0$. Therefore, $\bar{h} = 1$ implies that no cut is necessary.

Algorithm

STEP 1: (Initialization.) Begin with an all-integer tableau with $y_j \overset{L}{>} 0$ for all $j \in R$. (This step will be discussed later in this section.) Go to Step 2.

STEP 2: (Test for optimality.) If the solution is primal feasible, it is optimal to (1). If not, go to Step 3.

STEP 3: (Cutting and pivoting.) Choose a row ($i \neq 0$) in the tableau with $y_{i0} < 0$, say $i = r$. The topmost row with $y_{i0} < 0$ must be chosen at least periodically. Select the lexicographically smallest column with $y_{rj} < 0$, say $j = k$, as the pivot column. Compute $\bar{h}$ from (25). If $\bar{h} = 1$, execute one dual simplex iteration with pivot element y_{rk}. If $\bar{h} < 1$, adjoin the cut (15) with $h = \bar{h}$, to the bottom of the tableau. Execute a dual simplex iteration with s as the departing variable and x_k as the entering variable. In any case, if x_k is a slack from a cut, delete the x_k row. Return to Step 2.

When no cut is added at Step 3 and x_k is a slack from a cut, it was shown in Section 5.5 that deletion of the pivot row does not disturb $\hat{y}_j \overset{L}{>} 0$.

Theorem 2: *The algorithm just given terminates with an optimal solution to* (1) *after a finite number of dual simplex iterations.*

PROOF: The solution column is lexicographically decreasing at each iteration since, except for the pivot row,

$$y_0^{t+1} = y_0^t + [hy_{r0}^t]y_k^t \qquad \text{and} \qquad [hy_{r0}^t]y_k^t \overset{L}{<} 0$$

The pivot row, whether it is deleted or not, has no effect on $y_0^t \overset{L}{>} y_0^{t+1}$ (see Section 5.5). Thus the sequence $\{y_{00}^t\}$ is nonincreasing. Since it is bounded from below and all decreases are by integers, it must stop decreasing after a finite number of iterations, say q. But then $y_0^t \overset{L}{>} y_0^{t+1}$ implies that $\{y_{10}^t\}$, $t > q$, is nonincreasing. Furthermore, $\{y_{10}^t\}$, $t > q$, is nonnegative, since if $y_{10}^t < 0$ for some $t > q$, then a cut from row 1 and a pivot on the cut row would yield the contradiction $y_{10}^{t+1} > y_{10}^t$. Consequently, after $q^* \geq q$ iterations, $\{y_{10}^t\}$ must be a nonnegative constant. The same argument is applied to the remaining rows. Since the number of rows is bounded, an optimal solution is found after a finite number of iterations. ■

Finding an Initial Solution

To begin the calculations, it is necessary to have an all-integer tableau with $y_j \overset{L}{>} 0$ for all $j \in R$. Suppose the problem

$$\begin{aligned} \max x_0 &= cx \\ Ax + x_s &= b \\ x, x_s &\geq 0 \text{ integer} \end{aligned} \tag{26}$$

is feasible and bounded and all data are integer. If $c \leq 0$, then

$$x_B = x_s = b, \qquad x_N = x = 0 \tag{27}$$

is a dual feasible solution to (26), since $y_{0j} = -c_j \geq 0$ for all $j \in R$. If at least one $c_j = 0$, say c_k, it may be that $y_k \overset{L}{<} 0$. Then, if no obvious rearrangement of the rows yields $y_j \overset{L}{>} 0$, for all $j \in R$, we add the superfluous constraint

$$s = p - \sum_{j \in R} x_j \tag{28}$$

where p is a suitably large integer (see Section 5.5) as row 1 of the tableau.

If $c_j > 0$, for at least one j, the solution (27) is not dual feasible. However, there is a simple procedure for obtaining a dual feasible solution. Begin with the solution (27) and add the superfluous constraint (28) as row 1 (the row beneath the objective row).

Let

$$y_{0k} = \min_{j \in R} y_{0j} < 0$$

and pivot on column k and row 1. Since the pivot element $y_{1k} = 1$,

$$\hat{y}_{0j} = y_{0j} - y_{0k} \geq 0$$
$$\hat{y}_{0k} = -y_{0k} > 0$$

and the solution is dual feasible. Furthermore, $\hat{y}_{1j} = 1$ for all $j \in R$, so $\hat{y}_j \overset{L}{>} 0$, for all $j \in R$.

Example

For the problem given in Section 5.1, an obvious integer basic solution is shown in Table 6.

Table 6

		↓ $-x_1$	$-x_2$
x_0	0	-2	-1
← x_3	5	1	1
x_4	0	-1	1
x_5	21	6	2

The solution is not dual feasible. Although the x_3 row is not superfluous, it is of the form (28) and can serve the same purpose. Pivoting on x_3 and x_1 yields the dual feasible solution given in Table 7.

In Table 7, the x_5 row is the source and x_2 enters. Using (25) with $\overline{M}_2 = -1$ and $\overline{M}_3 = -1$ yields

$$h = \bar{h} = \min\{\tfrac{1}{4}, \tfrac{1}{6}\} = \tfrac{1}{6}$$

Table 7

			$-x_3$	$-x_2$ ↓
	x_0	10	2	1
	x_1	5	1	1
	x_4	5	1	2
Source	x_5	−9	−6	−4

Thus the cut from (15) is

$$s_1 = -2 + x_3 + x_2$$

as shown in Table 8.

Table 8

			$-x_3$	$-x_2$ ↓
	x_0	10	2	1
	x_1	5	1	1
	x_4	5	1	2
	x_5	−9	−6	−4
← Cut	s_1	−2	−1	−1

Pivoting on the cut row and x_2 yields Table 9.

Table 9

			$-x_3$ ↓	$-s_1$
	x_0	8	1	1
	x_1	3	0	1
	x_4	1	−1	2
Source	x_5	−1	−2	−4
	x_2	2	1	−1

In Table 9, the x_5 row is the source and x_3 enters, yielding $\overline{M}_3 = \overline{M}_{s_1} = -1$ and

$$h = \bar{h} = \min\{\tfrac{1}{2}, \tfrac{1}{4}\} = \tfrac{1}{4}$$

Thus the cut is

$$s_2 = -1 + x_3 + s_1$$

as shown in Table 10. Pivoting on this cut, we find the optimal solution $(x_0, x_1, x_4, x_5, x_2, x_3) = (7, 3, 2, 1, 1, 1)$ given in Table 11. The calculations are shown graphically in Figure 4.

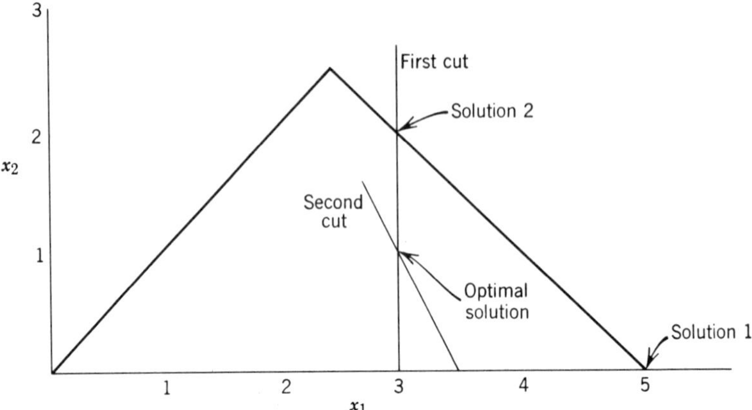

Figure 4

Table 10

			↓ $-x_3$	$-s_1$
	x_0	8	1	1
	x_1	3	0	1
	x_4	1	−1	2
	x_5	−1	−2	−4
	x_2	2	1	−1
← Cut	s_2	−1	−1	−1

Table 11

		$-s_2$	$-s_1$
x_0	7	1	0
x_1	3	0	1
x_4	2	−1	3
x_5	1	−2	−2
x_2	1	1	−2
x_3	1	−1	1

Although the dual all-integer algorithm resolves the roundoff problem, there is still a very desirable feature that is missing. Since there is no known upper bound on the number of iterations, in practice a problem may have to be terminated before optimality is achieved. In the dual all-integer algorithm, termination before optimality means that no primal feasible integer solution has been obtained. An obvious remedy is a primal all-integer algorithm, an algorithm that moves from one primal feasible integer solution to another until dual feasibility is reached.

5.10. PRIMAL ALL-INTEGER CUTS

Consider a primal feasible, all-integer tableau that is not dual feasible. If the pivot element in the primal simplex algorithm is 1, primal feasible, all-integer solutions can be maintained. Thus, if the natural pivot element is 1, an ordinary simplex iteration can be executed. In general, this will not be the case.

Assume that $y_{0k} < 0$ and

$$\frac{y_{r0}}{y_{rk}} = \min_{i=1,\ldots,m}\left(\frac{y_{i0}}{y_{ik}}, y_{ik} \geq 1\right) \tag{29}$$

so that, with x_k as the entering variable, x_{B_r} can be a departing variable in the primal simplex algorithm. Note that $y_{ik} > 0$ has been replaced by $y_{ik} \geq 1$ because the tableau is all-integer. If $y_{rk} = 1$, an ordinary simplex iteration can be executed. If $y_{rk} > 1$, let $h = 1/y_{rk} < 1$ in (6) and adjoin the cut

$$s + \sum_{j\in R}\left[\frac{y_{rj}}{y_{rk}}\right]x_j = \left[\frac{y_{r0}}{y_{rk}}\right], \qquad s \geq 0 \text{ integer} \tag{30}$$

The coefficient of x_k in (30) is $[y_{rk}/y_{rk}] = 1$. Since

$$\left[\frac{y_{r0}}{y_{rk}}\right]\bigg/\left[\frac{y_{rk}}{y_{rk}}\right] = \left[\frac{y_{r0}}{y_{rk}}\right] \leq \frac{y_{r0}}{y_{rk}}$$

the cut row can serve as the pivot row and the pivot element is 1. Although the x_{B_r} row can always serve as the source row, any row with $y_{qk} \geq 1$ and

$$\left[\frac{y_{q0}}{y_{qk}}\right] = \left[\frac{y_{r0}}{y_{rk}}\right] \leq \frac{y_{r0}}{y_{rk}} \tag{31}$$

will suffice, yielding the cut (30) with r replaced by q.

The cut (30) does not exclude the current feasible integer solution (it might be optimal), since $s = [y_{r0}/y_{rk}] \geq 0$. But, unless $y_{r0}/y_{rk} = [y_{r0}/y_{rk}]$, the solution that would have been obtained by entering x_k and removing x_{B_r} is eliminated. To see this, suppose we pivot on y_{rk}. In the new solution, $x_j = 0$ for all $j \in R - \{k\}$ and $x_k = y_{r0}/y_{rk}$. Substituting this solution into (30) yields

$$s = \left[\frac{y_{r0}}{y_{rk}}\right] - \frac{y_{r0}}{y_{rk}} \leq 0$$

and $s = 0$ if and only if y_{r0}/y_{rk} is an integer. When y_{r0}/y_{rk} is an integer, the solution obtained by pivoting on the x_{B_r} row or the cut row is

$$\hat{y}_0 = y_0 - y_k\frac{y_{r0}}{y_{rk}}$$

except for the x_{B_r} row, and $\hat{y}_{r0} = y_{r0}/y_{rk}$. This is an integer solution and cannot be eliminated.

5.11. A RUDIMENTARY PRIMAL ALL-INTEGER ALGORITHM

Ignoring finiteness for the moment, a primal all-integer algorithm is:

Algorithm

STEP 1: (Initialization.) Begin with an all-integer primal feasible tableau. (If there is not an obvious one, a Phase 1 method (see Exercise 13) can be used to find one.) Go to Step 2.

STEP 2: (Test for optimality.) If the solution is dual feasible, it is optimal to (1). If not, go to Step 3.

STEP 3: (Cutting and pivoting.) Choose a column $k \in R$ with $y_{0k} < 0$ and a row r from (29). If $y_{rk} = 1$, execute a primal simplex iteration with x_k as the entering variable and x_{B_r} as the departing variable. If $y_{rk} > 1$, adjoin the constraint (30) to the bottom of the tableau, with the source row chosen from (31). Execute a primal simplex iteration with s as the departing variable and x_k as the entering variable. In any case, if a slack variable from a cut becomes basic, its row is deleted. Return to Step 2.

Example

For the problem of Section 5.1, an initial primal feasible all-integer tableau is given in Table 6. Choosing x_1 as the entering variable in Table 6 yields the x_5 row as the source row, $h = \frac{1}{6}$, and the constraint

$$s_1 + x_1 = 3 \qquad (x_1 \leq 3)$$

shown in Table 12.

Table 12

			$\downarrow$ $-x_1$	$-x_2$
	x_0	0	-2	-1
	x_3	5	1	1
	x_4	0	-1	1
Source	x_5	21	6	2
$\leftarrow$ Cut	s_1	3	1	0

Transforming yields Table 13. The unique entering variable is x_2. The source row must be the x_5 row; thus $h = \frac{1}{2}$ and we have the constraint

$$s_2 + x_2 - 3s_1 = 1 \qquad (3x_1 + x_2 \leq 10)$$

This constraint is added to the bottom of Table 13, and transforming yields Table 14 (excluding the bottom row). The unique entering variable is s_1 and the source row is the x_4 row. Thus $h = \frac{1}{4}$ and the cut

$$s_3 + s_1 - s_2 = 0 \qquad (2x_1 + x_2 \leq 7)$$

is obtained. Pivoting on this cut yields dual feasibility and verifies the optimality of the solution given in Table 14.

The iterations are displayed graphically in Figure 5 and illustrate that the first cut excludes the extreme point (P) that would have been reached

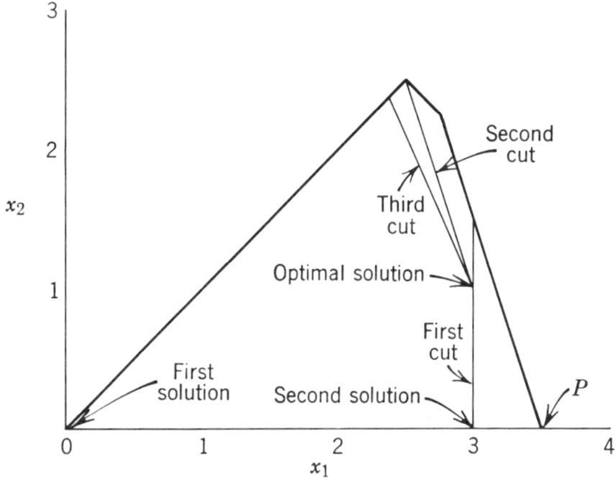

Figure 5

using an ordinary simplex iteration. By pivoting on the cut, we move from the present integer solution toward P and stop at the integer solution nearest to P.

Table 13

			$-s_1$	↓ $-x_2$
	x_0	6	2	-1
	x_3	2	-1	1
	x_4	3	1	1
Source	x_5	3	-6	2
	x_1	3	1	0
← Cut	s_2	1	-3	1

Table 14

			$-s_1$	$-s_2$
	x_0	7	-1	1
	x_3	1	-4	-1
Source	x_4	2	4	-1
	x_5	1	0	-2
	x_1	3	1	0
	x_2	1	-3	1
$\leftarrow$ Cut	s_3	0	1	-1

Discussion of Finiteness

If the cut row is never degenerate ($y_{r0}/y_{rk} \geq 1$), the sequence $\{y_{00}^t\}$ is increasing and the rudimentary algorithm is finite. The absence of degeneracy, while sufficient, is not necessary for finite termination. In the example just solved, the last cut was degenerate. To prove finiteness, we must show that an infinite sequence of degenerate cuts is impossible. This is accomplished in the next section by giving specific pivot column and source row selection rules.

5.12. A FINITE PRIMAL ALL-INTEGER ALGORITHM

Pivot column and source row selections are guided by a special row in the tableau

$$x_{B_*} = y_{*0} - \sum_{j \in R} y_{*j} x_j$$

satisfying the following properties:

(i) If $y_j \overset{L}{<} 0, j \in R$, then $y_{*j} \geq 1$.

(ii) Let $\alpha_j = y_j/y_{*j}, j \in R$, $R_* = \{j | j \in R, y_{*j} \geq 1\}$ and $\alpha_k = \operatorname{lexmin}_{j \in R^*} \alpha_j$

If $y_{*j} \leq -1$, then $\alpha_j \overset{L}{<} \alpha_k$.

In our description of the algorithm, the same row is the special row at each iteration, but this is not necessary (see Exercise 11).

If the initial tableau has at least one row satisfying properties (i) and (ii), any one of them can be designated to be the special row. However, any row with $y_{*j} \geq 1$ for all $j \in R$ satisfies (i) and (ii). Therefore if the initial tableau does not have a suitable row, add the superfluous constraint (28) as the special row. (Also see Exercise 12 for another possibility.)

We now give a pivot column selection rule for maintaining properties (i) and (ii) for all tableaus. Assume that (i) and (ii) hold and the solution is not dual feasible so that $R_* \neq \varnothing$. Then, if

$$\alpha_k = \operatorname*{lexmin}_{j \in R_*} \alpha_j \tag{32}$$

choose x_k as the entering variable. Note that $y_{0j} \leq -1$ implies $j \in R_*$ and $y_{0j} \geq 0$, $j \in R_*$ implies $\alpha_j \overset{L}{>} \alpha_k$. Thus $y_{0k} < 0$ and x_k could have been chosen in the rudimentary algorithm. Given the pivot column k, we adjoin the cut (30) using any source row q from (31). Pivoting on the cut row yields

$$\begin{aligned} y_j^1 &= y_j - \left[\frac{y_{qj}}{y_{qk}}\right] y_k, \qquad j \neq k \\ y_s^1 &= -y_k \text{ (the transformed pivot column)} \end{aligned} \tag{33}$$

except for the pivot (cut) row. The coefficients in the transformed pivot row are $[y_{qj}/y_{qk}]$ for $j = 0$ and all $j \in R$, and the pivot row is the bottom row of the tableau. If x_k is a slack variable from a cut the pivot row in deleted. (The details for this case are asked for in Exercise 30).

Proof of Property (i): Suppose that $y_j^1 \overset{L}{<} 0$ and $y_{*j}^1 \leq 0$. From (33) and the position of the pivot row

$$y_j \overset{L}{\leq} \left[\frac{y_{qj}}{y_{qk}}\right] y_k \tag{34}$$

Since

$$y_{*j}^1 = y_{*j} - \left[\frac{y_{qj}}{y_{qk}}\right] y_{*k} \leq 0$$

$y_{*k} \geq 1$ implies that

$$\frac{y_{*j}}{y_{*k}} \leq \left[\frac{y_{qj}}{y_{qk}}\right] \tag{35}$$

Since $y_k \overset{L}{<} 0$, combining (34) and (35) yields

$$y_j \overset{L}{\leq} \frac{y_{*j}}{y_{*k}} y_k \tag{36}$$

Relation (36) gives a contradiction. There are four cases (in (a), (b) and (c) the lexicographic inequality is assumed to be strict):

(a) $y_{*j} = 0$. Then (36) implies that $y_j \overset{L}{<} 0$, contradicting (i).

(b) $y_{*j} > 0$. Then (36) yields $\alpha_j = y_j/y_{*j} \overset{L}{<} y_k/y_{*k} = \alpha_k$, contradicting (32).

(c) $y_{*j} < 0$. Then (36) yields $\alpha_j \overset{L}{>} \alpha_k$, contradicting (ii).

(d) $y_j = (y_{*j}/y_{*k})y_k$ so that $\alpha_j = \alpha_k$. Then (34) and (35) imply $y_{*j}/y_{*k} = [y_{qj}/y_{qk}]$. Now (33) implies $y_j^1 = 0$ except for the pivot row. Thus the assumption $y_j^1 \overset{L}{<} 0$ and the condition $y_{qk} > 0$ imply $y_{qj} < 0$. Now $y_{*k} > 0$ implies $y_{*j} < 0$ or equivalently, $y_{*j} \leq -1$. Then $\alpha_j = \alpha_k$ contradicts (ii).

For the pivot column, $y_k \overset{L}{<} 0$ and $y_k^1 = -y_k \overset{L}{>} 0$, so that (i) is vacuous. ■

PROOF OF PROPERTY (ii): Let $\alpha_j^1 = y_j^1/y_{*j}^1$. Except for the pivot row

$$\alpha_j^1 - \alpha_k = \frac{y_j - [y_{qj}/y_{qk}]y_k}{y_{*j} - [y_{qj}/y_{qk}]y_{*k}} - \frac{y_k}{y_{*k}} = \frac{y_{*k}y_j - y_{*j}y_k}{y_{*j}^1 y_{*k}}, \tag{37}$$

$\alpha_s^1 = \alpha_k$ and $y_{*s}^1 \leq -1$.

Property (ii) is shown in two parts. We will prove that

(a) if $y_{*j}^1 \leq -1$, then $\alpha_j^1 \overset{L}{<} \alpha_k$, for $j \neq s$,

and then

(b) $\alpha_{k_1}^1 \overset{L}{>} \alpha_k$, where $\alpha_{k_1}^1 = \operatorname{lexmin}_{j \in R_*^1} \alpha_j^1$, $R_*^1 = \{j \mid j \in R^1, y_{*j}^1 \geq 1\}$.

Together (a) and (b) imply that $\alpha_{k_1}^1 \overset{L}{>} \alpha_j^1$ for all j such that $y_{*j}^1 \leq -1$, which is property (ii).

PROOF OF (a): Assume that it is false. From (37) we obtain

$$\frac{y_{*k}y_j - y_{*j}y_k}{y_{*j}^1 y_{*k}} \overset{L}{\geq} 0 \tag{38}$$

Since $y_{*j}^1 \leq -1$ and $y_{*k} \geq 1$, (38) yields

$$y_j \overset{L}{\leq} \frac{y_{*j}}{y_{*k}} y_k \tag{39}$$

But (39) is identical to (36), which we have already shown to be impossible. ■

PROOF OF (b): Assume that it is false. From (37) we obtain

$$\frac{y_{*k}y_{k_1} - y_{*k_1}y_k}{y_{*k_1}^1 y_{*k}} \overset{L}{\leq} 0 \tag{40}$$

Since $y_{*k_1}^1 \geq 1$ and $y_{*k} \geq 1$, (40) yields

$$y_{k_1} \overset{L}{\leq} \frac{y_{*k_1}}{y_{*k}} y_k \tag{41}$$

But (41) is identical to (36) with $j = k_1$. ■

Part (b) is important in its own right. It shows that the sequence $\{\alpha_{k_t}^t\}$ is lexicographically increasing, where

$$\alpha_{k_t}^t = \operatorname*{lexmin}_{j \in R_*^t} \alpha_j^t, \qquad R_*^t = \{j \mid y_{*j}^t \geq 1, j \in R^t\}$$

In the rudimentary algorithm no cut is required when $y_{rk} = 1$. However, in the finite algorithm, a cut is required in this case to maintain properties (i) and (ii) [see Exercise 30].

A rule for selecting the source row is

(a) If $y_{*k} > y_{*0}$, choose the row $*$ as the source.

(b) If $y_{*k} \leq y_{*0}$ and there exists a row $i \neq 0$ with $y_{ik} > 0$ and $y_{ik}/y_{*k} > y_{i0}$, choose the topmost of these rows, say $i = q$, as the source. After pivoting, if $y_{qk_1}^1/y_{*k_1}^1 > y_{q0}$, use q as the source again. The general rule is to use q as the source for t consecutive cuts, where

$$\frac{y_{qk_t}^t}{y_{*k_t}^t} \leq y_{q0} < \frac{y_{qk_{t-1}}^{t-1}}{y_{*k_{t-1}}^{t-1}} \tag{42}$$

When (42) applies, (a) is considered at the next iteration.

(c) If neither (a) nor (b), choose the source row from (31).

It must be shown that the rules for cases (a) and (b) yield source rows that satisfy (31). When $y_{*k} > y_{*0}$, $[y_{*0}/y_{*k}] = 0$ and row $*$ satisfies (31). Similarly, when $y_{qk}/y_{*k} > y_{q0}$, $y_{qk} > y_{q0}$ and $[y_{q0}/y_{qk}] = 0$ so that row q satisfies (31). If $y_{qk_1}^1/y_{*k_1}^1 > y_{q0}$, row q remains eligible, since $y_{q0}^1 = y_{q0}$. Note that when the source row is chosen by (a) or (b), the cut row is degenerate.

Theorem 3: *The primal all-integer algorithm with the pivot column and source row selection rule given above yields an optimal solution to* (1) *after a finite number of iterations.*

PROOF: If $y_{*k} > y_{*0}$, we will show that a finite number, say t, applications of (a) yields $y_{*k_t}^t \leq y_{*0}^t$. When (a) is applied, $y_{*0}^1 = y_{*0}$ and

$$y_{*k_1}^1 = y_{*k_1} - \left[\frac{y_{*k_1}}{y_{*k}}\right] y_{*k} \geq 0$$

Thus

$$0 \leq \frac{y_{*k_1}^1}{y_{*k}} = \frac{y_{*k_1}}{y_{*k}} - \left[\frac{y_{*k_1}}{y_{*k}}\right] < 1 \tag{43}$$

From (43) and $y_{*k} \geq 1$, we obtain $0 \leq y_{*k_1}^1 < y_{*k}$. Then integrality implies that $y_{*k_1}^1 \leq y_{*k} - 1$. Thus there exists t such that $y_{*k_t}^t \leq y_{*0}^t = y_{*0}$.

Also there can only be a finite number of consecutive applications of (b) with q as the source row. When (b) occurs, $y^1_{q0} = y_{q0}$ and, as above, $y^1_{qk_1} \leq y_{qk} - 1$. Thus there exists t such that $y^t_{qk_t} \leq y_{q0}$, and

$$y^t_{*k_t} > 0 \Rightarrow y^t_{qk_t}/y^t_{*k_t} \leq y_{q0}$$

Suppose that the algorithm does not terminate. Then there must be an infinite number of nonoptimal tableaus with

(i) $\{\alpha^t_{k_t}\}$ lexicographically increasing.
(ii) $0 < y^t_{*k_t} \leq y^t_{*0}$.

We have (ii) because every sequence of consecutive tableaus with $y^t_{*k_t} > y^t_{*0}$ is finite and is followed by a tableau with $y^t_{*k_t} \leq y^t_{*0}$. Consider an infinite subsequence of tableaus satisfying (ii). Since each component of y^t_0 is bounded, $\{y^t_{*k_t}\}$ is bounded in these tableaus, say by M. From (i) the ratio $y^t_{0k_t}/y^t_{*k_t}$ is nondecreasing. Because $y^t_{*k_t}$ can only assume the values $1, 2, \ldots, M$ and $y^t_{0k_t}$ is a negative integer, there can only be a finite number of values for the nondecreasing sequence $\{y^t_{0k_t}/y^t_{*k_t}\}$. Thus there exists p such that for all $t \geq p$, $\{y^t_{0k_t}/y^t_{*k_t}\}$ is constant. Then from (i), for $t > p$, $\{y^t_{1k_t}/y^t_{*k_t}\}$ is non-decreasing. Thus $y^t_{1k_t}/y^t_{*k_t} \leq y^t_{10}$, for $t > p$, since if $y^t_{1k_t}/y^t_{*k_t} > y^t_{10}$, (b) would be applied and in a finite number of steps $y^t_{1k_t}/y^t_{*k_t}$ would decrease. Since y^t_{10} is bounded from above and $y^t_{*k_t}$ can only assume a finite number of values, there can only be a finite number of values for the nondecreasing sequence $\{y^t_{1k_t}/y^t_{*k_t}\}$, $t > p$. Thus there exists p_1 such that, for all $t > p_1$, $\{y^t_{1k_t}/y^t_{*k_t}\}$ is constant. The same argument is then applied to $\{y^t_{2k_t}/y^t_{*k_t}\}$, and so forth. Since the number of rows is bounded and each component of $\alpha^t_{k_t}$ becomes constant after a finite number of iterations, an infinite number of tableaus contradicts $\{\alpha^t_{k_t}\}$ lexicographically increasing. ■

Example

In the tableau of Table 6, either the x_3 or x_5 row can be the special row. If we choose the x_3 row, since $y_{01}/y_{11} < y_{02}/y_{12}$, x_1 enters and the successive tableaus are given in Tables 12–14. If we designate the x_5 row to be the special row, since $y_{02}/y_{32} < y_{01}/y_{31}$, x_2 enters. We have $y_{30} > y_{32}$, $y_{22}/y_{32} > y_{20}$; using (b) of the selection rule, the x_4 row is the source. But $y_{22} = 1$, so that no cut is required and x_4 departs. The transformed tableau is given in Table 15.

The entering variable is x_1. Since $y_{30} > y_{31}$ and there is no row satisfying (b), the source row is chosen by (c). Thus the source row can be either the x_3 or x_5 row, since $[y_{10}/y_{11}] = [y_{30}/y_{31}] = 2$. Both rows yield the same cut

$$s_1 + x_1 - x_4 = 2 \qquad (x_2 \leq 2)$$

Table 15

			↓ $-x_1$	$-x_4$
	x_0	0	−3	1
Source	x_3	5	2	−1
	x_2	0	−1	1
Special	x_5	21	8	−2
← Cut	s_1	2	1	−1

The transformed tableau is given in Table 16. The entering variable is

Table 16

			$-s_1$	↓ $-x_4$
	x_0	6	3	−2
	x_3	1	−2	1
	x_2	2	1	0
Source, special	x_5	5	−8	6
	x_1	2	1	−1
← Cut	s_2	0	−2	1

x_4. Since $y_{30} < y_{34}$, the special row is the source row. This yields the degenerate cut

$$s_2 + x_4 - 2s_1 = 0 \qquad (x_1 + x_2 \leq 4)$$

The transformed tableau is given in Table 17. The entering variable is s_1. The source row is chosen by (c) and is the x_5 row. This yields the cut

$$s_3 + s_1 - 2s_2 = 1 \qquad (2x_1 + x_2 \leq 7)$$

By pivoting on this cut, the optimal solution is obtained.

Table 17

			$-s_1$	↓ $-s_2$
	x_0	6	−1	2
	x_3	1	0	−1
	x_2	2	1	0
Source, special	x_5	5	4	−6
	x_1	2	−1	1
	x_4	0	−2	1
← Cut	s_3	1	1	−2

The iterations are shown graphically in Figure 6. Note that a degenerate cut is required when there is no integer solution on the edge joining the current integer solution and the extreme point that would have been reached using an ordinary simplex iteration.

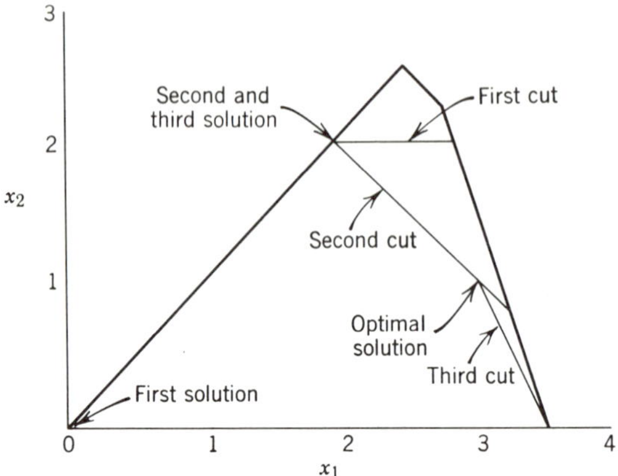

Figure 6

5.13. A SUMMARY OF THE BASIC ALGORITHMS

A tableau gives an optimal solution to (1) if the following three conditions hold:

(i) Primal feasibility, $y_{i0} \geq 0, i = 1, \ldots, m$.
(ii) Integrality, y_{i0} integer, $i = 1, \ldots, m$.
(iii) Dual feasibility, $y_{0j} \geq 0$, for all $j \in R$.

The first two conditions are, of course, necessary; (i)–(iii) are sufficient but not necessary. Each of the basic algorithms maintain two of the three conditions and adds cuts in an attempt to satisfy the third. The following general algorithm loosely describes the three basic algorithms.

Algorithm

STEP 1: (Initialization.) Begin with a solution that satisfies two of the three conditions. Go to Step 2.

STEP 2: (Optimality test.) If the third condition is satisfied, terminate. Otherwise, go to Step 3.

STEP 3: (Cutting and pivoting.) Add a cut from (6) with h chosen appropriately. Pivot to maintain the specified two conditions (more than one pivot may be necessary). Return to Step 2.

Stating which condition is not met until termination identifies the particular algorithm. If it is (ii), we have the method of integer forms; if (i), the dual all-integer algorithm; and if (iii), the primal all-integer algorithm.

Having described the basic algorithms, we now turn to the problem of how they might be improved by finding better cuts.

5.14. COMPOSITE CUTS IN THE METHOD OF INTEGER FORMS

In this section we combine several cuts from the method of integer forms into one cut called a *composite cut*. Pivoting on the composite cut has the advantage of bringing the solution "closer" to integrality, but in the process we may encounter tableaus that are neither primal nor dual feasible.

Suppose the cut (8)

$$0 = -f_{r0} + \sum_{j \in R} f_{rj} x_j - s_1 \tag{44}$$

is taken from row r. By pivoting on (44) and column k (k chosen by the dual simplex criterion), the system (3) is transformed to

$$x_{B_i} = \hat{y}_{i0} - \sum_{j \in R - \{k\}} \hat{y}_{ij} x_j - y_{is_1} s_1, \qquad i = 0, \ldots, m \tag{45}$$

The cut (8) with source row (45) and $i = r$ is

$$0 = -\hat{f}_{r0} + \sum_{j \in R - \{k\}} \hat{f}_{rj} x_j + f_{rs_1} s_1 - s_2 \tag{46}$$

If $f_{rs_1} > 0$, we can pivot on (46) and column s_1. The system (45) is transformed to

$$x_{B_i} = \hat{\hat{y}}_{i0} - \sum_{j \in R - \{k\}} \hat{\hat{y}}_{ij} x_j - y_{is_2} s_2, \qquad i = 0, \ldots, m \tag{47}$$

Let $RS(P)$ represent the right-hand side of Equation (P). The transformation from (3) to (47) is given by

$$RS(45) = RS(3) + \frac{y_{ik}}{f_{rk}} RS(44) \tag{48}$$

and

$$RS(47) = RS(45) + \frac{y_{is_1}}{f_{rs_1}} RS(46) \tag{49}$$

Thus

$$RS(47) = RS(3) + \frac{y_{ik}}{f_{rk}} RS(44) + \frac{y_{is_1}}{f_{rs_1}} RS(46) \tag{50}$$

Since $y_{is_1} = y_{ik}/f_{rk}$, (50) yields

$$RS(47) = RS(3) + \frac{y_{ik}}{f_{rk}}\left(RS(44) + \frac{RS(46)}{f_{rs_1}}\right) \tag{51}$$

Thus the coefficients $\hat{y}_{ij}$ in (47) can be obtained from (51).

There is a method for obtaining the coefficients in (47) that requires fewer calculations. Using (44) to eliminate s_1 in (46) yields the *composite cut*

$$\begin{aligned} 0 &= -\hat{f}_{r0} - f_{r0}f_{rs_1} + \sum_{j \in R - \{k\}} (\hat{f}_{rj} + f_{rj}f_{rs_1})x_j + f_{rk}f_{rs_1}x_k - s_2 \\ &= RS(46) + f_{rs_1}RS(44) \end{aligned} \tag{52}$$

By pivoting on (52) and column k, $RS(3)$ is transformed into

$$\begin{aligned} RS(3) + \frac{y_{ik}}{f_{rk}f_{rs_1}} RS(52) &= RS(3) + \frac{y_{ik}}{f_{rk}f_{rs_1}} (f_{rs_1}RS(44) + RS(46)) \\ &= RS(3) + \frac{y_{ik}}{f_{rk}}\left(RS(44) + \frac{RS(46)}{f_{rs_1}}\right) = RS(47) \end{aligned}$$

The point is that to obtain the coefficients in (47) using the composite cut, it is not necessary to transform the whole tableau twice. Instead (45) is obtained only for row r. If $f_{rs_1} \neq 0$, we find the cut (46) and the composite cut (52). The whole tableau is then transformed using (52) as the pivot row and k as the pivot column to obtain (47).

Equation (52) is the composite cut for a double transformation. Clearly, the procedure can be extended to composite cuts for p transformations. If in (47) with $i = r, f_{rs_2} > 0$, the cut from row r

$$0 = -\hat{\hat{f}}_{r0} + \sum_{j \in R - \{k\}} \hat{\hat{f}}_{rj}x_j + f_{rs_2}s_2 - s_3 \tag{53}$$

can be combined with (52) by eliminating s_2. This gives the composite cut

$$0 = RS(53) + f_{rs_2}RS(52) \tag{54}$$

Transforming (3) by pivoting on (54) and column k is equivalent to the three transformations: (a) pivot on (44) and column k, (b) pivot on (46) and column s_1, and (c) pivot on (53) and column s_2, but requires many fewer calculations.

Suppose that row r has been transformed p times and $x_k, s_1, \ldots, s_{p-1}$ have been eliminated yielding

$$x_{B_r} = y^p_{r0} - \sum_{j \in R - \{k\}} y^p_{rj}x_j - y_{rs_p}s_p \tag{55}$$

If $f_{rs_p} > 0$ we can, proceeding as above, eliminate s_p from (55). On the other hand, if $f_{rs_p} = 0$, we could not use the $(p + 1)$st cut to eliminate s_p from (55). Therefore, if we elect to continue the process described above as far as possible, we stop when $f_{rs_i} = 0$. Suppose this occurs after p transformations, so that $f_{rs_p} = 0$ and $f_{rs_i} > 0$, $i < p$. Instead of computing each successive composite cut, we want to find the pth one directly from (55). The pth composite cut is of the form

$$s_p = \alpha_{r0}^p - \sum_{j \in R} \alpha_{rj}^p x_j \tag{56}$$

Substituting (56) into (55) yields

$$x_{B_r} = (y_{r0}^p - y_{rs_p}\alpha_{r0}^p) - \sum_{j \in R - \{k\}} (y_{rj}^p - y_{rs_p}\alpha_{rj}^p)x_j + y_{rs_p}\alpha_{rk}^p x_k \tag{57}$$

Equating coefficients of $x_j, j \in R$ from row r of (3) and (57), we obtain

$$\begin{aligned} \alpha_{rj}^p &= \frac{y_{rj}^p - y_{rj}}{y_{rs_p}} \qquad \text{for } j \neq k \\ \alpha_{rk}^p &= -\frac{y_{rk}}{y_{rs_p}} \end{aligned} \tag{58}$$

The system (3) is transformed by pivoting on (56) and column k.

Example

From Table 1

$$x_2 = \frac{9}{4} - \frac{3x_3}{2} + \frac{x_5}{4} \tag{59}$$

and we obtain the cut

$$s_1 - \frac{x_3}{2} - \frac{3x_5}{4} = -\frac{1}{4} \tag{60}$$

Since the x_5 column is the dual simplex pivot column, we use (60) to eliminate x_5 from (59) and obtain

$$x_2 = \frac{7}{3} - \frac{5x_3}{3} + \frac{s_1}{3} \tag{61}$$

Because the coefficient of s_1 is not an integer in (61), a second cut

$$s_2 - \frac{2x_3}{3} - \frac{2s_1}{3} = -\frac{1}{3} \tag{62}$$

is obtained. Using (62) to eliminate s_1 from (61) yields

$$x_2 = \frac{5}{2} - 2x_3 + \frac{s_2}{2} \tag{63}$$

The cut from (63) is

$$s_3 - \frac{s_2}{2} = -\frac{1}{2} \tag{64}$$

Using (64) to eliminate s_2 from (63) yields

$$x_2 = 3 - 2x_3 + s_3 \tag{65}$$

Since the coefficient of s_3 in (65) is an integer, the procedure stops. The composite cut is then found from (58), using (59) and (65). We have $\alpha_{r0}^3 = (3 - \frac{9}{4})/(-1) = -\frac{3}{4}$, $\alpha_{r3}^3 = (2 - \frac{3}{2})/(-1) = -\frac{1}{2}$, and $\alpha_{r5}^3 = (0-(-\frac{1}{4}))/(-1) = -\frac{1}{4}$, and thus the composite cut

$$s_3 = -\frac{3}{4} + \frac{x_3}{2} + \frac{x_5}{4} \tag{66}$$

Note that (66) is a better cut than (60).

Adding the composite cut (66) to the bottom of Table 1 and pivoting by removing s_3 and entering x_5, we obtain Table 18.

Table 18

		↓ $-x_3$	$-s_3$
x_0	7	0	1
x_1	2	−1	1
x_2	3	2	−1
← x_4	−1	−3	2
x_5	3	2	−4

To restore primal feasibility, another dual simplex iteration is executed with x_4 leaving and x_3 entering, yielding Table 19.

Table 19

		$-x_4$	$-s_3$
x_0	7	0	1
x_1	$\frac{7}{3}$	$-\frac{1}{3}$	$\frac{1}{3}$
x_2	$\frac{7}{3}$	$\frac{2}{3}$	$\frac{1}{3}$
x_3	$\frac{1}{3}$	$-\frac{1}{3}$	$-\frac{2}{3}$
x_5	$\frac{7}{3}$	$\frac{2}{3}$	$-\frac{8}{3}$

All available source rows (x_1, x_2, x_3, x_5) now yield the same cut

$$\frac{2x_4}{3} + \frac{s_3}{3} \geq \frac{1}{3}$$

This cut is not sufficient to achieve the optimal solution, as can be seen by adding it and executing a dual simplex iteration. However, it can be improved by choosing x_1 as a source row and then finding the composite cut. With x_1 as the source row,

$$s_4 = -\frac{1}{3} + \frac{2x_4}{3} + \frac{s_3}{3}$$

as the first cut, and x_4 as the departing variable, the transformed x_1 row is

$$x_1 = \frac{5}{2} + \frac{s_4}{2} - \frac{s_3}{2}$$

This yields the next cut

$$s_5 = -\frac{1}{2} + \frac{s_4}{2} + \frac{s_3}{2}$$

and the transformed row

$$x_1 = 3 + s_5 - s_3$$

Thus the composite cut is

$$s_5 = -\frac{2}{3} + \frac{x_4}{3} + \frac{2s_3}{3} \tag{67}$$

When (67) is added to Table 19, one dual simplex iteration yields the optimal solution.

Composite Cuts and Determinants

In the example, the two transformations obtained using the composite cuts (66) and (67) yielded all-integer tableaus. These results were by no means coincidental.

Recall (Section 2.7) that $y_{ij} = q_{ij}/D$ for all i and j, where D is the absolute value of the determinant of the basis matrix and q_{ij} is an integer. Thus $D = 1$ is a sufficient condition for an all-integer tableau. We will show that the composite cuts systematically reduce D, frequently to 1.

When a cut is added from (8) and a pivot is executed on the cut row, the magnitude of the pivot element is $f_{rk} = e_{rk}/D < 1$. Thus the effect of the cut is to reduce the magnitude of the determinant to $D^1 = De_{rk}/D = e_{rk}$. Now if a second cut is permissible from row r, the magnitude of the pivot element is $f_{rs_1} = e_{rs_1}/D^1$, where e_{rs_1} is a positive integer smaller than D^1. By pivoting on it, we obtain $D^2 = e_{rs_1} < D^1$. In general,

$$1 \leq D^{p+1} = e_{rs_p} < D^p = e_{rs_{p-1}} < \cdots < D^1 = e_{rk} < D$$

As can be seen from (58), the magnitude of the pivot element from the composite cut (56) is $|y_{rk}/y_{rs_p}|$. Since $D^{p+1} = e_{rs_p}$, we have

$$D\left|\frac{y_{rk}}{y_{rs_p}}\right| = e_{rs_p}$$

Thus

$$|y_{rs_p}| = \frac{D|y_{rk}|}{e_{rs_p}}$$

Since, by hypothesis $f_{rs_p} = 0$, e_{rs_p} must be a divisor of $D|y_{rk}|$. Therefore, an upper bound on e_{rs_p} is the largest divisor of $D|y_{rk}|$ equal to or less than e_{rk}. In particular, if $D|y_{rk}|$ is a prime number, then $e_{rs_p} = 1$. In the example $D|y_{rk}| = 1$ for both composite cuts.

By using the composite cut, we may achieve a greater reduction in the determinant and hopefully produce an integer solution, but a complication arises. Column k was chosen as the pivot column based upon the dual simplex rule and the pivot row (44). Now, using column k and the pivot row (56), dual feasibility may be destroyed. (For no particular reason, this did not happen in the example.) Thus after the tableau is transformed by (56), the solution may be neither primal nor dual feasible nor integer. If the solution obtained after pivoting on the composite cut is not all-integer, another composite cut can be added. Since each composite cut reduces D by at least one, in a finite number of steps an all-integer solution is found.

When an all-integer solution is obtained, it may be either:

(a) Primal and dual feasible.
(b) Only dual feasible.
(c) Only primal feasible.
(d) Neither primal nor dual feasible.

If (a) occurs, the solution is optimal. If (b) or (c) occurs, dual (primal) simplex iterations can be executed until primal (dual) feasibility is achieved, at which point it may be necessary to generate another composite cut. Alternatively, a composite cut can be added at any noninteger tableau. When (d) occurs, Phase 1 simplex calculations are executed until primal or dual feasibility is achieved. The finiteness of this algorithm is an open question. Because dual feasibility can be destroyed, the lexicographic argument given for the method of integer forms does not carry over in a direct manner.

Example

Since the previous example did not illustrate case (d), consider the tableau in Table 20.

Table 20

		$-x_3$	$-x_4$
x_0	8	$\frac{1}{2}$	$\frac{1}{2}$
x_1	$\frac{5}{2}$	$\frac{1}{4}$	$-\frac{1}{4}$
x_2	$\frac{1}{2}$	$-\frac{1}{4}$	$\frac{5}{4}$

Using the x_1 row as a source, we obtain the cut

$$s_1 = -\frac{1}{2} + \frac{x_3}{4} + \frac{3x_4}{4}$$

Since the x_4 column is the pivot column by the dual simplex criterion, we transform the x_1 row to eliminate x_4 and then take another cut from the x_1 row. After two transformations, we obtain the composite cut

$$s_3 = -\frac{1}{2} + \frac{3x_3}{4} + \frac{x_4}{4}$$

The cut is shown in Table 21. Pivoting on s_3 and x_4 yields Table 22.

Table 21

			$-x_3$	$\downarrow$ $-x_4$
	x_0	8	$\frac{1}{2}$	$\frac{1}{2}$
Source	x_1	$\frac{5}{2}$	$\frac{1}{4}$	$-\frac{1}{4}$
	x_2	$\frac{1}{2}$	$-\frac{1}{4}$	$\frac{5}{4}$
← Cut	s_3	$-\frac{1}{2}$	$-\frac{3}{4}$	$-\frac{1}{4}$

Table 22

		$-x_3$	$-s_3$
x_0	7	-1	2
x_1	3	1	-1
x_2	-2	-4	5
x_4	2	3	-4

The tableau in Table 22 is all-integer but neither primal nor dual feasible (case d). We arbitrarily choose to restore primal feasibility first. This is done by adding a Phase 1 objective row of zeros, as shown in Table 23, and then applying the dual simplex algorithm. The departing variable is x_2 and x_3 enters. The resulting tableau (Table 24) is primal feasible, so that Phase 1 ends. Now we choose to execute a primal simplex iteration with x_2 entering and x_4 departing to drive toward dual feasibility. The tableau in Table 25 is primal and dual feasible but not all-integer. The x_0 row is chosen as the source yielding the cut

$$s_4 = -\frac{2}{3} + \frac{x_4}{3} + \frac{2s_3}{3} \tag{68}$$

According to the dual simplex criterion, either x_4 or s_3 can be the entering variable. However, regardless of which is chosen, the composite cut is (68). The cut (68) is appended to the tableau of Table 25. With s_3 as the entering variable, the optimal solution given in Table 26 is obtained. Note that since s_3 is a slack for a cut, its row is deleted.

Table 23

		↓ $-x_3$	$-s_3$
z_0	0	0	0
x_0	7	−1	2
x_1	3	1	−1
← x_2	−2	−4	5
x_4	2	3	−4

Table 24

		↓ $-x_2$	$-s_3$
x_0	$\frac{15}{2}$	$-\frac{1}{4}$	$\frac{3}{4}$
x_1	$\frac{5}{2}$	$\frac{1}{4}$	$\frac{1}{4}$
x_3	$\frac{1}{2}$	$-\frac{1}{4}$	$-\frac{5}{4}$
← x_4	$\frac{1}{2}$	$\frac{3}{4}$	$-\frac{1}{4}$

Table 25

			$-x_4$	↓ $-s_3$
Source	x_0	$\frac{23}{3}$	$\frac{1}{3}$	$\frac{2}{3}$
	x_1	$\frac{7}{3}$	$-\frac{1}{3}$	$\frac{1}{3}$
	x_3	$\frac{2}{3}$	$\frac{1}{3}$	$-\frac{4}{3}$
	x_2	$\frac{2}{3}$	$\frac{4}{3}$	$-\frac{1}{3}$
← Cut	s_4	$-\frac{2}{3}$	$-\frac{1}{3}$	$-\frac{2}{3}$

Table 26

		$-x_4$	$-s_4$
x_0	7	0	1
x_1	2	$-\frac{1}{2}$	$\frac{1}{2}$
x_3	2	1	−2
x_2	1	$\frac{3}{2}$	$-\frac{1}{2}$

5.15. CUTS WITH DIFFERENT VALUES OF h

For any integer h, (6) can be written

$$\sum_{j\in R} (hf_{ij} - [hf_{ij}])x_j \geq hf_{i0} - [hf_{i0}] \tag{69}$$

Note that when $h = 1$, $[hf_{ij}] = 0$ and (69) specializes to (7).

The cut

$$\sum_{j\in R} d_j x_j \geq d_0$$

is said to be stronger than, or to imply, the cut

$$\sum_{j\in R} p_j x_j \geq p_0$$

if $d_j \leq p_j$ for all $j \in R$ and $d_0 \geq p_0$, with at least one of the inequalities strict.

Sometimes $h > 1$ in (69) yields stronger cuts than (7). For example, consider the source row

$$x_1 = \frac{1}{4} - \frac{3x_2}{4} + \frac{x_3}{4}$$

The cuts obtainable from (69) with $h = (1, 2, 3)$ are

$$\frac{3x_2}{4} + \frac{3x_3}{4} \geq \frac{1}{4} \qquad (h = 1)$$

$$\frac{x_2}{2} + \frac{x_3}{2} \geq \frac{1}{2} \qquad (h = 2)$$

$$\frac{x_2}{4} + \frac{x_3}{4} \geq \frac{3}{4} \qquad (h = 3)$$

In this case, the $h = 3$ cut implies the $h = 2$ cut, which implies the $h = 1$ cut. In general no such order exists. However, given a specific criterion like maximizing $hf_{i0} - [hf_{i0}]$, it is possible to choose the best integer value of h (see Exercise 18). Furthermore, all cuts (69) with integer h can be generated by $0 \leq h \leq D - 1$. To show this, let $h = \rho D + h^*$, where ρ is an arbitrary integer and h^* is an integer satisfying $0 \le h^* \le D - 1$. Since $f_{ij} = e_{ij}/D$, $0 \le e_{ij} < D$, where e_{ij} is an integer satisfying

$$\begin{aligned} hf_{ij} - [hf_{ij}] &= (\rho D + h^*)\frac{e_{ij}}{D} - \left[(\rho D + h^*)\frac{e_{ij}}{D}\right], \qquad j \in R \cup \{0\} \\ &= \rho e_{ij} + \frac{h^* e_{ij}}{D} - \left[\rho e_{ij} + \frac{h^* e_{ij}}{D}\right] \\ &= \frac{h^* e_{ij}}{D} - \left[\frac{h^* e_{ij}}{D}\right] \\ &= h^* f_{ij} - [h^* f_{ij}] \end{aligned}$$

Additional cuts of the form (6) may be generated from integer linear combinations of source rows. Consider the two rows

$$x_1 = \frac{1}{4} + \frac{3x_2}{4} - \frac{x_3}{4}$$

$$x_4 = \frac{3}{2} - \frac{x_2}{2}$$

Since $x_1 + x_4$ must be an integer,

$$x_1 + x_4 = \frac{7}{4} + \frac{x_2}{4} - \frac{x_3}{4}$$

can be used as a source row in (69), yielding with $h = 1$ the cut

$$\frac{3x_2}{4} + \frac{x_3}{4} \ge \frac{3}{4}$$

which cannot be generated from either the x_1 row or the x_4 row.

A powerful result is that no more than D different cuts of the form (69) can be generated from a tableau, even permitting integer combinations of rows as sources. This can be derived from the fact that the set of cuts obtainable from (69) is isomorphic to a finite abelian group (see Exercise 24 of Chapter 7).

Noninteger h

It may be possible to obtain stronger cuts with noninteger h in (6). The coefficient of x_j in (6) is

$$d_j(h) = [h]y_{ij} - [hy_{ij}]$$

Let $h = [h] + f_h$. Note that $d_j([h] + f_h)$ is a nonincreasing (nondecreasing) step function of f_h for $y_{ij} > 0\ (<0)$. There is a jump of unit magnitude at h^o if $h^o y_{ij}$ is an integer, and we obtain

$$\left.\begin{aligned} d_j(h^o + \epsilon) &= d_j(h^o) \\ d_j(h^o - \epsilon) &= d_j(h^o) + 1 \end{aligned}\right\} y_{ij} > 0$$

$$\left.\begin{aligned} d_j(h^o + \epsilon) &= d_j(h^o) + 1 \\ d_j(h^o - \epsilon) &= d_j(h^o) \end{aligned}\right\} y_{ij} < 0$$

where $\epsilon > 0$ is arbitrarily small (see Figure 7).

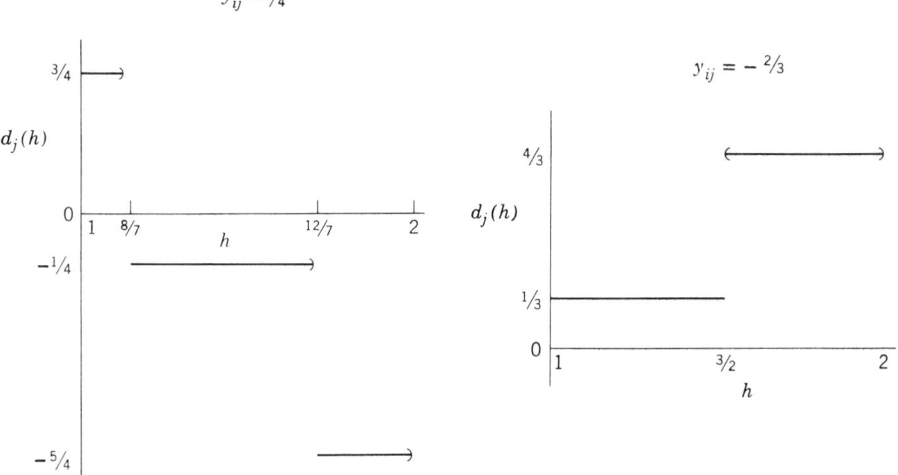

Figure 7

Given $[h]$, it is sometimes possible to choose $f_h > 0$ such that

$$\begin{aligned} d_j(h) &\leq d_j([h]) && \text{for all } j \in R \\ d_j(h) &< d_j([h]) && \text{for at least one } j \in R \\ d_0(h) &= d_0([h]) \end{aligned}$$

and thereby obtain a stronger cut. For example, using the source row,

$$x_1 = \frac{47}{10} - \frac{29}{10}x_2 - \frac{23}{10}x_3 + \frac{25}{10}x_4$$

the cut

$$-\frac{3}{10}x_2 - \frac{1}{10}x_3 + \frac{5}{10}x_4 \geq \frac{1}{10}$$

with $\frac{90}{29} \leq h < \frac{150}{47}$ is stronger than the cut

$$\frac{7}{10}x_2 + \frac{9}{10}x_3 + \frac{5}{10}x_4 \geq \frac{1}{10}$$

obtained with $h = 3$.

If a cut is to exclude the current solution, it is necessary to choose f_h so that

$$d_0(h) = [h]y_{i0} - [hy_{i0}] > 0$$

If $y_{i0} > 0$, $d_0(h)$ can decrease as f_h increases. Suppose that $d_0([h]) > 0$ so that $[h]f_{i0}$ is not integer. If there is no value of h, $[h] \leq h < [h] + 1$ such that hy_{i0} is an integer, then

$$d_0(h) = d_0([h]), \qquad [h] \leq h < [h] + 1$$

and any $f_h > 0$ can be used. On the other hand, suppose that h^* is the smallest h, $[h] \leq h < [h] + 1$ such that hy_{i0} is an integer. Then, since $[h]y_{i0}$ is not an integer,

$$h^*y_{i0} = \langle [h]y_{i0} \rangle$$

Since

$$d_0(h^*) = d_0([h]) - 1 < 0$$

it is necessary to choose $f_h < f_{h^*}$ for the cut to exclude the current solution.

A similar idea is applicable to improving cuts in the primal and dual all-integer algorithms. In the primal all-integer algorithm, we choose $h = 1/y_{qk}$ in (6), where the x_q row is the source, k is the pivot column, and

$$\left[\frac{y_{q0}}{y_{qk}}\right] \leq \frac{y_{r0}}{y_{rk}} = \min_i \left(\frac{y_{i0}}{y_{ik}},\ y_{ik} > 0\right)$$

This choice of h is sufficient but not necessary for obtaining a pivot element equal to 1. It may be possible to choose $h > 1/y_{qk}$. Necessary conditions for pivot element $[hy_{qk}] = 1$ are

(i) $1/y_{qk} \leq h < 2/y_{qk}$.
(ii) $[hy_{q0}]/1 \leq y_{q0}/y_{qk}$.

Note that (ii) is satisfied if

$$hy_{q0} < \left\langle \frac{y_{q0}}{y_{qk}} \right\rangle$$

Thus

$$\frac{1}{y_{qk}} \leq h < \min\left\{\frac{2}{y_{qk}}, \frac{\langle y_{q0}/y_{qk} \rangle}{y_{q0}}\right\} \tag{70}$$

Now as h increases from $1/y_{qk}$ in the cut,

$$\sum_{j \in R} p_j(h)x_j = \sum_{j \in R} [hy_{qj}]x_j \leq [hy_{q0}] = p_0(h) \tag{71}$$

positive coefficients increase in jumps and negative coefficients decrease in jumps. A cut (71) with $h > 1/y_{qk}$ such that

$$\begin{aligned} p_j(h) &\geq p_j(1/y_{qk}) && \text{for all } j \in R \\ p_j(h) &> p_j(1/y_{qk}) && \text{for at least one } j \in R \\ p_0(h) &= p_0(1/y_{qk}) \end{aligned}$$

is stronger than the cut (30) with $h = 1/y_{qk}$.

The following example illustrates a cut (71) with $h > 1/y_{qk}$, which is stronger than (30). The source row is

$$x_1 = 22 - 8x_2 - 15x_3 + 6x_4$$

and x_2 is the entering variable. The cut (30) obtained with $h = \frac{1}{8}$ is

$$x_2 + x_3 - x_4 \leq 2$$

From (70) we have $\frac{1}{8} \leq h < \min\{\frac{2}{8}, \frac{3}{22}\} = \frac{3}{22}$. Putting h arbitrarily close to $\frac{3}{22}$ yields the stronger cut

$$x_2 + 2x_3 - x_4 \leq 2$$

The improvement of dual all-integer cuts by using different values of h is asked for in Exercise 19.

5.16. DEEP CUTS

Suppose that the cut (7) is not satisfied as an equality by any $x \in S$, where S is the set of feasible solutions to (1). Then (7) can be made deeper, in the sense that there exists $t > f_{i0}$ such that the cut

$$\sum_{j \in R} f_{ij}x_j \geq t$$

is valid (excludes no feasible integer solution). Given f_{ij}, for all $j \in R$, the deepest valid cut,

$$\sum_{j \in R} f_{ij}x_j \geq t^*$$

is determined by finding the smallest value of t such that $\sum_{j\in R} f_{ij}x_j = t$ is satisfied by some $x \in S$. Thus

$$t^* = \min \sum_{j\in R} f_{ij}x_j, \qquad x \in S$$

Finding t^* is as difficult as solving (1). However, for any $Q \supset S$

$$t(Q) = \min \sum_{j\in R} f_{ij}x_j, \qquad x \in Q$$

yields a valid cut

$$\sum_{j\in R} f_{ij}x_j \geq t(Q)$$

For instance, let

$$Q_1 = \{x | Ax = b, x_j \geq 0 \text{ integer, for all } j \in R, x_{B_r} \text{ integer}\}$$

In Q_1 nonnegativity on all basic variables is dropped, as is the integrality on all basic variables except x_{B_r}. Requiring x_{B_r} integer implies that

$$\sum_{j\in R} f_{rj}x_j = f_{r0} + \rho, \qquad \rho \text{ integer} \tag{72}$$

Furthermore, $f_{rj} \geq 0$, $x_j \geq 0$, $j \in R$, imply that $\rho \geq 0$. Thus $t(Q_1) = f_{r0} + \rho^*$, where ρ^* is the smallest nonnegative integer such that (72) has a nonnegative integer solution.

Suppose the cut (7) is

$$\frac{6x_1}{10} + \frac{7x_2}{10} \geq \frac{8}{10} \tag{73}$$

Clearly, there is no solution to (73) of the form (72) with $\rho = 0$. However, with $\rho = 1$, we obtain the solution $x_1 = 3$, $x_2 = 0$. Thus $\rho^* = 1$, $t(Q_1) = \frac{18}{10}$ and the cut

$$\frac{6x_1}{10} + \frac{7x_2}{10} \geq \frac{18}{10}$$

is valid.

Let $Q_1 \supset Q_2 = \{x | Ax = b, x_j \geq 0 \text{ integer for all } j \in R \cup \{B_r\}\}$. We have

$$\begin{aligned} t(Q_2) = \min & \sum_{j\in R} f_{rj}x_j \\ & x_{B_r} + \sum_{j\in R} y_{rj}x_j = y_{r0} \\ & x_{B_r}, x_j \geq 0 \text{ integer}, j \in R \end{aligned} \tag{74}$$

Note that (74) is a one-constraint ILP.

Suppose that (73) was derived from the source row

$$x_3 = \frac{8}{10} - \frac{6x_1}{10} + \frac{13x_2}{10}$$

Then from (74) we obtain

$$\begin{aligned} t(Q_2) &= \min \frac{6x_1}{10} + \frac{7x_2}{10} \\ x_3 &+ \frac{6x_1}{10} - \frac{13x_2}{10} = \frac{8}{10} \\ x_1&, x_2, x_3 \geq 0 \text{ integer} \end{aligned} \tag{75}$$

By inspection the optimal solution to (75) is $x_1 = 0$, $x_2 = 4$, $x_3 = 6$ and $t(Q_2) = \frac{28}{10}$ which yields the cut

$$\frac{6x_1}{10} + \frac{7x_2}{10} \geq \frac{28}{10}$$

Cuts in the dual all-integer algorithm can also be made deeper using the same formulation. Consider at some iteration in the dual all-integer algorithm the equations

$$\begin{aligned} x_5 &= -1 + 2x_1 - x_2 + 5x_3 + x_4 \\ x_6 &= -10 - 3x_1 - 5x_2 + 6x_3 + 4x_4 \end{aligned}$$

With $h = \frac{1}{2}$, the x_5 row yields, from (15), the cut

$$x_1 + 3x_3 + x_4 \geq 1$$

Now if we require x_6 to be a nonnegative integer, consider

$$\begin{aligned} t(Q_2) &= \min x_1 + 3x_3 + x_4 \\ &- 3x_1 - 5x_2 + 6x_3 + 4x_4 - x_6 = 10 \\ &x_1, \ldots, x_4, x_6 \geq 0 \text{ integer} \end{aligned} \tag{76}$$

The solution to (76) is $x_1 = x_2 = x_3 = 0$, $x_4 = 3$, $x_6 = 2$ and $t(Q_2) = 3$. Hence

$$x_1 + 3x_3 + x_4 \geq 3$$

is a valid constraint.

Cuts used in the primal all-integer algorithm cannot be improved in this manner because they are always satisfied as an equality by a feasible integer solution. Improvements in primal all-integer cuts have to come from orientation changes instead of increased depth.

5.17. A CUTTING PLANE ALGORITHM FOR THE MILP

The MILP

$$\begin{aligned} \max x_0 = c_1x + c_2v& \\ A_1x + A_2v = b& \\ x \geq 0 \text{ integer}& \\ v \geq 0& \end{aligned} \tag{77}$$

can also be solved by a cutting plane algorithm.

Suppose we have a basic solution to the LP corresponding to (77) with the ith row given by

$$x_{B_i} = y_{i0} - \sum_{j \in R_1} y_{ij}x_j - \sum_{j \in R_2} y_{ij}v_j \tag{78}$$

where R_1 is the index set for the nonbasic variables restricted to be integers, $R_1 \cup R_2 = R$, and x_{B_i} is an integer variable. Assume that $f_{i0} > 0$. Our objective is to develop a cut that excludes the solution with $x_{B_i} = y_{i0}$, $x_j = 0$, $j \in R_1$, and $v_j = 0, j \in R_2$, yet excludes no feasible solutions to (77).

Taking fractional parts in (78) and rearranging terms yields

$$\sum_{j \in R_1} f_{ij}x_j + \sum_{j \in R_2} y_{ij}v_j - f_{i0} = [y_{i0}] - \sum_{j \in R_1} [y_{ij}]x_j - x_{B_i} \tag{79}$$

The right-hand side of (79) must be an integer; therefore, the left-hand side must be as well. In particular, either

$$\sum_{j \in R_1} f_{ij}x_j + \sum_{j \in R_2} y_{ij}v_j - f_{i0} \geq 0 \tag{80}$$

or

$$\sum_{j \in R_1} f_{ij}x_j + \sum_{j \in R_2} y_{ij}v_j - f_{i0} \leq -1 \tag{81}$$

Let

$$R_2^+ = \{j \mid j \in R_2, y_{ij} > 0\} \qquad \text{and} \qquad R_2^- = \{j \mid j \in R_2, y_{ij} < 0\}$$

Then, if (80) holds, so does

$$\sum_{j \in R_1} f_{ij}x_j + \sum_{j \in R_2^+} y_{ij}v_j \geq f_{i0} \tag{82}$$

and if (81) holds, then since $f_{ij} \geq 0$ for all $j \in R_1$

$$\sum_{j \in R_2^-} y_{ij}v_j \leq -1 + f_{i0} \tag{83}$$

Multiplying (83) by $f_{i0}/(-1 + f_{i0}) < 0$ yields

$$-\sum_{j \in R_2^-} \frac{f_{i0} y_{ij} v_j}{1 - f_{i0}} \geq f_{i0} \tag{84}$$

The left-hand sides of (82) and (84) are both nonnegative; since (82) or (84) must hold, we have

$$\sum_{j \in R_1} f_{ij} x_j + \sum_{j \in R_2^+} y_{ij} v_j - \sum_{j \in R_2^-} \frac{f_{i0} y_{ij} v_j}{1 - f_{i0}} \geq f_{i0} \tag{85}$$

For an ILP, $R_2^+ = R_2^- = \varnothing$ and (85) specializes to (8). It is important to note that when (85) is transformed into an equality, the slack variable is not necessarily integer.

If there exists $j \in R_1$ such that $f_{ij} > f_{i0}$, (85) can be improved. Partition R_1 as

$$R_1^+ = \{j \mid j \in R_1, f_{ij} \leq f_{i0}\}, \qquad R_1^- = \{j \mid j \in R_1, f_{ij} > f_{i0}\}$$

Since it is assumed that $f_{i0} > 0$, then $f_{ij} > 0$ for all $j \in R_1^-$. By subtracting $\sum_{j \in R_1^-} x_j$ from both sides of (79), we get

$$\sum_{j \in R_1^+} f_{ij} x_j + \sum_{j \in R_1^-} (f_{ij} - 1) x_j + \sum_{j \in R_2} y_{ij} v_j - f_{i0} =$$
$$[y_{i0}] - \sum_{j \in R_1^+} [y_{ij}] x_j - \sum_{j \in R_1^-} \langle y_{ij} \rangle x_j - x_{B_i} \tag{86}$$

From (86), analogous to (82) and (84), we have either

$$\sum_{j \in R_1^+} f_{ij} x_j + \sum_{j \in R_2^+} y_{ij} v_j \geq f_{i0} \tag{87}$$

or

$$-\frac{f_{i0}}{1 - f_{i0}} \left(\sum_{j \in R_1^-} (f_{ij} - 1) x_j + \sum_{j \in R_2^-} y_{ij} v_j \right) \geq f_{i0} \tag{88}$$

Combining (87) and (88) yields the cut

$$\sum_{j \in R_1^+} f_{ij} x_j + \sum_{j \in R_1^-} \frac{f_{i0}(1 - f_{ij}) x_j}{1 - f_{i0}} + \sum_{j \in R_2^+} y_{ij} v_j - \sum_{j \in R_2^-} \frac{f_{i0} y_{ij} v_j}{1 - f_{i0}} \geq f_{i0} \tag{89}$$

Note that (89) is the same as (85), except for the coefficients on x_j, $j \in R_1^-$. However, $f_{ij} > f_{i0}$ implies that

$$f_{ij} - f_{i0} f_{ij} > f_{i0} - f_{i0} f_{ij}$$

or equivalently

$$f_{ij} > \frac{f_{i0}(1 - f_{ij})}{1 - f_{i0}}$$

so that (89) is stronger than (85). If $R_2 = \varnothing$, (89) can be used for an ILP as well, but it may not be desirable to do so because the slack in (89) is not an integer variable.

The algorithm for the MILP (76) is identical to the algorithm for the method of integer forms, except that (89) is used when $f_{i0} > 0$ instead of (7). If it is assumed that x_0 is constrained to be an integer, then the proof of finiteness is also the same. However, if x_0 is not constrained to be an integer, the algorithm may not be finite.

Example

We return to the example of Table 1 but assume that only x_1 is constrained to be an integer. The solution in Table 1 is primal and dual feasible, but $f_{10} > 0$. Thus, using the x_1 row as a source, the cut from (89) is

$$\frac{x_5}{4} + \frac{3x_3}{2} \geq \frac{3}{4}$$

Adding the equation

$$s_1 = -\frac{3}{4} + \frac{x_5}{4} + \frac{3x_3}{2}$$

to Table 1 yields Table 27. Pivoting on x_3 and s_1 yields the optimal solution given in Table 28.

Table 27

			↓ $-x_3$	$-x_5$
	x_0	$\frac{31}{4}$	$\frac{1}{2}$	$\frac{1}{4}$
Source	x_1	$\frac{11}{4}$	$-\frac{1}{2}$	$\frac{1}{4}$
	x_2	$\frac{9}{4}$	$\frac{3}{2}$	$-\frac{1}{4}$
	x_4	$\frac{1}{2}$	-2	$\frac{1}{2}$
← Cut	s_1	$-\frac{3}{4}$	$-\frac{3}{2}$	$-\frac{1}{4}$

Table 28

		$-s_1$	$-x_5$
x_0	$\frac{30}{4}$	$\frac{1}{3}$	$\frac{1}{6}$
x_1	3	$-\frac{1}{3}$	$\frac{1}{3}$
x_2	$\frac{3}{2}$	1	$-\frac{1}{2}$
x_4	$\frac{3}{2}$	$-\frac{4}{3}$	$\frac{5}{6}$
x_3	$\frac{1}{2}$	$-\frac{2}{3}$	$\frac{1}{6}$

5.18. INTERSECTION CUTS

Intersection cuts are a class of cuts applicable to both ILP's and MILP's. Consider an ILP. Given a primal and dual feasible LP solution that is not all-integer, an intersection cut, like a cut in the method of integer forms, excludes the current solution but no feasible integer solutions. The cuts have appealing geometric properties that will motivate our arguments.

The ILP is given as

$$\max x_0 = \sum_{j=1}^{n} c_j x_j$$

$$\sum_{j=1}^{n} a_{ij} x_j + x_{n+i} = b_i, \qquad i = 1, \ldots, m \tag{90}$$

$$x_j \geq 0 \text{ integer}, \qquad j = 1, \ldots, n + m$$

Using a slightly different notation, the equations of (90) can be written as

$$x_i = y_{i0} - \sum_{j \in R} y_{ij} x_j, \qquad i = 1, \ldots, n + m \tag{91}$$

where if $k \in R$, (91) is the trivial equation $x_k = x_k$ ($y_{kk} = -1$ and $y_{kj} = 0$ otherwise). For the rows corresponding to basic variables, (91) is identical to (3). Note that R contains $(n + m) - m = n$ elements. In the following discussion, we only need the first n rows of (91). Dropping the m rows corresponding to slack variables, putting $y_j = (y_{1j}, \ldots, y_{nj})$ for all j and $x = (x_1, \ldots, x_n)$, we represent rows $1, \ldots, n$ of (91) as

$$x = y_0 - \sum_{j \in R} y_j x_j \tag{92}$$

It is assumed that $x = y_0$ yields an optimal solution to the LP corresponding to (90), but is not all-integer.

Let

$$S = \left\{ x \,\middle|\, \sum_{j=1}^{n} a_{ij} x_j \leq b_i,\ x_j \geq 0 \text{ integer}, j = 1, \ldots, n \right\}$$

and

$$T = \left\{ x \,\middle|\, x = y_0 - \sum_{j \in R} y_j x_j,\ x_j \geq 0, j \in R \right\}$$

Since T does not require either nonnegativity on the basic components of x or integrality, $S \subset T$. Assume T is of full dimension, so that T is an n-dimensional cone with apex at $x = y_0$ and edges given by the extreme rays

$$r_j = y_0 - \theta_j y_j, \quad \theta_j \geq 0, \qquad j \in R \tag{93}$$

(see Figure 8). The kth ray is obtained by putting $x_j = 0$ for $j \in R - \{k\}$ in (92). The vectors $\{y_j \mid j \in R\}$ can be shown to be linearly independent and therefore the rays have a unique intersection point at y_0 (see Exercise 32).

Given $\theta_j = \theta_j^* > 0$, let $p_j^* = y_0 - \theta_j^* y_j$ be the point on the jth ray determined by θ_j^* and consider the set of n distinct points $P^* = \{p_j^* | j \in R\}$. Since each point is on a different ray, P^* determines a hyperplane $d^*x = d_0^*$ as shown in Figure 8.

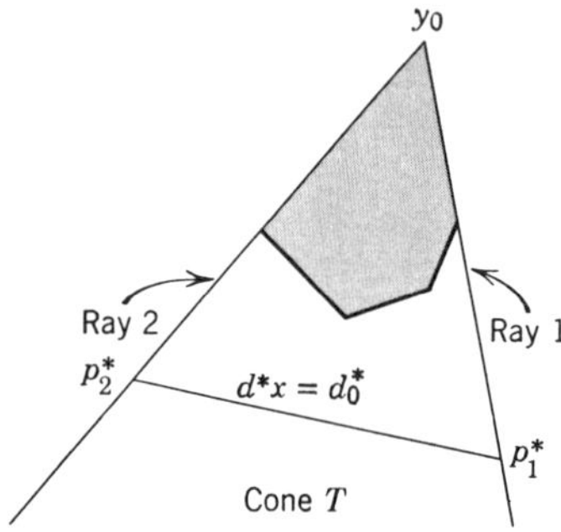

Figure 8

Now consider a closed and bounded convex set C in n-space containing y_0 in its interior; that is, $y_0 \in C - C_b$, where C_b is the boundary of C. Thus (see Exercise 33), for each $j \in R$, it can be shown that there exists a unique $\theta_j^\circ > 0$ such that

$$y_0 - \theta_j^\circ y_j = p_j^\circ \in C_b$$

Consider the hyperplane

$$d^\circ x = d_0^\circ \tag{94}$$

determined by $P^\circ = \{p_j^\circ | j \in R\}$ (see Figure 9). Since r_j has a unique inter-

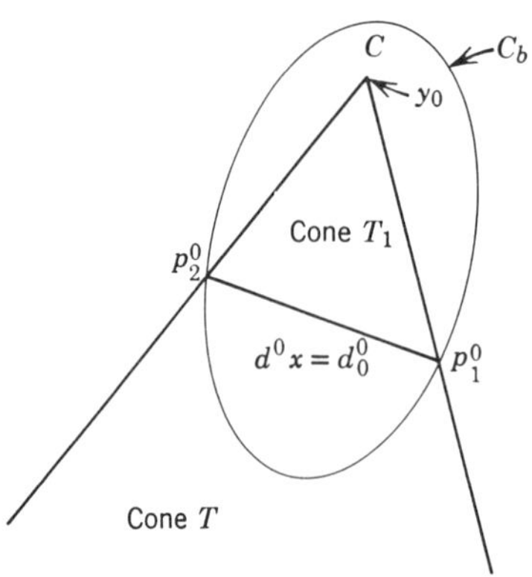

Figure 9

section point ($p_j^o \neq y_0$) with the hyperplane, y_0 is not on the hyperplane. Specifically, assume

$$d^o y_0 > d_0^o$$

and partition T as

$$T_1 = T \cap \{x | d^o x > d_0^o\}$$
$$T_2 = T \cap \{x | d^o x \leq d_0^o\}$$

Note that $T_1 \subset T_3 = T \cap \{x | d^o x \geq d_0^o\}$ and T_3 has extreme points $\{P^o, y_0\}$. Since each of these extreme points belongs to C, $T_3 \subset C$. But the only elements of T_3 that are in C_b are the elements of P^o, and each element of P^o is on the hyperplane (94). Thus $T_1 \subset C - C_b$. Since $S \subset T$, every $x \in S$ is either in T_1 or T_2. If

$$(C - C_b) \cap S = \varnothing$$

then $T_1 \cap S = \varnothing$ so that $S \subset T_2$. But $y_0 \in T_1$, so that the constraint

$$d^o x \leq d_0^o \tag{95}$$

excludes y_0 but no integer points. The cut (95) is an *intersection cut*.

The crux of the problem is to find a suitable convex set C. It is sufficient that C satisfy

(i) $y_0 \in C - C_b$.
(ii) $(C - C_b) \cap S = \varnothing$.

but for computational purposes, C should be chosen so that the coefficients (d^o, d_0^o) can easily be determined.

Choosing C

Let $\alpha^t = (\alpha_1^t, \ldots, \alpha_n^t)$, $t = 1, \ldots, 2^n$, where $\alpha_i^t = [y_{i0}]$ or $[y_{i0}] + 1$. The convex hull of these 2^n points defines an n-dimensional hypercube H containing y_0. Since y_0 is not all-integer, $y_0 \neq \alpha^t$ for any t. The center of the hypercube is defined to be $\rho = ([y_{10}], \ldots, [y_{n0}]) + (\frac{1}{2}, \ldots, \frac{1}{2})$.

We now propose to choose C as the Euclidean hypersphere Q with center ρ and radius $\sqrt{n}/2$, that is

$$Q = \{x | (x - \rho)(x - \rho) \leq n/4\} \tag{96}$$

Note that $H \subset Q$ and $H \cap Q_b = \{\alpha^t | t = 1, \ldots, 2^n\}$ as shown in Figure 10. Therefore, $y_0 \in Q - Q_b$ and $Q - Q_b$ contains no integer points, which implies that Q satisfies properties (i) and (ii). There is a complex formula for determining the n points P^o, where Q_b intersects the n extreme rays (see Exercise 27), which, in turn, yields the coefficients in (95). In general, these co-

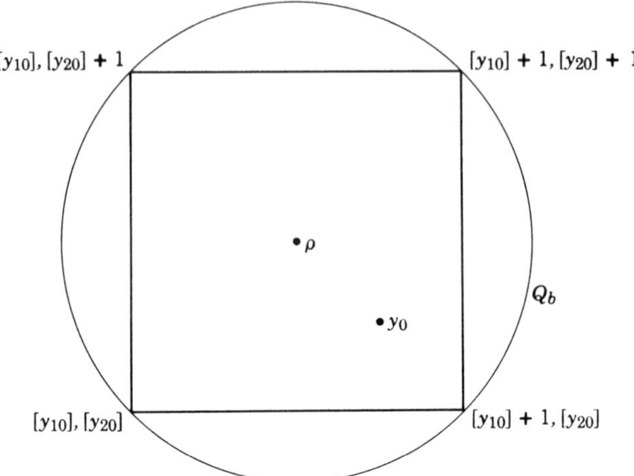

Figure 10

efficients are irrational and would have to be approximated for digital computation. Furthermore, it is doubtful if an algorithm based exclusively on the cuts (95) could, in general, be finite.

Example

Starting from the optimal LP solution in Table 1, the extreme rays, the circle Q_b, and the resultant cut (95) are shown in Figure 11. Reoptimizing

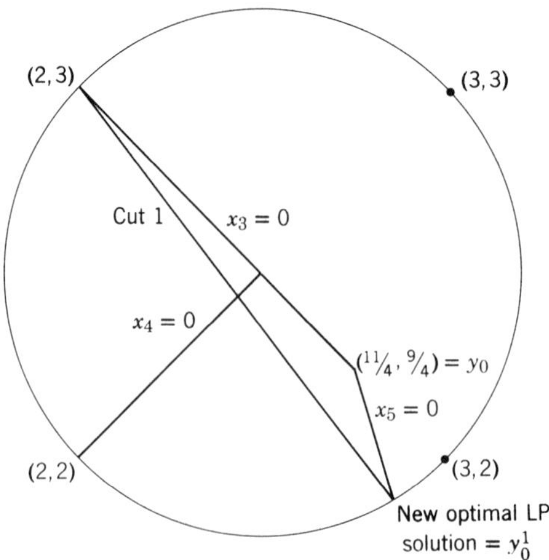

Figure 11

over the new feasible region yields the optimal LP solution shown in Figure 11. The process is repeated in Figure 12, yielding a second cut and a new optimal LP solution, which is not integer. The completion of the problem is left as Exercise 26.

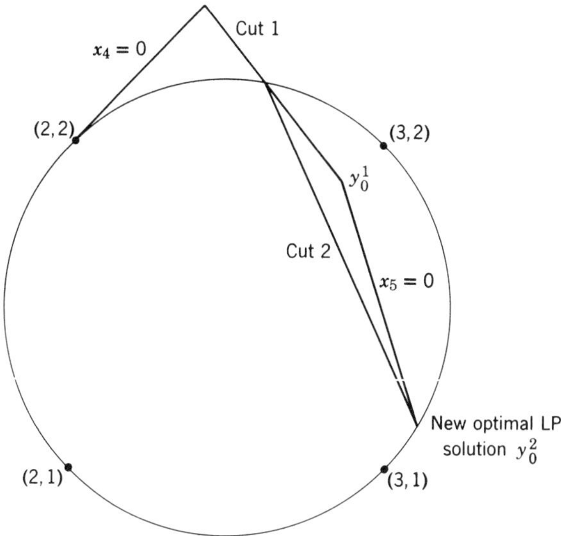

Figure 12

5.19. EXERCISES

1. For Exercise 1 of Chapter 2, assume that the variables are restricted to be integers and

(a) Solve graphically.
(b) Construct the convex hull of feasible integer solutions.
(c) Solve by the method of integer forms.
(d) Solve by the dual all-integer algorithm.
(e) Solve by the primal all-integer algorithm.
(f) Solve by the algorithm of Section 5.14.

2. Do Parts (c), (d), (e), and (f), of Exercise 1 for the problem

$$\begin{aligned} \max x_0 = 4x_1 + 5x_2 + x_3 & \\ 3x_1 + 2x_2 &\leq 10 \\ x_1 + 4x_2 &\leq 11 \\ 3x_1 + 3x_2 + x_3 &\leq 13 \\ x_1, x_2, x_3 &\geq 0 \text{ integer} \end{aligned}$$

3. What happens in each of the three basic algorithms if

(a) There is no feasible integer solution.

(b) The corresponding LP is unbounded.

4. Prove or give a counterexample to the following theorem. If the corresponding LP is unbounded, the ILP has a feasible solution.

5. Let $R_i^+ = \{j \mid j \in R, y_{ij} \geq 0\}$ and $R_i^- = \{j \mid j \in R, y_{ij} < 0\}$. A row

$$x_{B_i} = y_{i0} - \sum_{j \in R} y_{ij} x_j$$

is said to be monotone if $R_i^- = \varnothing$ or if $j \in R_i^-$ implies $x_j = 0$ in an optimal ILP solution.

Consider the following source row selection rule in the method of integer forms. If there exists at least one monotone row in the tableau, choose one as the source. Otherwise make an arbitrary choice. Prove that if the columns are maintained lexicographically positive, this rule yields a finite algorithm. How can the rule be implemented?

6. Suppose that $f_{i0} > 0$ and let $R_1 = \{j \mid j \in R, f_{ij} > 0\}$. For each $j \in R_1 \cup \{0\}$, let $f_{ij} = a_{ij}/b_{ij}$, where the fraction has been reduced as much as possible (a_{ij} and b_{ij} are relatively prime; their only common divisor is 1). Let $R_2 = \{j \mid j \in R_1, b_{i0}$ and b_{ij} are not relatively prime$\}$. Prove that $\sum_{j \in R_2} x_j \geq 1$ is a valid cut.

7. Give heuristic source row selection rules for the dual all-integer algorithm, analogous to the ones given in Section 5.3 for the method of integer forms.

8. Assume an all-integer tableau with $x_{B_i} = y_{i0} - \sum_{j \in R} y_{ij} x_j$ and $y_{ij} \geq 0$ for all $j \neq 0, k$, and $y_{ij} < 0$ for $j = 0, k$. Show that a valid cut is

$$x_k \geq \left\langle \frac{y_{i0}}{y_{ik}} \right\rangle$$

Furthermore, show constructively that any all-integer tableau with a primal infeasibility in row i can be transformed such that row i has the properties given above (assuming a feasible solution exists). Give an algorithm based on these cuts and a source row selection rule that will assure finite termination.

9. Show that the source row selection rule given in Section 5.12 is a special case of the rule: Choose any row that assures $y_{*k} \leq y_{*0}$ periodically and periodically reduces y_{ik}/y_{*k} for the smallest i such that $y_{ik}/y_{*k} > y_{i0}$. Prove that this rule yields a finite algorithm.

10. In the primal all-integer algorithm, prove that finite termination will be assured by the source row selection rule: Choose any row that will result in

$$y_{0k_{t+1}}^{t+1} > y_{0k_t}^{t}$$

if possible; otherwise choose a row according to the rudimentary algorithm.

11. Show that in the primal all-integer algorithm, the row designated to be the "special" row can be changed from one iteration to the next. If more than one row is available in a tableau, which one would you choose?

12. Prove that the surrogate constraint based on the optimal dual variables (Section 4.7) can be used as the special row in the primal all-integer algorithm. Why might this be a good choice?

13. Describe a Phase 1 ILP for obtaining an all-integer primal feasible tableau.

14. Let

$$x_1 = \frac{29}{11} - \frac{5}{11}x_2 + \frac{2}{11}x_3$$

Find the composite cut assuming that x_3 is the entering variable.

15. Generate all cuts (69) with integer h from the equation

$$x_1 = \frac{7}{10} + \frac{9}{10}x_4 + \frac{3}{10}x_5$$

Now consider the tableau

		$-x_4$	$-x_5$
x_0	$\frac{14}{10}$	$\frac{2}{10}$	$\frac{4}{10}$
x_1	$\frac{7}{10}$	$-\frac{9}{10}$	$-\frac{3}{10}$
x_2	$\frac{13}{10}$	$-\frac{1}{10}$	$\frac{3}{10}$
x_3	$\frac{18}{10}$	$\frac{4}{10}$	$-\frac{2}{10}$

containing the row given above. Can any additional cuts be generated from (69) using this tableau, integer h, and integer multiples of rows?

16. Eliminate all dominated cuts from the set of cuts generated in Exercise 15.

17. Give an explicit formula for the cut in Section 5.15 with $h = h^* - \epsilon$, $\epsilon > 0$ and arbitrarily small. For the cut of Exercise 15 with $[h] = 4$, find h^* and the corresponding cut. Show that the cut with $h = h^* - \epsilon$ maximizes the number of negative coefficients $d_j, j \in R$. Discuss the advantages and liabilities of choosing $h = h^* - \epsilon$.

18. Given positive integers a and b. Let $(a, b) =$ the greatest common divisor of a and b. Show that

(1) There exist integers p and q such that

$$(a, b) = pa + qb$$

(2) The integer value of h that maximizes $hf_{i0} - [hf_{i0}]$ in (69) is given by $h = D - p_0$, where $f_{i0} = e_{i0}/D$ and $(e_{i0}, D) = p_0 e_{i0} + q_0 D$

(3) The nonzero integer value of h that minimizes $hf_{ij} - [hf_{ij}]$ in (69) is given by $h = p_j$, where $f_{ij} = e_{ij}/D$ and $(e_{ij}, D) = p_j e_{ij} + q_j D$

19. In Section 5.9, it was suggested that $h = \bar{h}$ be used in the cut (15). Show by example that it is sometimes possible to obtain a stronger cut with $h < \bar{h}$. In general how should such an h be chosen?

20. Prove that the cut

$$\frac{3}{10}x_1 + \frac{2}{10}x_2 \geq \frac{1}{10}$$

in the method of integer forms, implies the deeper cut

$$\frac{3}{10}x_1 + \frac{2}{10}x_2 \geq \frac{11}{10}$$

21. Show that the equations

$$x_0 = \frac{11}{4} - \frac{1}{4}x_2 - \frac{1}{2}x_3 + 2x_4$$

$$x_1 = \frac{3}{4} - \frac{1}{2}x_2 + \frac{3}{4}x_3 + 7x_4$$

imply the cut

$$x_2 + x_3 \geq 2$$

22. Solve the problem of Exercise 2, assuming that only x_2 is required to be integer.

23. Solve the problem of Exercise 2, assuming that only x_2 and x_3 are required to be integer.

24. Discuss the computational and theoretical implications of dropping rows that correspond to slack variables from cuts in the method of integer forms, the dual all-integer algorithm, and the primal all-integer algorithm.

25. When $R_2 = \varnothing$, the cut (85) specializes to (7). However, (7) is a special case of (69). Develop the counterpart of (69) for the MILP by generalizing (85).

26. Complete the graphical solution of the example given in Section 5.18.

27. Given $y_0 \in Q$ of (96), show that the intersection of the ray (93) with Q_b is given by

$$\theta_j = \frac{(y_0 - \rho) + z^{1/2}}{y_j^2}$$

where $z = (\rho a_j)^2 + f(1 - f)y_j^2$, f is the vector of the fractional parts of y_0, and 1 is the vector of ones.

28. Use the results of Exercise 27 to solve the example of Exercise 26 numerically.

29. Is the hypercube given in Section 5.18 the only one that would work? Under what circumstances would other Euclidean hyperspheres be suitable? What about hyperspheres defined by norms other than the Euclidean norm?

30. In the primal all-integer algorithm of Section 5.12, show that the pivot row can be deleted when a slack variable from a cut enters the basis. Also, when $y_{rk} = 1$, consider the possibility of executing a pivot without adding a cut. Give an example that shows that such a pivot can destroy property (i) or property (ii).

31. Prove that the column corresponding to a slack variable from a cut cannot be a scalar multiple of another column. Then, using the results of Exercise 15 of Chapter 2, prove that k determined by (9) is unique if the entering variable is a slack variable from a cut.

32. Prove that the n extreme rays given by (93) are linearly independent if T is of full dimension. Modify the procedure of Section 5.18 to treat the case where the dimension of T is smaller than n.

33. Prove that an extreme ray of (93) has a unique intersection point with a closed and bounded convex set C containing y_0 in its interior.

5.20. NOTES

5.1. As far as we know, implied linear constraints were first used in Dantzig, Fulkerson, and Johnson (1954). An elaboration of this paper is contained in Dantzig, Fulkerson, and Johnson (1959). Also of historical interest is Markowitz and Manne (1957).

Gomory, in 1958, developed a systematic way of generating implied constraints and a corresponding finite algorithm. His 1958 paper is more readily available as Gomory (1963a). Abbreviated versions of this paper are Gomory (1958) and (1960a).

5.2. The fundamental constraint (6) is from Gomory (1963b).

5.3–5.5. The method of integer forms is developed in Gomory (1963a). Srinivasan (1965) investigates different rules for choosing the source row. The result that there is no upper bound on the number of iterations as a function of problem size appears in Jeroslow and Kortanek (1969) and (1971). Related papers are Jeroslow (1969), (1971) and Rubin (1970), which show that no upper bound exists on the number of faces defining a convex polyhedron of integer points in two variable, one constraint ILP's.

5.6. The cut (11) is from Dantzig (1959). It was not known whether cuts from (11) could provide a finite algorithm until the question was answered negatively by Gomory and Hoffman (1963). The cut (12) was suggested in Charnes and Cooper (1961). A finite algorithm based on the cuts (12) or (13) is given in Bowman and Nemhauser (1970). Related cuts are discussed in Rubin and Graves (1972).

5.7. Computer aspects of linear programming, including reinversion techniques and roundoff problems, are discussed in Orchard-Hays (1968). Woolsey (1971) discusses their effect on cutting plane algorithms.

5.8–5.9. These two sections are based on Gomory (1963b). It is shown in Finkelstein (1970) that there is no upper bound on the number of iterations as a function of problem size for the dual all-integer algorithm. Pnueli (1971) gives methods for determining different initial integer, dual feasible solutions.

Glover (1965a) develops a dual all-integer algorithm using a cut different from (15) (see Exercise 8). Two variants of this algorithm are given in Glover (1965b) and (1967a).

5.10–5.12. The rudimentary algorithm first appeared in Ben-Israel and Charnes (1962). This paper also contains a finite algorithm, but it is not practical for computing because it involves solving simultaneous equations in integers when the cut (30) is degenerate. Mathis (1971) shows, by examples, that the rudimentary algorithm can cycle or fail to converge without cycling. The first finite algorithm based exclusively on the cuts (30) is given in Young (1965). Young's original pivot column and source row selection rules and finiteness proof are very complicated. Simplified rules and proofs are given in Glover (1968a) and Young (1968a). Our presentation follows Glover's. Arnold and Bellmore (1971a), (1971b), and (1971c) have modified Glover's algorithm in an attempt to eliminate long sequences of degenerate pivots.

5.14. The material of this section is from Martin (1963). Bowman (1969) gives a different development of Martin's procedure.

5.15. The discussion of integer h is from Gomory (1963a). The cuts based on noninteger h are derived in Glover (1966). For finding better values of h in the dual all-integer algorithm, see Wilson (1967). Glover (1967c) notes that Wilson's result follows from Glover's results on noninteger h. Better values of h can also be found for primal all-integer cuts, as noted by Bowman and Nemhauser (1971).

5.16. This section is based on Bowman and Nemhauser (1971). Similar ideas are proposed in Kianfar (1971); Pollatschek and Avi-Itzhak (1971); Mitra, Richards, and Wolfenden (1970); and Salkin and Breining (1971).

5.17. This section is based on Gomory (1960b). White (1961) has shown that if the objective function is not required to be integer, the algorithm may not be finite. Another mixed integer cut is developed in Beale (1958).

An extension of Gomory's algorithm to the case in which the variables are nonuniformly discrete (say $x = 0, 5, 8, 14, 35$) is given in Dalton and Llewellyn (1966). An alternate approach to general discrete variables is to transform them to nonnegative integer variables, as shown in Section 1.4.

Rutledge (1967) gives a simplex-type method for MILP's with binary variables. Salkin (1971) points out that (89) is stronger than (85) for ILP's and gives a two-variable problem for which (89) requires fewer iterations. However, in larger problems, it would appear that (89) is undesirable for ILP's because continuous slack variables are required.

5.18. The basic reference is Balas (1971a). Extensions are contained in Glover (1969b); Burdet (1970); Balas et al. (1971); and Balas (1970d), (1970e), (1971c), and (1971e). Intersection cuts for ILP's in 0–1 variables are given by Young (1968b) and (1971) and Taha (1971).

6 The Knapsack Problem

6.1. PROBLEM FORMULATION

Suppose that n different types of scientific equipment are being considered for inclusion on a space vehicle going to the moon. For $j = 1, \ldots, n$, let $c_j > 0$ be the scientific value per unit and $a_j > 0$ the weight per unit of the jth type. If the total weight limitation is b, the problem of maximizing the total value of the equipment taken is

$$\max x_0 = \sum_{j=1}^{n} c_j x_j$$

$$\sum_{j=1}^{n} a_j x_j \leq \text{b} \tag{1}$$

$$x_j \geq 0 \text{ integer}, \qquad j = 1, \ldots, n$$

where x_j is the number of the jth type included. All data in (1) is assumed to be integer. Changing the interpretation from space trip to hike explains the name *knapsack problem*.

The knapsack problem can be solved by any of the general ILP algorithms but, because it has only one constraint, more efficient algorithms are possible. Certainly, the problem of optimally loading a space vehicle is far too complicated to be described by our simple model, and it is doubtful that any hiker would take the knapsack model seriously. Nevertheless, the knapsack problem is worth studying. There are algorithms for more complex models that use the knapsack problem as a subroutine and require the solution of a very large number of knapsack problems. A general ILP in bounded variables can be transformed into a knapsack problem (see Section 6.12). Also, insights derived from studying the knapsack problem will be very helpful in developing another approach to the general ILP to be given in Chapter 7.

6.2. RECURSIVE EQUATIONS I

Let $f_k(y)$ be the maximum value that can be achieved in (1) from a knapsack of size or weight limitation y using only the first k items $(y = 0, 1, \ldots, b; k = 1, 2, \ldots, n)$. Note that $f_n(b) = x_0^0$ is the optimal value of (1). Our objective is to calculate recursively the sequence of functions $\{f_k(y)\}$, $k = 1, \ldots, n$. The formulation and computational aspects of recursive optimization are called *dynamic programming*.

By definition,

$$
\begin{aligned}
f_k(y) = \max & \sum_{j=1}^{k} c_j x_j \\
& \sum_{j=1}^{k} a_j x_j \le y \\
& x_j \ge 0 \text{ integer}, \quad j = 1, \ldots, k
\end{aligned} \tag{2}
$$

Isolating x_k in (2) yields, for $k = 2, \ldots, n$ and $y = 0, 1, \ldots, b$,

$$
f_k(y) = \max_{x_k = 0, 1, \ldots, [y/a_k]} c_k x_k + \left[\begin{aligned} \max & \sum_{j=1}^{k-1} c_j x_j \\ & \sum_{j=1}^{k-1} a_j x_j \le y - a_k x_k \\ x_j \ge 0 & \text{ integer}, \quad j = 1, \ldots, k-1 \end{aligned} \right] \tag{3}
$$

By definition, the problem in the brackets has an optimal value of $f_{k-1}(y - a_k x_k)$. Thus, for $k = 2, \ldots, n$ and $y = 0, 1, \ldots, b$, (3) can be written as

$$
f_k(y) = \max_{x_k = 0, 1, \ldots, [y/a_k]} (c_k x_k + f_{k-1}(y - a_k x_k)) \tag{4}
$$

To begin the recursive computation of (4), note that

$$f_1(y) = \max_{x_1=0,1,\ldots,[y/a_1]} c_1 x_1$$
$$= c_1\left[\frac{y}{a_1}\right]$$

since $c_1 > 0$. In fact, (4) is easily extended to include $k = 1$, by using the initial condition $f_0(y) = 0$, $y = 0, 1, \ldots, b$.

Equation (4) says that in optimizing over the first k types, regardless of the number of type k items chosen, the remaining space, $y - a_k x_k$, must be allocated optimally over the first $k - 1$ types. This explanation is a statement of the dynamic programming *principle of optimality*.

In some applications of the knapsack problem, the constraint is an equality, $\sum_{j=1}^{n} a_j x_j = b$. Recursion (4) can be applied to the equality problem by changing the initial condition. After all n types have been placed in the knapsack, no space can remain. Thus the proper initial condition is $f_0(0) = 0$ and $f_0(y) = -\infty$, $y > 0$.

We will not give the computational details of solving (4), since no use is made of the linearity of the objective function and constraint. It is important only because it is easily generalized to treat problems with a nonlinear objective function and/or constraint, and also problems with bounded variables.

6.3. RECURSIVE EQUATIONS II

If, for a given y and k, there is an optimal solution to (4) with $x_k = 0$, then $f_k(y) = f_{k-1}(y)$. On the other hand, if $x_k > 0$ in some optimal solution to (4), then an optimal knapsack of size y over the first k items can be put together by combining one item of type k with an optimal knapsack of size $y - a_k$ over the first k items. Thus, for a given y,

$$x_k = 0 \Rightarrow f_k(y) = f_{k-1}(y) \tag{5}$$

$$x_k > 0 \Rightarrow f_k(y) = c_k + f_k(y - a_k) \tag{6}$$

Combining (5) and (6) with $f_0(y) = 0$, $y = 0, 1, \ldots, b$, yields

$$f_k(y) = \max\{f_{k-1}(y);\, c_k + f_k(y - a_k),\, a_k \le y\} \tag{7}$$
$$y = 0, 1, \ldots, b,\ k = 1, \ldots, n$$

If the knapsack constraint is changed to an equality, (7) still applies with the initial condition $f_0(0) = 0$, and $f_0(y) = -\infty$ otherwise. Recursion (7) is easier to execute than (4) because the maximization in (7) involves the comparison of only two numbers. To solve (7) for $f_k(y)$, it is necessary to know $f_{k-1}(y)$ and $f_k(y - a_k)$. Thus $f_k(y)$ must be calculated in the order $y = 0, 1, \ldots, b$.

An indicator $p_k(y)$ is computed for all y and k and is used to find an optimal solution after $f_n(b)$ has been calculated. The indicator is defined as

$$p_k(y) = \begin{cases} 0 & \text{if } f_k(y) = f_{k-1}(y) \\ 1 & \text{if } f_k(y) > f_{k-1}(y) \end{cases}$$

Algorithm

STEP 1: Initialize $k = 0, f_0(y) = 0$, for $y = 0, 1, \ldots, b$. Go to Step 2.

STEP 2: Set $k = k + 1$. For all $y < a_k$, set $f_k(y) = f_{k-1}(y)$ and $p_k(y) = 0$. Let $y = a_k$. Go to Step 3.

STEP 3: Compute $V = c_k + f_k(y - a_k)$. If $V > f_{k-1}(y)$, set $f_k(y) = V$ and $p_k(y) = 1$. Otherwise put $f_k(y) = f_{k-1}(y)$ and $p_k(y) = 0$. Go to Step 4.

STEP 4: If $y < b$, let $y = y + 1$ and return to Step 3. If $y = b$, go to Step 5.

STEP 5: If $k < n$, return to Step 2. If $k = n$, set $z = 0$ and go to Step 6 (z is a counter for the optimal value of x_k, $k = 1, \ldots, n$).

STEP 6: If $p_k(y) = 1$, set $z = z + 1$ and go to Step 7. If $p_k(y) = 0$, go to Step 8.

STEP 7: Set $y = y - a_k$ and return to Step 6.

STEP 8: Set $x_k = z$ and $z = 0$. If $k > 1$, put $k = k - 1$ and return to Step 6. If $k = 1$, terminate.

When Step 5 has been executed for the last time, $f_n(b)$ has been computed. Steps 6–8 find an optimal set of x_k's. If $p_n(b) = 0$, then $x_n = 0$; if $p_n(b) = 1$, $x_n \geq 1$. When a case is reached with $p_k(y) = 0$, $x_k = z$. Note that the algorithm gives a parametric solution, since it finds $f_n(y)$ for $y = 0, 1, \ldots, b$.

When executing the algorithm on a computer, three lists, each of size $b + 1$, are kept in rapid access storage. Specifically, when computing f_k, lists for f_{k-1}, f_k, and p_k are needed. When the f_k list is complete, the f_{k-1} list is discarded and the p_k list can be put in auxiliary storage. If rapid access storage is scarce, it is not necessary to maintain a p_k list. A minus sign can be attached to f_k to indicate $p_k = 1$. Since $f_k \geq 0$, obviously the minus signs do not mean negativity. Following this convention, the f_{k-1} list is placed in auxiliary storage when the f_k list is complete.

Example

$$\max x_0 = 11x_1 + 7x_2 + 5x_3 + x_4$$

$$6x_1 + 4x_2 + 3x_3 + x_4 \leq 25$$

$$x_1, \ldots, x_4 \geq 0 \text{ integer}$$

The calculations are shown in Table 1. To keep the table compact, the minus sign convention has been used to eliminate the $p_k(y)$ lists.

Table 1

y	$f_1(y)$	$V(y)$	$f_2(y)$	$V(y)$	$f_3(y)$	$V(y)$	$f_4(y)$
0	0	—	0	—	0	—	0
1	0	—	0	—	0	1 + 0	1−
2	0	—	0	—	0	1 + 1	2−
3	0	—	0	5 + 0	5−	1 + 2	5
4	0	7 + 0	7−	5 + 0	7	1 + 5	7
5	0	7 + 0	7−	5 + 0	7	1 + 7	8−
6	11−	7 + 0	11	5 + 5	11	1 + 8	11
7	11−	7 + 0	11	5 + 7	12−	1 + 11	12
8	11−	7 + 7	14−	5 + 7	14	1 + 12	14
9	11−	7 + 7	14−	5 + 11	16−	1 + 14	16
10	11−	7 + 11	18−	5 + 12	18	1 + 16	18
11	11−	7 + 11	18−	5 + 14	19−	1 + 18	19
12	22−	7 + 14	22	5 + 16	22	1 + 19	22
13	22−	7 + 14	22	5 + 18	23−	1 + 22	23
14	22−	7 + 18	25−	5 + 19	25	1 + 23	25
15	22−	7 + 18	25−	5 + 22	27−	1 + 25	27
16	22−	7 + 22	29−	5 + 23	29	1 + 27	29
17	22−	7 + 22	29−	5 + 25	30−	1 + 29	30
18	33−	7 + 25	33	5 + 27	33	1 + 30	33
19	33−	7 + 25	33	5 + 29	34−	1 + 33	34
20	33−	7 + 29	36−	5 + 30	36	1 + 34	36
21	33−	7 + 29	36−	5 + 33	38−	1 + 36	38
22	33−	7 + 33	40−	5 + 34	40	1 + 38	40
23	33−	7 + 33	40−	5 + 36	41−	1 + 40	41
24	44−	7 + 36	44	5 + 38	44	1 + 41	44
25	44−	7 + 36	44	5 + 40	45−	1 + 44	45

After having completed the f_4 list, $x_0^0 = f_4(25) = 45$. Since $f_4(25)$ does not have a minus sign, $x_4 = 0$. Letting $k = 3$ and noting the minus sign attached to $f_3(25)$, we have $x_3 \geq 1$. Thus y is reduced to $25 - 3 = 22$ and $f_3(22)$ is examined for a minus sign. Since it does not have one, $x_3 = 1$. Continuing in this manner, an optimal solution $x_1 = 3$, $x_2 = 1$, $x_3 = 1$, $x_4 = 0$ is found. Note that there are alternative optima that we have not attempted to identify; in particular $x_1 = 4$, $x_2 = x_3 = 0$, $x_4 = 1$ is another optimal solution.

6.4. RECURSIVE EQUATIONS III

Another recursive algorithm is derived that, for many data sets, is more efficient than the algorithm of Section 6.3.

Let the value of an optimal knapsack of size y be denoted by $g(y)$ and

$$D(y) = \{k \mid x_k > 0 \text{ in some optimal solution to the problem with knapsack size } y\}$$

Suppose that $g(y) \neq 0$. Then

$$g(y) = c_k + g(y - a_k), \qquad k \in D(y) \tag{8}$$

and

$$g(y) > c_j + g(y - a_j), \qquad j \notin D(y),\ y - a_j \geq 0 \tag{9}$$

Let $S(y) = \{j \mid y - a_j \geq 0\}$. Since, without loss of generality, it can be assumed that $j \neq k \Rightarrow a_j \neq a_k$, it follows that $|S(y)| \leq \min(y, n)$. Now $D(y) \subseteq S(y)$ and (8) and (9) imply that

$$g(y) = \begin{cases} \max\limits_{j \in S(y)} c_j + g(y - a_j) & S(y) \neq \varnothing \\ 0 & S(y) = \varnothing \end{cases} \tag{10}$$

Our objective is to calculate $g(y)$ recursively for $y = 1, \ldots, b$. Given the initial condition $g(0) = 0$, we can calculate $g(1)$ from

$$g(1) = \begin{cases} c_k & \text{if } S(1) = \{k\}\text{; i.e., } a_k = 1 \\ 0 & \text{if } S(1) = \varnothing \end{cases}$$

In the general case, given $g(0), \ldots, g(y - 1)$, $g(y)$ can be calculated from (10), since $y - a_j < y$ for all j. Note that $|S(y)|$ additions and comparisons are required to calculate $g(y)$, $y = 1, \ldots, b$. However, there is a very simple improvement of (10) that can drastically reduce the number of calculations.

Consider any $Q(y)$ such that $Q(y) \subseteq S(y)$ and

$$Q(y) \cap D(y) = \varnothing \Rightarrow D(y) = \varnothing$$

Then (10) can be written

$$g(y) = \begin{cases} \max\limits_{j \in Q(y)} c_j + g(y - a_j) & Q(y) \neq \varnothing \\ 0 & Q(y) = \varnothing \end{cases} \tag{11}$$

Obviously, solving (11) requires no more, and generally fewer, computations than solving (10). We will show that an appropriate $Q(y)$ can be found.

Let

$$d(y) = \begin{cases} \min\{j \mid j \in D(y)\} & D(y) \neq \varnothing \\ n & D(y) = \varnothing \end{cases} \tag{12}$$

Note that $j \in D(y)$ implies that $D(y - a_j) \subseteq D(y)$ and from (12)

$$d(y) \leq d(y - a_j)$$

Thus an appropriate $Q(y)$ is

$$Q(y) = \{j \mid j \in S(y), j \leq d(y - a_j)\} \tag{13}$$

To apply (13) in the solution of (11), suppose that $g(0), \ldots, g(y - 1)$ and $d(0), \ldots, d(y - 1)$ are known. For every $j \in S(y)$, $y - a_j$ is calculated, but the addition $c_j + g(y - a_j)$ is done only if $j \leq d(y - a_j)$. [See Exercise 21 for a modified recursion that eliminates the need for calculating $y - a_j$ for some $j \in S(y)$.]

This procedure will be most effective when the $d(y)$ are small. Thus it is desirable to index the variables such that those with small indices are likely to be positive in an optimal solution. A reasonable order is obtained by letting $\rho_j = c_j/a_j$ and indexing the variables such that $\rho_1 \geq \rho_2 \geq \cdots \geq \rho_n$.

Algorithm

STEP 1: Order the variables such that $\rho_1 \geq \rho_2 \geq \cdots \geq \rho_n$. Initialize $g(y) = 0$ and $d(y) = n$, $y = 0, 1, \ldots, b$. Let $y = 1, j = 1$. Go to Step 2.

STEP 2: If $y - a_j \geq 0$ and $j \leq d(y - a_j)$, go to Step 3; otherwise go to Step 4.

STEP 3: Compute $V = g(y - a_j) + c_j$. If $V \leq g(y)$, go to Step 4. Otherwise let $g(y) = V$, $d(y) = j$, and go to Step 4.

STEP 4: If $j < n$, let $j = j + 1$ and return to Step 2. Otherwise go to Step 5.

STEP 5: If $y < b$, let $j = 1$, $y = y + 1$, and return to Step 2. If $y = b$, go to Step 6.

STEP 6: Initialize $y = b$ and $x_j = 0, j = 1, \ldots, n$. Go to Step 7.

STEP 7: If $g(y) = 0$ terminate; otherwise go to Step 8.

STEP 8: Set $x_{d(y)} = x_{d(y)} + 1$ and $y = y - a_{d(y)}$. Return to Step 7.

When Step 5 is executed for the last time, $g(b)$ has been calculated. Steps 6–8 find an optimal solution. Only two lists, one for $g(y)$ and another for $d(y)$, must be kept in rapid access memory during the execution of Steps 2–6. If at least some of the $d(y)$ are small, this algorithm can require many fewer calculations than the algorithm of Section 6.3. In the following example, we see that there is a y^* such that $d(y) = 1$, for all $y \geq y^*$.

Example

The example of Section 6.3 is solved, by the algorithm just given, in Table 2. Note that the variables are already indexed appropriately. The first

Table 2

y	g(y)			d(y)		
0	0			4		
1	0	1		4		
2	0	2		4		
3	0	5		4	3	
4	0	7		4	2	
5	0	8		4	2	
6	0	11		4	1	
7	0	12		4	1	
8	0	13	14	4	1	2
9	0	16		4	1	
10	0	18		4	1	
11	0	19		4	1	
12	0	22		4	1	
13	0	23		4	1	
14	0	25		4	1	
15	0	27		4	1	
16	0	29		4	1	
17	0	30		4	1	
18	0	33		4	1	
19	0	34		4	1	
20	0	36		4	1	
21	0	38		4	1	
22	0	40		4	1	
23	0	41		4	1	
24	0	44		4	1	
25	0	45		4	1	

columns of the $g(y)$ and $d(y)$ lists represent the initialization. Changes, if any, are shown in subsequent columns. After having computed $g(25) = 45$, from $d(25) = d(19) = d(13) = d(7) = 1$, and $d(1) = 4$, we obtain an optimal solution, $x_1 = 4$, $x_2 = x_3 = 0$, and $x_4 = 1$.

In the example, for all $y \geq 9$, $d(y) = 1$. Having a y^* such that $d(y) = 1$ for all $y \geq y^*$ is by no means peculiar to this example. The existence, identification, and use of y^* is the subject of Section 6.5.

6.5. A PERIODIC PROPERTY

Assume, as before, that the variables are indexed such that $\rho_1 \geq \rho_2 \geq \cdots \geq \rho_n$. Our objective is to show that if $\rho_1 > \rho_2$, there exists a smallest integer y^* such that for all $y \geq y^*$, an optimal knapsack of size y has $x_1 \geq 1$. For a knapsack of size y, a feasible solution is $x_1 = [y/a_1]$ and $x_j = 0$, $j = 2, \ldots, n$. Letting the value of this solution be denoted by θ_1, yields

$$\theta_1 = c_1\left[\frac{y}{a_1}\right] \geq \frac{c_1 y}{a_1} - c_1 = \rho_1(y - a_1)$$

An upper bound on the value of a knapsack of size y with $x_1 = 0$ is

$$\theta_2 = \frac{c_2 y}{a_2} = \rho_2 y$$

Thus a sufficient condition for an optimal knapsack of size y to have $x_1 \geq 1$ is $\rho_1(y - a_1) \geq \rho_2 y$ or equivalently $y \geq c_1/(\rho_1 - \rho_2)$. Now, since y is an integer, we obtain

$$y^* \leq \left\langle \frac{c_1}{\rho_1 - \rho_2} \right\rangle \tag{14}$$

For the case $\rho_1 = \rho_2$, (14) is meaningless, but it can still be shown that for large enough y, there exists a solution having $x_1 \geq 1$ (see Exercise 25).

Consider a knapsack of size $y + ka_1$, $k = 1, 2, \ldots$, where $y \geq y^* - a_1$. Since $d(y) = 1$ for all $y \geq y^*$, it follows from (8) that

$$g(y + ka_1) = kc_1 + g(y), \qquad k = 1, 2, \ldots \tag{15}$$

Thus, if $x_1^o, x_2^o, \ldots, x_n^o$ is an optimal solution for the knapsack of size y, $(x_1^o + k), x_2^o, \ldots, x_n^o$ is an optimal solution for the knapsack of size $y + ka_1$. Except for the linear term kc_1, the optimal value is periodic with period a_1 for all $y \geq y^*$. By knowing y^* and an optimal solution for all $y \leq y^*$, from (15), it is trivial to calculate the optimal solution for $y > y^*$.

The bound on y^* given by (14) is generally very weak. However, the algorithm of Section 6.4 yields y^*. Let $a_k = \max_j a_j$ and suppose $d(y) = 1$, $y^o \leq y < y^o + a_k$. Then from (10),

$$g(y^o + a_k) = \max_j c_j + g(y^o + a_k - a_j)$$

where for all j, $y^o + a_k - a_j \geq y^o$. Thus $d(y^o + a_k) = 1$ and it follows that $y^o \geq y^*$. If $d(y^o - 1) > 1$, $y^o = y^*$. Thus when a_k successive values of y have $d(y) = 1$, the algorithm can be terminated. This periodic property can save an enormous number of calculations, particularly for problems with large values of b.

In our example, (14) yields $y^* \leq 132$, but from Table 2, $y^* = 9$. Calculations can be terminated at $y = 9 + 6 - 1 = 14$. An optimal solution for $y = 25$ can be obtained from an optimal solution for $y = 13 = 25 - 2a_1$. We find that $g(13) = 23$, and $x_1 = 2$, $x_2 = x_3 = 0$, and $x_4 = 1$ is an optimal solution for $y = 13$. Thus $g(25) = 23 + 2c_1 = 45$ and $x_1 = 4$, $x_2 = x_3 = 0$, and $x_4 = 1$ is an optimal solution for $y = 25$.

In Chapter 7, the periodic property for knapsack problems will be extended to general ILP's. In preparation for this, we consider a knapsack problem associated with finite abelian groups.

6.6. MODULO ARITHMETIC AND ABELIAN GROUPS

Let p be a real number, δ a positive integer, and x the remainder when p is divided by δ; that is, $p = m\delta + x$, where m is an integer and $0 \leq x < \delta$. We say that x equals p modulo δ and write this relation as $x = p \pmod{\delta}$. For example, $7 = 43 \pmod{12}$. Note that $p \pmod 1$ is the fractional part of p. A cut of the method of integer forms is obtained by taking the coefficients of a row of a simplex tableau modulo 1.

If $p \pmod{\delta} = q \pmod{\delta}$, we say that p is congruent to q modulo δ and write this relation as

$$p \equiv q \pmod{\delta}$$

The congruence relation is linear, in the sense that

$$a \pmod{\delta} + b \pmod{\delta} \equiv (a + b) \pmod{\delta}$$

$$n(a \pmod{\delta}) \equiv (na) \pmod{\delta}$$

The important properties of modulo arithmetic can be obtained by introducing the concept of a finite abelian group. A *finite abelian group* is a finite set $S = \{g_0, g_1, \ldots, g_{|S|-1}\}$ and a binary operation $\oplus$ that satisfies the following properties:

1. Closure: $g_i \oplus g_j \in S$, for all $g_i, g_j \in S$.
2. Associativity: $(g_i \oplus g_j) \oplus g_k = g_i \oplus (g_j \oplus g_k) = g_i \oplus g_j \oplus g_k$, for all $g_i, g_j, g_k \in S$.
3. Commutivity: $g_i \oplus g_j = g_j \oplus g_i$, for all $g_i, g_j \in S$.
4. Identity: g_0 is such that $g_0 \oplus g_i = g_i$, for all $g_i \in S$.
5. Inverse: there exists $g_{j(i)} \in S$ such that $g_{j(i)} \oplus g_i = g_0$, for all $g_i \in S$ ($g_{j(i)}$ is denoted by $-g_i$).

The group is denoted by $G = (S, \oplus)$ and its *order* by $|G| = |S|$.

Let $S = \{g_0, g_1, \ldots, g_{\delta-1}\}$ and

$$g_i \oplus g_j = g_{(i+j) \,(\mathrm{mod}\ \delta)} \tag{16}$$

The pair $(S, \oplus)$, where $\oplus$ is given by (16), defines a group since

1. $0 \leq (i + j) \,(\mathrm{mod}\ \delta) < \delta$.
2. $((i + j) \,(\mathrm{mod}\ \delta) + k) \,(\mathrm{mod}\ \delta) = (i + (j + k) \,(\mathrm{mod}\ \delta)) \,(\mathrm{mod}\ \delta) = (i + j + k) \,(\mathrm{mod}\ \delta)$.
3. $(i + j) \,(\mathrm{mod}\ \delta) = (j + i) \,(\mathrm{mod}\ \delta)$.
4. $(0 + i) \,(\mathrm{mod}\ \delta) = i$.
5. $((\delta - i) + i) \,(\mathrm{mod}\ \delta) = 0$.

The group defined by (16) is denoted by $G(\delta)$. In applying group theory to integer programming, only $G(\delta)$, $\delta = 1, 2, \ldots$, and certain groups derived from $G(\delta)$ need to be considered.

Let $G = (S, \oplus)$ and $T \subseteq S$. If $H = (T, \oplus)$ is a group, then H is said to be a *subgroup* of G. For example, consider $G(8)$ and $T = \{g_0, g_2, g_4, g_6\}$. Then H as defined above is a subgroup of $G(8)$.

Define

$$mg_i = g_i \oplus g_i \oplus \cdots \oplus g_i \qquad (m \text{ terms})$$

and

$$0g_i = g_0$$

Note that in $G(\delta)$,

$$mg_i = g_{(mi) \,(\mathrm{mod}\ \delta)}$$

Let $T(g_i) = \{g_j | g_j = mg_i$ for some $m\}$. It can be shown that $H(g_i) = (T(g_i), \oplus)$ is a subgroup of $G = (S, \oplus)$; $H(g_i)$ is called *the subgroup generated by* g_i. The order of the subgroup is called the order of g_i and is denoted by $|g_i|$.

In $G(6)$,

$$mg_2 = \begin{cases} g_0 & m = 3k, \ k \text{ a nonnegative integer} \\ g_2 & m = 3k + 1 \\ g_4 & m = 3k + 2 \end{cases}$$

Thus $T(g_2) = \{g_0, g_2, g_4\}$ and $|g_2| = 3$.

In $G(\delta)$, it follows from (16) that $|g_i| = m_i$, where m_i is the smallest positive integer such that $(im_i) \,(\mathrm{mod}\ \delta) = 0$. The generalization to arbitrary finite abelian groups is that $|g_i| = m_i$, where m_i is the smallest positive integer such that $m_i g_i = g_0$. It can be shown that m_i is a divisor of $|G|$, which implies that $|G| g_i = g_0$ for all elements of the group.

If there exists $g_i \in S$, such that $|g_i| = |G|$, G is called *cyclic*. In $G(\delta)$, $m_1 = \delta$; thus g_1 generates $G(\delta)$, which implies that $G(\delta)$ is cyclic for all δ.

Direct Sum Groups

Let

$$S = \{(a_i, b_k) | i = 0, \ldots, \delta_1 - 1, k = 0, \ldots, \delta_2 - 1\}$$

and

$$(a_i, b_k) \oplus (a_j, b_l) = (a_{(i+j) \pmod{\delta_1}}, b_{(k+l) \pmod{\delta_2}})$$

Then it can be shown that $G(\delta_1, \delta_2) = (S, \oplus)$ defines a group of order $\delta_1\delta_2$, where $G(\delta_1, \delta_2)$ is said to be the *direct sum* of $G(\delta_1)$ and $G(\delta_2)$. In exactly the same way a direct sum group $G(\delta_1, \delta_2, \ldots, \delta_m)$ of order $\prod_{i=1}^{m} \delta_i$ can be constructed.

For example, consider $G(2, 3)$. The elements of $G(2, 3)$ are the ordered pairs $g_{0,0} = (a_0, b_0)$, $g_{1,0} = (a_1, b_0)$, $g_{0,1} = (a_0, b_1)$, $g_{1,1} = (a_1, b_1)$, $g_{0,2} = (a_0, b_2)$, and $g_{1,2} = (a_1, b_2)$. Note that in $G(2, 3)$,

$$g_{i,k} \oplus g_{j,l} = g_{(i+j) \pmod{2}, (k+l) \pmod{3}}$$

A $\oplus$ table for $G(2, 3)$ is given in Table 3.

Table 3

	$g_{0,0}$	$g_{1,0}$	$g_{0,1}$	$g_{1,1}$	$g_{0,2}$	$g_{1,2}$
$g_{0,0}$	$g_{0,0}$					
$g_{1,0}$	$g_{1,0}$	$g_{0,0}$				
$g_{0,1}$	$g_{0,1}$	$g_{1,1}$	$g_{0,2}$			
$g_{1,1}$	$g_{1,1}$	$g_{0,1}$	$g_{1,2}$	$g_{0,2}$		
$g_{0,2}$	$g_{0,2}$	$g_{1,2}$	$g_{0,0}$	$g_{1,0}$	$g_{0,1}$	
$g_{1,2}$	$g_{1,2}$	$g_{0,2}$	$g_{1,0}$	$g_{0,0}$	$g_{1,1}$	$g_{0,1}$

A table for $mg_{i,k}$, $m = 0, 1, \ldots, |G| - 1$, can also be constructed. Note that if $(m - 1)g_{i,k} = g_{j,l}$ then $mg_{i,k} = g_{i,k} \oplus g_{j,l}$. Thus the $mg_{i,k}$ table for $G(2, 3)$, Table 4, can be constructed from Table 3. The group $G(2, 3)$ is cyclic, since it is generated by both g_{11} and g_{12}.

In general, the elements of $G(\delta_1, \ldots, \delta_m)$ are denoted as $g_{i_1, \ldots, i_m}$, where $0 \leq i_l < \delta_l$, $l = 1, \ldots, m$. If

$$g_{i_1, \ldots, i_m} \oplus g_{j_1, \ldots, j_m} = g_{k_1, \ldots, k_m}$$

Table 4

m	0	1	2	3	4	5	$\lvert g_{ij}\rvert$
$g_{0,0}$	$g_{0,0}$	$g_{0,0}$	$g_{0,0}$	$g_{0,0}$	$g_{0,0}$	$g_{0,0}$	1
$g_{1,0}$	$g_{0,0}$	$g_{1,0}$	$g_{0,0}$	$g_{1,0}$	$g_{0,0}$	$g_{1,0}$	2
$g_{0,1}$	$g_{0,0}$	$g_{0,1}$	$g_{0,2}$	$g_{0,0}$	$g_{0,1}$	$g_{0,2}$	3
$g_{1,1}$	$g_{0,0}$	$g_{1,1}$	$g_{0,2}$	$g_{1,0}$	$g_{0,1}$	$g_{1,2}$	6
$g_{0,2}$	$g_{0,0}$	$g_{0,2}$	$g_{0,1}$	$g_{0,0}$	$g_{0,2}$	$g_{0,1}$	3
$g_{1,2}$	$g_{0,0}$	$g_{1,2}$	$g_{0,1}$	$g_{1,0}$	$g_{0,2}$	$g_{1,1}$	6

then

$$k_l = (i_l + j_l) \pmod{\delta_l}, \qquad l = 1, \ldots, m$$

Thus a group addition in $G(\delta_1, \ldots, \delta_m)$ involves m modulo additions.

Sometimes it is convenient to indicate the elements of $G(\delta_1, \ldots, \delta_m)$ with single subscripts, that is, as $g_0, g_1, \ldots, g_{\Pi\delta_i - 1}$. When this convention is used, the correspondence between the single and multiple subscript representations must be known in order to perform group additions.

Isomorphism

Two groups G and G' (of the same order) are said to be *isomorphic* if there is a one-to-one correspondence ($a \leftrightarrow a'$) between their elements that preserves their respective binary operations; that is, if $a \leftrightarrow a'$ and $b \leftrightarrow b'$, then $a \oplus b \leftrightarrow a' \oplus' b'$. The concept of isomorphism is important because it allows us to work with the most convenient group representation. For example, $G(2, 3)$ is isomorphic to $G(6)$. It is easy to check that a correspondence between elements is given by $g_{0,0} \leftrightarrow g_0$, $g_{1,0} \leftrightarrow g_3$, $g_{0,1} \leftrightarrow g_2$, $g_{1,1} \leftrightarrow g_5$, $g_{0,2} \leftrightarrow g_4$, and $g_{1,2} \leftrightarrow g_1$. Two isomorphic groups are the same mathematical object, only their representations are different. Thus $G(6)$ cyclic implies $G(2, 3)$ cyclic. Note that the group $G(1, \ldots, 1, \delta_1, \ldots, \delta_m)$ is isomorphic to $G(\delta_1, \ldots, \delta_m)$.

The group $G(2, 2)$ represented in Tables 5 and 6 is not cyclic, since no element generates the group. It can be concluded that $G(2, 2)$ is not isomorphic to $G(4)$.

Table 5

	$g_{0,0}$	$g_{1,0}$	$g_{0,1}$	$g_{1,1}$
$g_{0,0}$	$g_{0,0}$			
$g_{1,0}$	$g_{1,0}$	$g_{0,0}$		
$g_{0,1}$	$g_{0,1}$	$g_{1,1}$	$g_{0,0}$	
$g_{1,1}$	$g_{1,0}$	$g_{0,1}$	$g_{1,0}$	$g_{0,0}$

Table 6

m	0	1	2	3	$\lvert g_{ij}\rvert$
$g_{0,0}$	$g_{0,0}$	$g_{0,0}$	$g_{0,0}$	$g_{0,0}$	1
$g_{1,0}$	$g_{0,0}$	$g_{1,0}$	$g_{0,0}$	$g_{1,0}$	2
$g_{0,1}$	$g_{0,0}$	$g_{0,1}$	$g_{0,0}$	$g_{0,1}$	2
$g_{1,1}$	$g_{0,0}$	$g_{1,1}$	$g_{0,0}$	$g_{1,1}$	2

An important group theoretic result for integer programming is that every direct sum group $G(\rho_1, \ldots, \rho_m)$ of order greater than 1 is isomorphic to a unique direct sum group $G(\delta_1, \ldots, \delta_k)$ such that

(a) $\prod_{i=1}^{m} \rho_i = \prod_{i=1}^{k} \delta_i$.
(b) $1 < \delta_1 \leq \cdots \leq \delta_k$.
(c) δ_i is a divisor of δ_{i+1}, $i = 1, \ldots, k - 1$.
(d) $k \leq m$.

The group $G(\delta_1, \ldots, \delta_k)$ is called the *canonical* representation of all groups isomorphic to it. A method for obtaining the canonical form is given in Section 7.4. Note that, from Property (d), group additions using the canonical form involve the minimum number of modulo additions over all isomorphic representations.

The above result greatly restricts the number of nonisomorphic groups of a given order. For example, if $\prod_{i=1}^{m} \rho_i = \rho$ and ρ is prime, then $G(\rho_1, \ldots, \rho_m)$ must be $G(1, \ldots, 1, \rho)$ isomorphic to $G(\rho)$. Note that, from property (c), all direct sum groups of order 6 are isomorphic to $G(6)$, even though 6 is not prime.

To find the set of canonical representations of the isomorphically different groups of order $|G|$, all of the factorizations of $|G|$ that satisfy properties (b) and (c) are needed. For example, for $|G| = 135$, there are exactly three canonical representations: $G(135)$, $G(3, 45)$, and $G(3, 3, 15)$.

Cosets

The set

$$T_g(g_i) = \{h | h = g \oplus mg_i, 0 \leq m < |g_i|\}$$

is called *the coset of $T(g_i)$ that contains g*. Since $g_0 \oplus mg_i = mg_i$, we have $T_{g_0}(g_i) = T(g_i)$. It can be shown that for $g \neq g'$ either $T_g(g_i) = T_{g'}(g_i)$ or $T_g(g_i) \cap T_{g'}(g_i) = \varnothing$. Furthermore,

$$\bigcup_{g \in S} T_g(g_i) = S$$

and the number of distinct cosets of $T(g_i)$ is $|G|/|g_i|$. Thus the distinct cosets of $T(g_i)$ yield a partition of S into $|G|/|g_i|$ subsets. Each of these subsets has cardinality $|g_i|$.

For example, in $G(2, 2)$,

$$T_{g_{0,0}}(g_{1,0}) = T_{g_{1,0}}(g_{1,0}) = \{g_{0,0}, g_{1,0}\}$$

$$T_{g_{0,1}}(g_{1,0}) = T_{g_{1,1}}(g_{1,0}) = \{g_{0,1}, g_{1,1}\}$$

6.7. A GROUP KNAPSACK PROBLEM

Let $G = (S, \oplus)$ be a finite abelian group and $S' = \{g_{i_1}, g_{i_2}, \ldots, g_{i_n}\} \subseteq S$; $(S', \oplus)$ is not necessarily a subgroup of G, nor is G necessarily cyclic. Given $g^* \in S$, and $Q = \{i_1, i_2, \ldots, i_n\}$, we consider the problem of finding non-negative integers t_{i_j}, $j = 1, \ldots, n$, such that

$$\bigoplus_{j \in Q} t_j g_j = t_{i_1} g_{i_1} \oplus t_{i_2} g_{i_2} \oplus \cdots \oplus t_{i_n} g_{i_n} = g^* \tag{17}$$

Now for integers α and m such that $\alpha + m|g_k| \geq 0$,

$$(\alpha + m|g_k|)g_k = \alpha g_k \oplus m|g_k|g_k = \alpha g_k \oplus m g_0 = \alpha g_k \oplus g_0 = \alpha g_k$$

Thus, if (17) has a solution with $t_k = \alpha$, there are solutions with $t_k = \alpha + m|g_k|$ for all m such that $\alpha + m|g_k| \geq 0$.

In the approach to integer linear programming discussed in Chapter 7, there arises the *group knapsack* problem

$$\begin{aligned} f_n(g^*) = \min \sum_{j \in Q} d_j t_j & \\ \bigoplus_{j \in Q} t_j g_j = g^* & \\ t_j \geq 0 \text{ integer}, j \in Q & \end{aligned} \tag{18}$$

where $d_j \geq 0$ are given for all $j \in Q$. Note that $d_j \geq 0$ implies that if (18) has a solution, it has an optimal solution with $t_j < |g_j|$ for all $j \in Q$. However, if there exists j^* such that $d_{j*} < 0$ and (18) has a solution, then it is unbounded.

It is our intent to show that the dynamic programming recursions for (1) can also be applied to solve (18).

6.8. GROUP RECURSION I

Let $Q_k = \{i_1, \ldots, i_k\}$, $k = 1, \ldots, n$, and for all g

$$f_k(g) = \min \sum_{j \in Q_k} d_j t_j$$

$$\bigoplus_{j \in Q_k} t_j g_j = g \tag{19}$$

$$t_j \geq 0 \text{ integer}, j \in Q_k$$

Note that (19) has a solution with $t_{i_k} = \alpha$ if and only if

$$\bigoplus_{j \in Q_{k-1}} t_j g_j = g \oplus -\alpha g_{ik}, \; t_j \geq 0 \text{ integer}, j \in Q_{k-1} \tag{20}$$

has a solution.

Combining (19) and (20) yields the recursion for all g and $k = 2, \ldots, n$,

$$f_k(g) = \min_{t_{i_k} = 0, \ldots, |g_{t_{i_k}}| - 1} (d_{i_k} t_{i_k} + f_{k-1}(g \oplus -t_{i_k} g_{i_k})) \tag{21}$$

If g is in the subgroup generated by g_{i_1}, then

$$f_1(g) = d_{i_1} t^*_{i_1}$$

where $t^*_{i_1}$ is the smallest nonnegative integer such that $t_{i_1} g_{i_1} = g$. Otherwise

$$f_1(g) = M \text{ (arbitrarily large)}$$

To extend (21) to $k = 1$, initialize $f_0(g_0) = 0$; $f_0(g) = M$ otherwise (g_0 is the identity element). As in the ordinary knapsack problem, (19) can be solved for all g by finding $f_n(g)$ for all g.

The recursion (21) is analogous to (4) for the knapsack problem. For a group knapsack problem, the efficiency of an algorithm depends on the group structure. Recursion (21) is useful if $|g_j|$, $j \in Q$, are small, as in $G(2, 2, \ldots, 2)$. For the group $G(2, 2, \ldots, 2)$, $g_j \oplus g_j = g_0$ or $g_j = -g_j$, and (21) specializes to

$$f_k(g) = \min \{ f_{k-1}(g), d_{i_k} + f_{k-1}(g \oplus g_{i_k}) \} \tag{22}$$

for $k = 1, \ldots, n$.

Example

To illustrate the solution of (22), consider the problem

$$\min 5t_{1,0} + 3t_{0,1} + t_{1,1}$$

$$t_{1,0} g_{1,0} \oplus t_{0,1} g_{0,1} \oplus t_{1,1} g_{1,1} = g_{1,0}$$

$$t_{i,j} \geq 0 \text{ integer}$$

over $G(2, 2)$.

The problem is solved in Table 7, yielding $f_3(g_{1,0}) = 4$ and $t_{1,1} = 1$. Since $g_{1,0} \oplus g_{1,1} = g_{0,1}$, $t_{0,1} = 1$. Then $g_{0,1} \oplus g_{0,1} = g_{0,0}$ and $t_{1,0} = 0$, resulting in an optimal solution $t_{1,0} = 0$, and $t_{0,1} = t_{1,1} = 1$.

Table 7

k	1	2			3		
Q_k	$\{(1, 0)\}$	$\{(1, 0), (0, 1)\}$			$\{(1, 0), (0, 1), (1, 1)\}$		
g	$f_1(g)$	$g \oplus g_{0,1}$	$d_{0,1} + f_1(g \oplus g_{0,1})$	$f_2(g)$	$g \oplus g_{1,1}$	$d_{1,1} + f_2(g \oplus g_{1,1})$	$f_3(g)$
$g_{0,0}$	0	$g_{0,1}$	$3 + M$	0	$g_{1,1}$	$1 + 8$	0
$g_{1,0}$	5	$g_{1,1}$	$3 + M$	5	$g_{0,1}$	$1 + 3$	4
$g_{0,1}$	M	$g_{0,0}$	$3 + 0$	3	$g_{1,0}$	$1 + 5$	3
$g_{1,1}$	M	$g_{1,0}$	$3 + 5$	8	$g_{0,0}$	$1 + 0$	1

6.9. GROUP RECURSION II

If $t_{i_k} = 0$ solves (21) for a given k and $g \in S$, then $f_k(g) = f_{k-1}(g)$. On the other hand, if $t_{i_k} > 0$ in an optimal solution to (21), then

$$f_k(g) = d_{i_k} + f_k(g \oplus -g_{i_k})$$

Thus, analogous to (7) for the knapsack problem, for all g and $k = 1, \ldots, n$,

$$f_k(g) = \min \{f_{k-1}(g), d_{i_k} + f_k(g \oplus -g_{i_k})\} \tag{23}$$

The initial condition, as in (22), is

$$f_0(g) = \begin{cases} 0 & g = g_0 \\ M & \text{otherwise} \end{cases}$$

To compute $f_k(g)$ from (23), $f_k(g \oplus -g_{i_k})$ must be known and the group elements must be ordered such that $g \oplus -g_{i_k}$ precedes g. Although, in general, an ordering for all $g \in S$ is not possible, one can order the elements of a coset. The computation of $f_k(g)$ is done separately for each distinct coset and can be divided into two cases.

Case 1: $g \in T(g_{i_k})$. There exists m, $0 \leq m < |g_{i_k}|$ such that $g = mg_{i_k}$. Thus

$$g \oplus -g_{i_k} = mg_{i_k} \oplus -g_{i_k} = (m - 1)g_{i_k}$$

and for all $g \in T(g_{i_k})$, (23) can be written as

$$f_k(mg_{i_k}) = \min \{f_{k-1}(mg_{i_k}), d_{i_k} + f_k((m - 1)g_{i_k})\} \tag{24}$$

Since $f_k(g_0) = 0$ for all k, (24) can be used to calculate $f_k(g)$ for all $g \in T(g_{i_k})$, beginning with $m = 1$ and terminating with $m = |g_{i_k}| - 1$. If $|g_{i_k}| = |G|$, (24) yields $f_k(g)$ for all g.

CASE 2: $g \notin T(g_{i_k})$. Consider the coset

$$T_g(g_{i_k}) = \{h | h = g \oplus mg_{i_k}, 0 \le m < |g_{i_k}|\}$$

For all $h \in T_g(g_{i_k})$

$$h \oplus -g_{i_k} = g \oplus mg_{i_k} \oplus -g_{i_k} = g \oplus (m - 1)g_{i_k}$$

and for $m = 1, \ldots, |g_{i_k}| - 1$, (23) can be written as

$$f_k(g \oplus mg_{i_k}) = \min \{f_{k-1}(g \oplus mg_{i_k}), d_{i_k} + f_k(g \oplus (m - 1)g_{i_k})\} \quad (25)$$

In order to start the recursion of (25), $f_k(g \oplus mg_{i_k})$ must be known for some m. It will now be shown that

Lemma 1: If

$$f_{k-1}(g \oplus m^*g_{i_k}) = \min_{0 \le m < |g_{i_k}|} f_{k-1}(g \oplus mg_{i_k}) \quad (26)$$

then

$$f_k(g \oplus m^*g_{i_k}) = f_{k-1}(g \oplus m^*g_{i_k}) \quad (27)$$

PROOF: From (21)

$$f_k(g \oplus m^*g_{i_k}) = \min_{t_{i_k} = 0, \ldots, |g_{i_k}| - 1} (d_{i_k}t_{i_k} + f_{k-1}(g \oplus m^*g_{i_k} \oplus -t_{i_k}g_{i_k}))$$

From (26) and $m^*g_{i_k} \oplus -t_{i_k}g_{i_k} \in T_g(g_{i_k})$, we have

$$f_{k-1}(g \oplus m^*g_{i_k}) \le f_{k-1}(g \oplus m^*g_{i_k} \oplus -t_{i_k}g_{i_k}), \qquad t_{i_k} = 0, \ldots, |g_{i_k}| - 1 \quad (28)$$

Then $d_{i_k} \ge 0$ and (28) imply that (27) holds. ■

The obvious way to determine m^* is by solving (26). Then (25) could be solved for $m = m^* + 1, \ldots, m^* + |g_{i_k}| - 1$. This would also yield $f_k(g \oplus mg_{i_k})$, $m = 0, \ldots, |g_{i_k}| - 1$, since

$$f_k(g \oplus mg_{i_k}) = f_k(g \oplus (m + |g_{i_k}|)g_{i_k})$$

However, it is not necessary to solve (26) to determine m^*. The following procedure determines m^*, while simultaneously calculating values of $f_k(g \oplus mg_{i_k})$.

Assume that $f_k(g) = f_{k-1}(g)$ and then use (25) to calculate estimated values of $f_k(g \oplus mg_{i_k})$, denoted by $f_k'(g \oplus mg_{i_k})$. This yields

$$f_k'(g \oplus g_{i_k}) = \min \{f_{k-1}(g \oplus g_{i_k}), d_{i_k} + f_{k-1}(g)\} \quad (29)$$

and for $m = 2, \ldots, |g_{i_k}| - 1$,

$$f'_k(g \oplus mg_{i_k}) = \min \{f_{k-1}(g \oplus mg_{i_k}), d_{i_k} + f'_k(g \oplus (m-1)g_{i_k})\} \tag{30}$$

Since $f_k(g) \le f_{k-1}(g)$, it follows that for all m

$$f'_k(g \oplus mg_{i_k}) \ge f_k(g \oplus mg_{i_k})$$

and then from (30) that

$$f'_k(g \oplus m^*g_{i_k}) = f_{k-1}(g \oplus m^*g_{i_k}) \tag{31}$$

Note that (31) implies for $m \ge m^*$

$$f'_k(g \oplus mg_{i_k}) = f_k(g \oplus mg_{i_k}) \tag{32}$$

and, in particular,

$$f'_k(g \oplus |g_{i_k}|g_{i_k}) = f'_k(g \oplus g_0) = f'_k(g) = f_k(g) \tag{33}$$

From (32) and (33) it follows that a procedure for calculating $f_k(g \oplus mg_{i_k})$, $0 \le m < |g_{i_k}|$, is

1. Use (29) and (30) to calculate $f'_k(g \oplus mg_{i_k})$ for $1 \le m \le |g_{i_k}|$.
2. Let

$$f_k(g) = f'_k(g \oplus |g_{i_k}|\, g_{i_k})$$

If $f_k(g) = f_{k-1}(g)$, let

$$f_k(g \oplus mg_{i_k}) = f'_k(g \oplus mg_{i_k}), \quad 1 \le m \le |g_{i_k}| - 1.$$

If $f_k(g) < f_{k-1}(g)$, use (25) to calculate $f_k(g \oplus mg_{i_k})$ for $m = 1, \ldots, m^\circ$, where m° is the first m such that

$$f_k(g \oplus m^\circ g_{i_k}) = f'_k(g \oplus m^\circ g_{i_k})$$

For $m = m^\circ + 1, \ldots, |g_{i_k}| - 1$, let

$$f_k(g \oplus mg_{i_k}) = f'_k(g \oplus mg_{i_k})$$

Algorithm

This algorithm is similar to the one given in Section 6.3. The indicator $p_k(g)$ is defined as

$$p_k(g) = \begin{cases} 0 & \text{if } f_k(g) = f_{k-1}(g) \\ 1 & \text{if } f_k(g) < f_{k-1}(g) \end{cases}$$

Again, instead of keeping a list for $p_k(g)$, a minus sign can be attached to $f_k(g)$ to indicate that $p_k(g) = 1$.

STEP 1: Initialize $f_0(g) = M$ for $g \ne g_0$, $f_0(g_0) = 0$. Let $k = 0$ and go to Step 2.

STEP 2: Set $k = k + 1$, $m = 1, p_k(g_0) = 0$, and $f_k(g_0) = 0$. Go to Step 3.

STEP 3: Compute $V = d_{i_k} + f_k((m - 1)g_{i_k})$. If $V < f_{k-1}(mg_{i_k})$, set $f_k(mg_{i_k}) = V$ and $p_k(mg_{i_k}) = 1$. Otherwise put $f_k(mg_{i_k}) = f_{k-1}(mg_{i_k})$ and $p_k(mg_{i_k}) = 0$. Go to Step 4.

STEP 4: If $m < |g_{i_k}| - 1$, let $m = m + 1$ and return to Step 3. If $m = |g_{i_k}| - 1$, go to Step 5.

STEP 5: If $f_k(g)$ has not been computed for some g, go to Step 6. Otherwise go to Step 10.

STEP 6: Choose any g such that $f_k(g)$ has not been computed. Put $f_k(g) = f_{k-1}(g)$, $p_k(g) = 0$, $m = 1$. Go to Step 7.

STEP 7: Set $m = m + 1$. If $m < |g_{i_k}|$, go to Step 8. If $|g_{i_k}| \leq m < m^* + |g_{i_k}|$, go to Step 9.

STEP 8: Compute $V = d_{i_k} + f_k(g \oplus (m - 1)g_{i_k})$. If $V < f_{k-1}(g \oplus mg_{i_k})$, set $f_k(g \oplus mg_{i_k}) = V$ and $p_k(g \oplus mg_{i_k}) = 1$. Otherwise put $f_k(g \oplus mg_{i_k}) = f_{k-1}(g \oplus mg_{i_k})$, $p_k(g \oplus mg_{i_k}) = 0$. Return to Step 7.

STEP 9: Let $m' = m - |g_{i_k}|$. Compute $V = d_{i_k} + f_k(g \oplus (m' - 1)g_{i_k})$. If $V < f_k(g \oplus m'g_{i_k})$, set $f_k(g \oplus m'g_{i_k}) = V$ and $p_k(g \oplus m'g_{i_k}) = 1$ and return to Step 7. Otherwise put $f_k(g \oplus m'g_{i_k}) = f_{k-1}(g \oplus m'g_{i_k})$ and $p_k(g \oplus m'g_{i_k}) = 0$ and Return to Step 5.

STEP 10: If $k < n$, return to Step 2. If $k = n$, go to Step 11.

STEP 11: Put $g = g^*$ and $z = 0$ (z is a counter for the optimal value of t_{i_k}). Go to Step 12.

STEP 12: If $p_k(g) = 1$, set $z = z + 1$, and go to Step 13. Otherwise go to Step 14.

STEP 13: Let $g' = g \oplus -g_{i_k}$. Put $g = g'$ and return to Step 12.

STEP 14: Set $t_{i_k} = z$ and $z = 0$. If $k > 1$, put $k = k - 1$ and return to Step 12. If $k = 1$, terminate.

Example

Consider the following problem over $G(6)$.

$$\begin{gathered} \min 3t_3 + 10t_2 + 5t_5 \\ t_3g_3 \oplus t_2g_2 \oplus t_5g_5 = g_1 \\ t_3, t_2, t_5 \geq 0 \text{ integer} \end{gathered} \tag{34}$$

The calculations are shown in Table 8. Let $Q_1 = \{3\}$, $Q_2 = \{3, 2\}$, and $Q_3 = \{3, 2, 5\}$. Note that g_5 generates the group, but g_3 and g_2 do not. Minus signs are used to indicate $p_k(g) = 1$.

Table 8

g	$f_0(g)$	$V(g)$	$f_1(g)$	$V(g)$	$f_2(g)$	$V(g)$	$f_3(g)$
g_0	0	—	0	—	0	—	0
g_1	M	M	M	$10 + 13$	$23-$	$5 + 8$	$\underline{13}-$
g_2	M	M	M	$10 + 0$	$10-$	$5 + 3$	$\underline{8}-$
g_3	M	$3 + 0$	$\underline{3}-$	$10 + M$	$\underline{3}$	$5 + 10$	$\underline{3}$
g_4	M	M	M	$10 + 10$	$20-$	$5 + 5$	$10-$
g_5	M	M	M	$10 + 3$	$13-$	$5 + 0$	$5-$

The subgroup generated by g_3 is $\{g_0, g_3\}$, and we obtain

$$f_1(g_3) = 3 + f_1(g_0) = 3$$

But for $g \neq g_0$ or g_3, $g \oplus mg_3 \neq g_0$ for any m; thus $f_1(g_i) = M$, $i \neq 0, 3$.

The subgroup generated by g_2 is $\{g_0, g_2, g_4\}$. This yields

$$f_2(g_2) = 10 + f_2(g_0) = 10$$

and

$$f_2(g_4) = 10 + f_2(g_2) = 20$$

Now let $g = g_1$ and assume that $f_2(g_1) = M$. Then calculate

$$f_2'(g_1 \oplus g_2) = f_2'(g_3) = \min\{10 + f_2(g_1), f_1(g_3)\} = \min\{10 + M, 3\} = 3$$

Therefore, $m^* = 1$ and $f_2(g_3) = 3$. Thus

$$f_2(g_1 \oplus 2g_2) = f_2(g_5) = 10 + f_2(g_3) = 13$$

Finally,

$$f_2(g_1 \oplus 3g_2) = f_2(g_1) = 10 + f_2(g_5) = 23$$

Since g_5 generates the group, the calculation of $f_3(g)$ is straightforward using (24). This yields $f_2(g_1) = 13$ and $t_5 \geq 1$ because of the minus sign on $f_3(g_1)$. Next $g_2 = g_1 \oplus -g_5$ so that $f_3(g_2)$ is checked for a minus sign; it has one, so $t_5 \geq 2$. Since $g_3 = g_2 \oplus -g_5$, $f_3(g_3)$ is examined for a minus sign. It does not have one, so $t_5 = 2$; nor does $f_2(g_3)$, so $t_2 = 0$. The minus sign on $f_1(g_3)$ indicates that $t_3 = 1$. The optimal solution to (34) is $t_3 = 1$, $t_2 = 0$, $t_5 = 2$ with cost 13.

For cyclic groups, the recursion of Section 6.4 also can be modified to treat the group knapsack problem. These recursions, both for the ordinary and group knapsack problems, are closely related to the shortest path algorithm given in Section 3.5, as will be discussed in Section 6.11.

6.10. A RELATION BETWEEN THE KNAPSACK AND GROUP KNAPSACK PROBLEMS

Stating (1) with an equality constraint by adding the slack x_{n+1} yields

$$\begin{aligned} \max x_0 &= \sum_{j=1}^{n+1} c_j x_j \\ &\sum_{j=1}^{n+1} a_j x_j = b \\ &x_j \geq 0 \text{ integer}, \quad j = 1, \ldots, n+1 \end{aligned} \tag{35}$$

where $a_{n+1} = 1$ and $c_{n+1} = 0$.

Assume that the variables in (35) are ordered so that $\rho_1 \geq \rho_2 \geq \cdots \geq \rho_{n+1}$. An optimal solution to the LP corresponding to (35) is given by $x_1 = b/a_1$, $x_0 = \rho_1 b$ and $x_j = 0$ otherwise. Writing the basic variables in terms of the nonbasic variables, yields

$$x_0 = \rho_1 b - \sum_{j=2}^{n+1} (\rho_1 a_j - c_j) x_j \tag{36}$$

$$x_1 = \frac{b}{a_1} - \sum_{j=2}^{n+1} \frac{a_j x_j}{a_1} \tag{37}$$

where $\rho_1 a_j - c_j \geq 0, j = 2, \ldots, n+1$. Assume that b/a_1 is not an integer; otherwise the linear programming solution solves (35). From (36) and (37) it follows that (35) can be restated as

$$\begin{aligned} \min &\sum_{j=2}^{n+1} (c_1 a_j - c_j a_1) x_j \\ &\frac{b}{a_1} - \sum_{j=2}^{n+1} \frac{a_j x_j}{a_1} \geq 0 \text{ integer} \\ &x_j \geq 0 \text{ integer}, \quad j = 2, \ldots, n+1 \end{aligned} \tag{38}$$

Now if b is large enough, so that x_1 is certain to be positive in an optimal solution (see Section 6.5), the nonnegativity on x_1 can be dropped.

Equation (37) and x_1 integer are equivalent to

$$\sum_{j=2}^{n+1} \frac{a_j x_j}{a_1} \equiv \frac{b}{a_1} \pmod 1$$

or

$$\sum_{j=2}^{n+1} a_j x_j \equiv b \pmod{a_1}$$

or

$$\sum_{j=2}^{n+1} (a_j \pmod{a_1}) x_j \equiv b \pmod{a_1}$$

or

$$\sum_{j=2}^{n+1} \alpha_j x_j \equiv \alpha_0 \pmod{a_1} \tag{39}$$

where $\alpha_j = a_j \pmod{a_1}$, $j = 2, \ldots, n + 1$, and $\alpha_0 = b \pmod{a_1}$.

Note that (39) can be written as a group equation over the group $G(a_1)$. Let

$$Q = \{i \mid i = \alpha_j, \text{ for some } j,\ 2 \le j \le n + 1\}$$

and $g^* = g_{\alpha_0}$. Then (39) can be written as

$$\bigoplus_{i \in Q} t_i g_i = g^*$$

where

$$t_i = \sum_{j \in P_i} x_j, \qquad i \in Q \tag{40}$$

and

$$P_i = \{j \mid \alpha_j = i, j = 2, \ldots, n + 1\}$$

Unless $|P_i| = 1$ for all i, which is only the case if $j \neq j'$ implies that $\alpha_j \neq \alpha_{j'}$, the t_i of (40) do not uniquely determine the x_j. However, for $j, j' \in P_i$, $c_1 a_j - c_j a_1 < c_1 a_{j'} - c_{j'} a_1$ implies $x_{j'} = 0$ in every optimal solution to (38) for large b. Let

$$d_i = \min_{j \in P_i} (c_1 a_j - c_j a_1) = c_1 a_{j(i)} - c_{j(i)} a_1$$

Then, in the variables t_i, (38) for large b can be written as the group knapsack problem

$$\min \sum_{i \in Q} d_i t_i$$

$$\bigoplus_{i \in Q} t_i g_i = g^* \tag{41}$$

$$t_i \ge 0 \text{ integer}, \qquad i \in Q$$

If t^* is an optimal solution to (41), then

$$x^*_{j(i)} = t^*_i, \qquad i \in Q$$
$$x^*_j = 0, \qquad \text{otherwise}$$

is an optimal solution to (38) for large b.

Thus (35) (for large b) is equivalent to a group knapsack problem. The advantage in representing (35) as a group problem is that the order of the group is a_1, where $a_1 \leq b$. In (35) the number of calculations is proportional to b, in (41) to a_1.

Example

Consider the example of Section 6.3

$$\begin{aligned} \max x_0 &= 11x_1 + 7x_2 + 5x_3 + x_4 \\ & 6x_1 + 4x_2 + 3x_3 + x_4 \leq 25 \\ & x_1, \ldots, x_4 \geq 0 \text{ integer} \end{aligned} \tag{42}$$

Since $a_4 = 1$, the problem is unchanged by changing the constraint to an equality. The optimal solution to the LP corresponding to (42) is $x_1 = \frac{25}{6}$, $x_0 = \frac{275}{6}$, and $x_j = 0$ otherwise. Writing x_0 and x_1 in terms of the remaining variables, we obtain

$$\begin{aligned} x_0 &= \frac{275}{6} - \frac{2x_2}{6} - \frac{3x_3}{6} - \frac{5x_4}{6} \\ x_1 &= \frac{25}{6} - \frac{4x_2}{6} - \frac{3x_3}{6} - \frac{x_4}{6} \end{aligned} \tag{43}$$

Thus for large enough b (we know from Section 6.5 that 25 is large enough), (42) can be written

$$\begin{aligned} \min 2x_2 &+ 3x_3 + 5x_4 \\ & 4x_2 + 3x_3 + \ \ x_4 \equiv 1 \pmod 6 \\ & x_2, x_3, x_4 \geq 0 \text{ integer} \end{aligned} \tag{44}$$

The group minimization problem corresponding to (44) is

$$\begin{aligned} \min 2t_4 &+ 3t_3 + 5t_1 \\ & t_4 g_4 \oplus t_3 g_3 \oplus t_1 g_1 = g_1 \\ & t_4, t_3, t_1 \geq 0 \text{ integer} \end{aligned} \tag{45}$$

where t_3 corresponds to x_3, t_4 to x_2, and t_1 to x_4.

By inspection, an optimal solution to (45) is $t_3 = t_4 = 0$ and $t_1 = 1$ with cost 5. Translating back to the x's yields $x_2 = x_3 = 0$ and $x_4 = 1$. Then (43) gives $x_1 = \frac{25}{6} - \frac{1}{6} = 4$ and $x_0 = \frac{275}{6} - \frac{5}{6} = 45$. Another optimal solution to (45) is $t_3 = t_4 = 1$ and $t_1 = 0$. In terms of the x's, $x_2 = x_3 = 1$ and $x_4 = 0$, which results in $x_1 = \frac{25}{6} - \frac{4}{6} - \frac{3}{6} = 3$ and $x_0 = \frac{275}{6} - \frac{2}{6} - \frac{3}{6} = 45$.

The example was solved more easily as a group knapsack problem than as an ordinary knapsack problem. But, it must be remembered, the group knapsack formulation only guarantees a feasible solution to (35) for suitably large values of b.

6.11. SHORTEST PATH FORMULATIONS OF KNAPSACK PROBLEMS

Both the knapsack and group knapsack problems can be represented as shortest path problems.

The Knapsack Problem

Consider the formulation of the knapsack problem given by (35). Let $\Gamma = (V, E)$ be a directed graph with vertex set $V = \{0, 1, \ldots, b\}$, and edge set $E = \{(i, k) | k - i = a_j \text{ for some } j, j = 1, \ldots, n + 1\}$. The graph Γ contains no cycles, since $a_j > 0$ for all j implies that if $(i, k) \in E$, then $k > i$. The edge (i, k) with $k - i = a_j$ is assigned a length $-c_j$. Let $L(p)$ be the length of path p.

Theorem 1: *There is a correspondence between paths in Γ from vertex 0 to vertex b and feasible solutions to (35) such that if the path p^* corresponds to the solution $(x_1^*, \ldots, x_{n+1}^*)$, then $L(p^*) = -\sum_{j=1}^{n+1} c_j x_j^*$.*

PROOF: Consider any feasible solution to (35), $(x_1, \ldots, x_{n+1}) = (x_1^*, \ldots, x_{n+1}^*)$. Clearly, there is a path p^* in Γ with vertex sequence $(0, a_1, \ldots, x_1^* a_1, x_1^* a_1 + a_2, \ldots, x_1^* a_1 + x_2^* a_2, \ldots, \sum_{j=1}^{k} x_j^* a_j, \ldots, \sum_{j=1}^{n+1} x_j^* a_j = b^*)$, and $L(p) = -\sum_{j=1}^{n+1} c_j x_j^*$.

Now consider $p = (0, i_1, \ldots, i_k = b)$. From the definition of E, there exists r such that $i_1 = a_r$; that is, $i_1 = \sum_{j=1}^{n+1} a_j \alpha_{j1}$, where $\alpha_{r1} = 1$ and $\alpha_{j1} = 0$, $j \neq r$. Proceeding by induction, suppose that

$$i_{k-1} = \sum_{j=1}^{n+1} a_j \alpha_{j,k-1}, \qquad \alpha_{j,k-1} \geq 0 \text{ integer}, \qquad j = 1, \ldots, n + 1$$

It then follows from the definition of E that there exists r such that

$$i_k = a_r + i_{k-1} = \sum_{j=1}^{n+1} a_j \alpha_{jk}$$

where $\alpha_{rk} = \alpha_{r,k-1} + 1$ and $\alpha_{jk} = \alpha_{j,k-1}, j \neq r$. Since $i_k = b, x_j = \alpha_{jk}$, $j = 1, \ldots, n + 1$, is a feasible solution to (35) with $x_0 = \sum_{j=1}^{n+1} c_j \alpha_{jk} = -L(p)$. ■

As a consequence of Theorem 1, finding a shortest path in Γ from vertex 0 to vertex b solves the knapsack problem. The negative edge lengths do not create any difficulties because Γ is acyclic. In fact, Γ has a highly structured topology, since the vertices can be ordered as $\{0, 1, \ldots, b\}$, with the length of a shortest path to i equal to or less than the length of a shortest path to k, for $i < k$.

The algorithm of Section 6.4 actually finds a shortest path in Γ, although the graph is not explicitly considered. In fact, this algorithm is related to the algorithm for finding shortest paths given in Section 3.5 (see Exercise 24).

The Group Knapsack Problem

Now consider the group knapsack problem given by (18). Let $\Gamma' = (V', E')$ be a directed graph with vertex set $V' = \{0, 1, \ldots, |G| - 1\}$, where vertex i corresponds to g_i and vertex i^* corresponds to g^*. Let $E' = \{(i, k) | g_i \oplus g_j = g_k \text{ for some } j \in Q\}$. Associated with edge (i, k) is the length d_j. In general, this graph contains cycles, but $d_j \geq 0$ implies that all cycles are of nonnegative length.

Theorem 2: *There is a correspondence between paths in Γ' from vertex 0 to vertex i^* and solutions to (18). If the path p^* corresponds to the solution $\{t_j^*\}, j \in Q$, then $\sum_{j \in Q} d_j t_j^* = L(p^*)$*

The proof of Theorem 2 is similar to the proof of Theorem 1, and is asked for in Exercise 22. From Theorem 2, a shortest path from vertex 0 to vertex i^* in Γ' solves the group knapsack problem. Since the existence of a shortest path from vertex 0 to vertex i^* implies that there exists a shortest path that is acyclic, edges directed to vertex 0 and edges directed away from vertex i^* can be omitted from Γ'.

Example

Consider the problem (34). The graph topology and edge lengths are given in Table 9. Vertex i corresponds to g_i; the (i, j)th element in the table is the length of the edge from vertex i to vertex j; $(-)$ indicates no edge. It is desired to find a shortest path from vertex 0 to vertex 1.

Applying the algorithm of Section 3.5 yields a shortest path (0, 3, 2, 1) of length 13. This path translates into an optimal solution $t_3 = 1$, $t_2 = 0$, and $t_5 = 2$ with cost 13.

Table 9

	1	2	3	4	5
0	—	10	3	—	5
2	5	—	—	10	3
3	—	5	—	—	10
4	3	—	5	—	—
5	10	3	—	5	—

6.12. TRANSFORMING AN ILP IN BOUNDED VARIABLES TO A KNAPSACK PROBLEM IN BOUNDED VARIABLES

Consider the ILP constraint set

$$S = \{x \mid Ax = b, 0 \leq x \leq u, x \text{ integer}\}$$

The purpose of this section is to establish the existence (by construction) of weights $w = (w_1, \ldots, w_m)$ such that

$$T = \{x \mid wAx = wb, 0 \leq x \leq u, x \text{ integer}\} = S$$

Since, except for the upper bounds, T is described by one constraint

$$\sum_j \left(\sum_i w_i a_{ij}\right) x_j = \sum_j a_j x_j = b_0 = \sum_i w_i b_i$$

it follows that an ILP in bounded variables can be solved as an equality constrained knapsack problem in bounded variables. Unfortunately, it is usually the case that the weights are large numbers and yield large coefficients $(a_1, \ldots, a_n, b_0)$. For this reason, it has not yet been established that transforming an ILP to a knapsack problem is desirable for computation.

A knapsack problem in bounded variables can be solved by a slight modification of (4). In particular, (4) is modified to

$$f_k(y) = \max_{x_k = 0, 1, \ldots, \min\{u_k, [y/a_k]\}} (c_k x_k + f_{k-1}(y - a_k x_k)) \tag{46}$$

For an equality constraint, the initial condition is $f_0(0) = 0$ and $f_0(y) = -\infty$, otherwise. Note that for a bounded variable knapsack problem with an equality constraint, it is meaningful to allow $c_k < 0$. Negative objective coefficients do not cause any difficulty in solving (46) and, in fact, may arise in the knapsack problem to be derived.

It is important to realize that Recursions II and III cannot be used to solve knapsack problems in bounded variables, since their derivations depend upon the variables not having explicit upper bounds.

Combining Two Constraints

We will show that two constraints can be combined into one without changing the set of feasible solutions. Then, by combining the constraints two at a time, it is clear that m constraints can be combined into one.

Consider the two constraints

$$\begin{aligned} \sum_{j=1}^{n} d_j x_j - b_1 &= 0 \\ \sum_{j=1}^{n} f_j x_j - b_2 &= 0 \end{aligned} \tag{47}$$

and assume that the coefficients d_j and f_j, $j = 1, \ldots, n$, are integers.

Let

$$\lambda^+ = \max \sum_{j=1}^{n} d_j x_j - b_1, \qquad 0 \le x_j \le u_j \text{ integer}, j = 1, \ldots, n$$

$$\lambda^- = \min \sum_{j=1}^{n} d_j x_j - b_1, \qquad 0 \le x_j \le u_j \text{ integer}, j = 1, \ldots, n$$

Defining $d_j^+ = \max\{0, d_j\}$ and $d_j^- = \min\{0, d_j\}$, we obtain

$$\lambda^+ = \sum_{j=1}^{n} d_j^+ u_j - b_1 \qquad \text{and} \qquad \lambda^- = \sum_{j=1}^{n} d_j^- u_j - b_1$$

Finally, define $\lambda = \max\{\lambda^+, |\lambda^-|\}$. Note that

$$\lambda = \max \left| \sum_{j=1}^{n} d_j x_j - b_1 \right|, \qquad 0 \le x_j \le u_j \text{ integer}, j = 1, \ldots, n$$

Theorem 3: *The integer vector x^0, $0 \le x^0 \le u$ is a solution to* (47) *if and only if*

$$\sum_{j=1}^{n} (d_j + \alpha f_j) x_j^0 - b_1 - \alpha b_2 = 0 \tag{48}$$

where α is any integer satisfying $|\alpha| > \lambda$.

PROOF: ($\Rightarrow$) From the fact that x^o is a solution to (47), it follows that

$$\alpha \sum_{j=1}^{n} f_j x_j^o - \alpha b_2 = 0$$

and

$$\sum_{j=1}^{n} d_j x_j^o - b_1 + \alpha \sum_{j=1}^{n} f_j x_j^o - \alpha b_2 = \sum_{j=1}^{n} (d_j + \alpha f_j) x_j^o - b_1 - \alpha b_2 = 0$$

($\Leftarrow$): Suppose that x^o solves (48) and

$$\sum_{j=1}^{n} f_j x_j^o = b_2 + k \tag{49}$$

where k is an arbitrary integer. It will be shown that $|\alpha| > \lambda$ implies that $k = 0$. Multiply (49) by α and subtract the result from (48), which yields

$$\sum_{j=1}^{n} d_j x_j^o - b_1 + k\alpha = 0 \tag{50}$$

Since

$$\left| \sum_{j=1}^{n} d_j x_j^o - b_1 \right| \le \lambda < |\alpha|$$

(50) implies that $k = 0$. Putting $k = 0$ into (49) and (50) yields the result that x^o solves (47). ■

Example

Consider the problem of Section 5.1

$$\max x_0 = 2x_1 + x_2$$

$$x_1 + x_2 + x_3 = 5 \tag{51}$$

$$-x_1 + x_2 + x_4 = 0 \tag{52}$$

$$6x_1 + 2x_2 + x_5 = 21 \tag{53}$$

$$x_1 \ldots, x_5 \ge 0 \text{ integer}$$

Equation (53) implies that $x_1 \le 3$, and then (52) yields $x_2 \le 3$. Also, the bounds $x_3 \le 5$, $x_4 \le 3$, and $x_5 \le 21$ are obvious.

For (53)

$$\lambda^+ = 6(3) + 2(3) + 1(21) - 21 = 24$$
$$\lambda^- = 0(3) + 0(3) + 0(21) - 21 = -21$$

and $\lambda = 24$. Thus (51) or (52) can be weighted by $|\alpha| > 24$ and combined with (53). Choosing (52) and $\alpha = 25$ yields

$$6x_1 + 2x_2 + x_5 + 25(-x_1 + x_2 + x_4) = 21$$

or

$$-19x_1 + 27x_2 + 25x_4 + x_5 = 21 \tag{54}$$

For (51), $\lambda^+ = 6$, $\lambda^- = -5$, and $\lambda = 6$. Weighting (54) by 7 and combining it with (51), we obtain

$$x_1 + x_2 + x_3 + 7(-19x_1 + 27x_2 + 25x_4 + x_5) = 5 + 7(21)$$

or

$$-132x_1 + 190x_2 + x_3 + 175x_4 + 7x_5 = 152 \tag{55}$$

Thus $w = (1, 25(7), 7) = (1, 175, 7)$.

Note that the coefficient of x_1 in (55) is negative. To obtain nonnegative coefficients, as we have required in knapsack problems, substitute $y_j = u_j - x_j$ whenever a negative coefficient appears. Substituting $y_1 = 3 - x_1$ in (55) yields

$$132y_1 + 190x_2 + x_3 + 175x_4 + 7x_5 = 548 \tag{56}$$

Thus the original ILP is equivalent to

$$\begin{aligned} \max x_0 = 6 - 2y_1 + \quad x_2 & \\ 132y_1 + 190x_2 + x_3 + 175x_4 + 7x_5 = 548 & \\ 0 \le y_1 \le 3,\ 0 \le x_2 \le 3,\ 0 \le x_3 \le 5,\ 0 \le x_4 \le 3,\ 0 \le x_5 \le 21 & \\ y_1, x_2, \ldots, x_5 \text{ integer} & \end{aligned} \tag{57}$$

The problem (57) can be solved using the recursion (46).

6.13. EXERCISES

1. Solve

$$\begin{aligned} \max x_0 = 11x_1 + 9x_2 + 5x_3 + 3x_4 & \\ 6x_1 + 5x_2 + 3x_3 + 2x_4 \le 45 & \\ x_1, \ldots, x_4 \ge 0 \text{ integer} & \end{aligned}$$

by Recursion I.

2. Solve the problem of Exercise 1 by Recursion II.

3. Solve the problem of Exercise 1 by Recursion III.

4. Find the smallest right-hand side of the inequality in Exercise 1, say y^*, such that there exists an optimal solution with $x_1 \geq 1$ for all $y \geq y^*$.

5. Construct group addition tables for $G(8)$, $G(2, 4)$, and $G(2, 2, 2)$. Using these tables, show that none of these groups is isomorphic to another.

6. For an arbitrary finite abelian group G, prove that the distinct cosets of $T(g)$ yield a partition of the group elements into $|G|/|g|$ subsets.

7. Solve

$$\begin{aligned} \min\ & 5t_1 + 6t_2 + t_4 + 3t_5 + 2t_6 \\ & t_1 g_1 \oplus t_2 g_2 \oplus t_4 g_4 \oplus t_5 g_5 \oplus t_6 g_6 = g_7 \\ & t_j \geq 0 \text{ integer, all } j \end{aligned}$$

over $G(8)$ using Group Recursion I.

8. Solve the problem of Exercise 7 using Group Recursion II.

9. Solve

$$\begin{aligned} \min\ & 3t_{1,0} + 2t_{0,1} + 4t_{1,1} + t_{2,1} + 5t_{2,0} \\ & t_{1,0}g_{1,0} \oplus t_{0,1}g_{0,1} \oplus t_{1,1}g_{1,1} \oplus t_{2,1}g_{2,1} \oplus t_{2,0}g_{2,0} = g_{3,0} \\ & t_{i,j} \geq 0 \text{ integer, all } i, j \end{aligned}$$

over $G(4, 2)$ by Group Recursion I.

10. Solve the problem of Exercise 9 using Group Recursion II.

11. Reduce the problem of Exercise 1 to a group knapsack problem by dropping $x_1 \geq 0$ and then solve using Group Recursion II.

12. Prove that if the group knapsack problem given by (18) has a feasible solution, it has an optimal solution t^* with

$$\sum_{j=1}^{n} t_j^* \leq |G| - 1$$

13. Draw the graph associated with shortest path formulation of the problem in Exercise 7. Solve it by finding a shortest path.

14. Show that

$$\begin{aligned} & \sum_{j=1}^{n} c_j x_j \geq z, \qquad c_1 > 0 \\ & \sum_{j=1}^{n} a_j x_j \leq b, \qquad a_1 > 0 \\ & x_j \geq 0 \text{ integer}, \qquad j = 1, \ldots, n \end{aligned}$$

has a solution $(x_1^*, \ldots, x_n^*)$ if and only if

$$\sum_{j=2}^{n} (c_j a_1 - c_1 a_j) x_j \geq z a_1 - c_1 b$$
$$\sum_{j=2}^{n} a_j x_j \leq b$$
$$x_j \geq 0 \text{ integer}, \quad j = 2, \ldots, n$$

has a solution $(x_2^*, \ldots, x_n^*)$ and x_1^* integer satisfies

$$\max\left\{0, \frac{z - \sum_{j=2}^{n} c_j x_j^*}{c_1}\right\} \leq x_1^* \leq \frac{b - \sum_{j=2}^{n} a_j x_j^*}{a_1}$$

15. Show that

$$\sum_{j=1}^{n} c_j x_j \geq z, \qquad c_1 < 0$$
$$\sum_{j=1}^{n} a_j x_j \leq b, \qquad a_1 > 0$$
$$x_j \geq 0 \text{ integer}, \quad j = 1, \ldots, n$$

has a solution $(x_1^*, \ldots, x_n^*)$ if and only if

$$\sum_{j=2}^{n} c_j x_j \geq z$$
$$\sum_{j=2}^{n} a_j x_j \leq b$$
$$x_j \geq 0 \text{ integer}, \quad j = 2, \ldots, n$$

has a solution $(x_2^*, \ldots, x_n^*)$ and x_1^* integer satisfies

$$0 \leq x_1^* \leq \min\left\{\frac{b - \sum_{j=2}^{n} a_j x_j}{a_1}, \frac{z - \sum_{j=2}^{n} c_j x_j}{c_1}\right\}$$

16. Use the results of Exercises 14 and 15 to develop an iterative scheme that eliminates one variable at a time and will construct a solution to

$$\sum_{j=1}^{n} c_j x_j \geq z$$
$$\sum_{j=1}^{n} a_j x_j \leq b$$
$$x_j \geq 0 \text{ integer}, \quad j = 1, \ldots, n$$

or show that none exists.

17. Apply the results of Exercise 16 to obtain an enumeration algorithm for the knapsack problem (i.e., if there is no solution for $z = z^*$, try $z = z^* - 1$, etc.).

18. Solve the problem of Exercise 1 by the algorithm developed in Exercise 17.

19. Show that x solves

$$\begin{aligned} 3x_1 + 2x_2 \quad + x_4 \quad\quad &= 10 \\ x_1 + 4x_2 \quad\quad + x_5 \quad &= 11 \\ 3x_1 + 3x_2 + x_3 \quad\quad + x_6 &= 13 \\ x_1, \ldots, x_6 \geq 0 \text{ integer} \end{aligned}$$

if and only if it solves

$$521x_1 + 562x_2 + 168x_3 + x_4 + 14x_5 + 168x_6 = 2348$$

$$0 \leq x_1 \leq 3, 0 \leq x_2 \leq 2, 0 \leq x_3 \leq 13, 0 \leq x_4 \leq 10, 0 \leq x_5 \leq 11, 0 \leq x_6 \leq 13$$
$$x_1, \ldots, x_6 \geq 0 \text{ integer}$$

20. Use the results of Exercise 19 to solve the example of Chapter 5, Exercise 2, as a knapsack problem.

21. In the algorithm of Section 6.4, if $j > d(y - a_j)$, then $j + 1$ is investigated next. An algorithm can be devised which has the property that if $j > d(y - a_j)$, then no larger j's have to be investigated. The algorithm is based on the following recursion for $i = 1, \ldots, b$, and $y = i, \ldots, b$

$$h_i(y) = \begin{cases} \max\{h_{i-1}(y); c_k + g(y - a_k), y - a_k = i - 1, k \leq d(i-1)\} \\ h_{i-1}(y), \qquad \text{if } Q_i(y) = Q_{i-1}(y) \end{cases}$$

where $Q_i(y) = \{j \mid 0 \leq y - a_j < i, j \leq d(y - a_j)\}$, g and d are defined in Section 6.4 and $h_0(y) = 0$, for all y.

(a) Prove that $h_i(i) = g(i)$, $i = 1, \ldots, b$.
(b) Translate this recursion into algorithmic form.
(c) Solve the problem of Section 6.2 using this algorithm.

22. Prove Theorem 2 of Section 6.11.

23. Extend Theorem 3 of Section 6.12 so that both equations of (47) can be multiplied by weights greater than 1 in absolute value.

24. Compare Recursion III with the shortest path algorithm of Section 3.5, using the shortest route formulation of the knapsack problem given in Section 6.11.

25. Use the results of Exercise 12 to prove that an alternative to the upper bound given by (14) is

$$y^* \leq a_1(1 + a_k)$$

where $a_k = \max_j a_j$.

6.14. NOTES

6.1. The name knapsack problem appears to have been suggested in Dantzig (1957). An alternate name, the "loading" problem, was given by Bellman (1957).

The "cutting stock" problem is an LP having a very large number of variables. Gilmore and Gomory have shown that in solving the cutting stock problem by a simplex method, the column to enter the basis can be found by solving a knapsack problem. Many of the advances in techniques for the knapsack problem have come from the work of Gilmore and Gomory in connection with their research on the cutting stock problem. Specifically, see Gilmore and Gomory (1961), (1963), (1965), and (1966).

6.2. The recursion of this section is given in Dantzig (1957) and Bellman (1957). Efficient dynamic programming procedures for computing (4) in the case of bounded variables and/or multiple constraints are given in Weingartner and Ness (1967) and Nemhauser and Ullmann (1969). For more general treatments of finite stage dynamic programming, see Bellman and Dreyfus (1962) and Nemhauser (1966).

6.3. The recursion of this section is from Gilmore and Gomory (1965).

6.4. The recursion of this section is based on Gilmore and Gomory (1966). See Exercise 21 for a variation that will be more efficient for some data sets. A more general knapsack problem containing a discount factor is treated by a related recursion in Shapiro and Wagner (1967).

6.5. The periodic result is given in Gilmore and Gomory (1966) and Shapiro and Wagner (1967). The upper bound (14) appears in Hu (1969). Another upper bound is discussed in Exercise 25.

6.6. A method for solving small ILP's by hand using the concepts of modulo arithmetic is given in Wilson (1970).

There are a variety of textbooks on abstract algebra containing the elements of group theory. A classic text is Birkhoff and Maclane (1941).

6.7. The group knapsack problem was introduced in Gomory (1965), which has been reprinted in Dantzig and Veinott (1968).

6.8. The recursion of this section is a straightforward application of dynamic programming.

6.9. The recursion of this section is from Gomory (1965).

6.10. The results of this section are a specialization of the results of Gomory (1965) to knapsack problems and have been included to motivate the development in Chapter 7.

6.11. General discussions of shortest route formulations of finite state, dynamic programming problems can be found in Shapiro (1968c) and Goldman and Nemhauser (1967).

Dijkstra's algorithm (see Section 3.5) has been adapted to the shortest route formulation of the group knapsack problem in Hu (1970). Shapiro (1968a) gives a different shortest route algorithm for the group knapsack problem.

6.12. This section is based on Bradley (1971c). However, Bradley gives a considerably generalized version of Theorem 3, which allows much greater flexibility in choosing weights. Elmaghraby and Wig (1970), Glover and Woolsey (1970), and Padberg (1970) give results related to Bradley's.

Flexibility in choosing weights is important because it allows one to exercise some control over the magnitude of the coefficients in the derived knapsack equation. Some results on choosing "good" weights are given in Bradley (1971b) where it is also shown that it is possible to derive a knapsack problem from a bounded ILP with the property $c_j = a_j$ for all j. Faaland (1971) presents an algorithm for knapsack problems having this property.

Further Notes

Other algorithms for the knapsack problem are given by Pandit (1962), Kolesar (1967), Greenberg (1969a), Greenberg and Hegererich (1970), and Cabot (1970). Cabot's algorithm is discussed in Exercises 14–18. Other algorithms for the group knapsack problem are given by White (1966) and Glover (1969a).

7 Integer Programming over Cones

7.1. PROBLEM FORMULATION AND BASIC ANALYSIS

By dropping certain nonnegativity restrictions from an ILP, a related problem, called an integer program over a cone (ILPC), is obtained. ILPC's can be formulated as group knapsack problems and are considerably easier to solve than the corresponding ILP's. ILPC's are of practical value because they are useful in solving ILP's and of theoretical interest because of their group knapsack structure.

Consider the ILP

$$\begin{aligned} \max v_0 &= c'v \\ A'v &\leq b \\ v &\geq 0 \text{ integer} \end{aligned} \tag{1}$$

and the corresponding LP

$$\begin{aligned} \max v_0 &= c'v \\ A'v &\leq \mathrm{b} \\ v &\geq 0 \end{aligned} \tag{2}$$

Problem (2) can be written

$$\begin{aligned} \max v_0 &= c'v \\ A'v + Is &= b \\ v, s &\geq 0 \end{aligned} \tag{3}$$

where s is a vector of slack variables. Let $x = (v, s)$, $x_0 = v_0$, $A = (A', I)$, and $c = (c', 0)$, and write (3) as

$$\begin{aligned} \max x_0 &= cx \\ Ax &= b \\ x &\geq 0 \end{aligned} \tag{4}$$

Let B be a basis matrix for the LP (4), and partition $x = (x_B, x_N)$, $A = (B, N)$, and $c = (c_B, c_N)$. In partitioned form (4) is

$$\begin{aligned} \max x_0 &= c_B x_B + c_N x_N \\ & B x_B + N x_N = b \\ & x_B, x_N \geq 0 \end{aligned} \tag{5}$$

Expressing x_0 and x_B in terms of x_N yields

$$\begin{aligned} \max x_0 &= c_B B^{-1} b - (c_B B^{-1} N - c_N) x_N \\ x_B &= B^{-1}(b - N x_N) \\ & x_B, x_N \geq 0 \end{aligned} \tag{6}$$

If $B^{-1}b \geq 0$, B is called a *primal feasible basis* (matrix); if $c_B B^{-1} N - c_N \geq 0$, B is called a *dual feasible basis* (matrix). If B is primal and dual feasible, it is called an *optimal basis* (matrix).

Consider the relaxation of (6) in which the constraints $x_B \geq 0$ are omitted

$$\begin{aligned} \max x_0 &= c_B B^{-1} b - (c_B B^{-1} N - c_N) x_N \\ x_B &= B^{-1}(b - N x_N) \\ & x_N \geq 0 \end{aligned} \tag{7}$$

In x_N space, the feasible solutions to (7) correspond to the *cone* given by the nonnegative orthant. For this reason, LP's of the form (7) are called *LP's over cones.*

If x_{B_i} is a slack variable s_i in (7), dropping $s_i \geq 0$ is equivalent to dropping the ith constraint of (2). Thus, in v-space, precisely those constraints that define the extreme point specified by B are kept. A two-dimensional case is shown in Figure 1. Every point in the shaded polyhedron is a feasible solution to (2). Every point within the two rays that meet at the extreme point defined by B is a feasible solution to (7). The feasible solutions to (7) are seen to be an unbounded convex set, which is a cone translated from the origin. A specific algebraic description of the translated cone constraints has been given in Section 5.18, but will not be needed here.

Assuming that the data in (1) is integer so that the slack variables are integer, corresponding to (4), we can write (1) as

$$\begin{aligned} \max x_0 &= cx \\ & Ax = b \\ & x \geq 0 \text{ integer} \end{aligned} \tag{8}$$

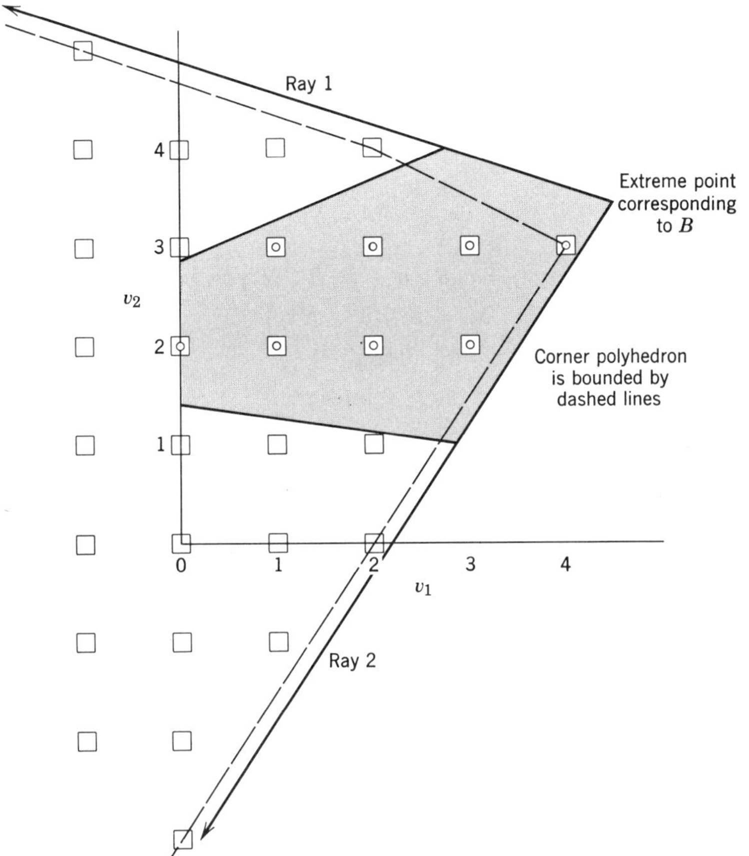

Figure 1

For any basis B, (8) can be stated as

$$\begin{aligned} \max x_0 &= c_B x_B + c_N x_N \\ B x_B &+ N x_N = b \\ x_B, &\; x_N \geq 0 \text{ integer} \end{aligned} \tag{9}$$

or

$$\begin{aligned} \max x_0 &= c_B B^{-1} b - (c_B B^{-1} N - c_N) x_N \\ x_B &= B^{-1}(b - N x_N) \\ x_B, &\; x_N \geq 0 \text{ integer} \end{aligned} \tag{10}$$

Again dropping the nonnegativity constraints on x_B, we obtain from (9) and (10), respectively

$$\begin{aligned} \max x_0 = c_B x_B + c_N x_N & \\ B x_B + N x_N = b & \\ x_B \text{ integer} & \\ x_N \geq 0 \text{ integer} & \end{aligned} \tag{11}$$

$$\begin{aligned} \max x_0 &= c_B B^{-1} b - (c_B B^{-1} N - c_N) x_N \\ x_B &= B^{-1}(b - N x_N) \\ & x_B \text{ integer} \\ & x_N \geq 0 \text{ integer} \end{aligned} \tag{12}$$

The equivalent problems (11) and (12) are ILP's in which the corresponding LP's are over cones. Thus (11) and (12) are called *ILP's over cones* or *ILPC's*. An ILPC is a relaxation of the corresponding ILP in which, for a given B, the nonnegativity restrictions on x_B are omitted.

In the v-space of Figure 1, the feasible solutions to (1) are indicated by circles and the feasible solutions to (11) or (12) by squares. If an ILPC has a feasible solution, it has an infinite number of feasible solutions. The convex hull of these solutions is an unbounded polyhedron, called the *corner polyhedron*. (The convex hull of S is the smallest convex set containing S.)

It will be seen in Section 7.3 that an ILPC is considerably easier to solve than the corresponding ILP. In fact, an ILPC can be solved as a group knapsack problem over a direct sum group of order $D = |\det B|$.

If (x_B^*, x_N^*) is an optimal solution to (12) and $x_B^* \geq 0$ then, because (12) is a relaxation of (8), (x_B^*, x_N^*) solves (8). A sufficient condition for $x_B^* \geq 0$ in (12) will be given in Section 7.5. It turns out that this condition will be met if each component of b is suitably large. In this sense an ILPC asymptotically solves the corresponding ILP. Thus an ILPC is sometimes called an *asymptotic integer linear program*.

Even when no optimal ILPC solution is feasible to the ILP, a branch and bound algorithm in which the subproblems are ILPC's can be used to solve the ILP. This will be discussed in Section 7.6.

In Section 7.7, the properties of corner polyhedra will be investigated. These polyhedra are much simpler, both in number and characterization of extreme points, than general convex hulls of integer points. A study of their properties constitutes the beginning of an analysis of convex hulls of integer points. Pursuing this line of work, hopefully, will lead to a better fundamental understanding of integer programming.

For a given ILP, each basis for the corresponding LP defines an ILPC. Thus there is a family of ILPC's corresponding to an ILP. However, only the dual feasible subfamily is useful. In particular, consider the ILPC given by (12) and assume that B is not dual feasible. Let $N = (a_1, \ldots, a_r)$ and let (x_B^*, x_N^*) be a feasible solution to (12). Any component of $B^{-1}a_j$ can be written as e_{ij}/D, where e_{ij} is an integer. Thus

$$\begin{aligned} x_j^o &= x_j^*, \qquad j = 1, \ldots, r \qquad j \neq k, \\ x_k^o &= x_k^* + mD \\ x_B^o &= B^{-1}(b - Nx_N^o) \end{aligned}$$

is a feasible solution to (12) for any nonnegative integer m. Since B is not dual feasible, suppose that $c_B B^{-1} a_k - c_k < 0$. Then

$$-(c_B B^{-1} a_k - c_k)(x_k^* + mD) \to \infty \qquad \text{as } m \to \infty$$

and (12) is unbounded.

Thus only dual feasible bases will be considered throughout this chapter. We will also assume that $B^{-1}b$ is not all-integer; otherwise the ILPC is solved by $x_B = B^{-1}b$ and $x_N = 0$.

Eliminating the constant term from the objective function of (12), changing from max to min, and replacing x_B integer by

$$B^{-1}(b - Nx_N) \equiv 0 \pmod 1$$

where (mod 1) means that each element of the vector is taken modulo 1, we obtain the ILPC statement

$$\begin{aligned} \min z_0 &= (c_B B^{-1} N - c_N) x_N \\ B^{-1} N x_N &\equiv B^{-1} b \pmod 1 \\ x_N &\geq 0 \text{ integer} \end{aligned} \tag{13}$$

where $c_B B^{-1} N - c_N \geq 0$.

Assume that x_N^* is an optimal solution to (13) so that the corresponding value of x_B is

$$x_B^* = B^{-1}(b - Nx_N^*) \tag{14}$$

We can think of $B^{-1}Nx_N^*$ as a minimum cost correction to $B^{-1}b$ that yields x_B^* integer. If the correction is such that $x_B^* \geq 0$, then (x_B^*, x_N^*) solves the ILP (8). For this reason, it is intuitively appealing to choose B such that $B^{-1}b \geq 0$. Thus we generally work with an ILPC for which B is an optimal LP basis. However, the theory and algorithms to be developed apply to any ILPC generated by a dual feasible basis.

Example

Consider the ILP introduced in Section 5.1

$$\begin{aligned} \max x_0 = 2x_1 + \ \ x_2 \qquad & \\ x_1 + \ \ x_2 + x_3 \qquad\qquad & = 5 \\ -x_1 + \ \ x_2 \qquad + x_4 \qquad & = 0 \\ 6x_1 + 2x_2 \qquad\qquad + x_5 & = 21 \\ x_1, \ldots, x_5 \geq 0 \text{ integer} & \end{aligned}$$

The optimal solution to the corresponding LP (see Section 2.4) is $x_B = (x_1, x_2, x_4) = (\frac{11}{4}, \frac{9}{4}, \frac{1}{2})$ and $x_N = (x_3, x_5) = (0, 0)$.

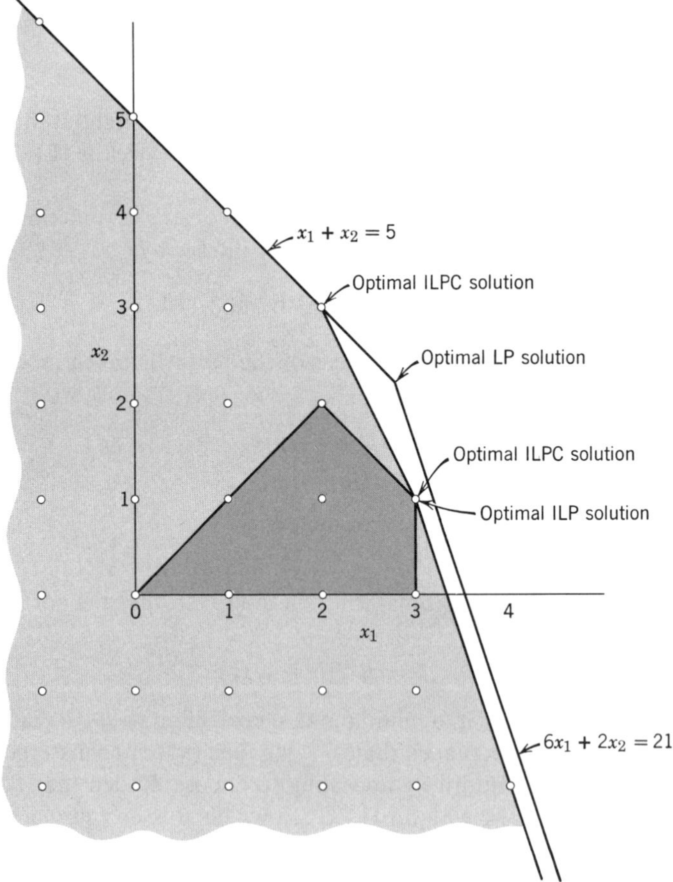

Figure 2

The ILPC (11) corresponding to the optimal LP basis is

$$
\begin{aligned}
\max x_0 = 2x_1 + \ \ x_2 & \\
x_1 + \ \ x_2 + x_3 \qquad\qquad & = \ 5 \\
-x_1 + \ \ x_2 \qquad + x_4 \qquad & = \ 0 \\
6x_1 + 2x_2 \qquad\qquad + x_5 & = 21 \\
x_3, x_5 \geq 0 \text{ integer} & \\
x_1, x_2, x_4 \text{ integer} &
\end{aligned}
$$

Since x_4 is a slack variable, in (x_1, x_2)-space (corresponding to the v-space of Figure 1), the ILPC is

$$
\begin{aligned}
\max x_0 = 2x_1 + \ \ x_2 & \\
x_1 + \ \ x_2 & \leq \ 5 \\
6x_1 + 2x_2 & \leq 21 \\
x_1, x_2 \text{ integer} &
\end{aligned}
$$

The LP cone constraints $x_1 + x_2 \leq 5$ and $6x_1 + 2x_2 \leq 21$ are shown in Figure 2. They define a translated cone with apex at the optimal LP solution $x_1 = \frac{11}{4}$, $x_2 = \frac{9}{4}$. The corner polyhedron is lightly shaded in Figure 2. Within this polyhedron is the darkly shaded convex hull of feasible solutions to the ILP. The set of optimal solutions to the ILP is contained in the corner polyhedron. The optimal ILP solution is $x_1 = 3$ and $x_2 = 1$. It is easy to see that $x_1 = 3$, $x_2 = 1$, and $x_1 = 2$, $x_2 = 3$, are alternate optimal solutions to the ILPC. Consequently, in the example, there is an optimal solution to the ILPC that is feasible to the ILP. This is not always the case.

Now we look at the example in (x_3, x_5)-space. The representation of the ILPC (13) is (see Table 5 of Section 2.4)

$$
\begin{aligned}
\min z_0 = \ & \frac{x_3}{2} + \frac{x_5}{4} \\
& \frac{3x_3}{2} - \frac{x_5}{4} \equiv \frac{9}{4} \pmod 1 \\
& -2x_3 + \frac{x_5}{2} \equiv \frac{1}{2} \pmod 1 \\
& -\frac{x_3}{2} + \frac{x_5}{4} \equiv \frac{11}{4} \pmod 1 \\
& x_3, x_5 \geq 0 \text{ integer}
\end{aligned}
\tag{15}
$$

It will be shown in Section 7.2 that the constraints of (15) are equivalent to

$$\begin{aligned} 2x_3 + 3x_5 &\equiv 1 \pmod 4 \\ x_3, x_5 &\geq 0 \text{ integer} \end{aligned} \tag{16}$$

The feasible solutions to (16) are designated by circles in the (x_3, x_5)-space of Figure 3. The two optimal ILPC solutions, $(x_3, x_5) = (1, 1)$ and $(x_3, x_5) = (0, 3)$, are indicated with squares around the circles.

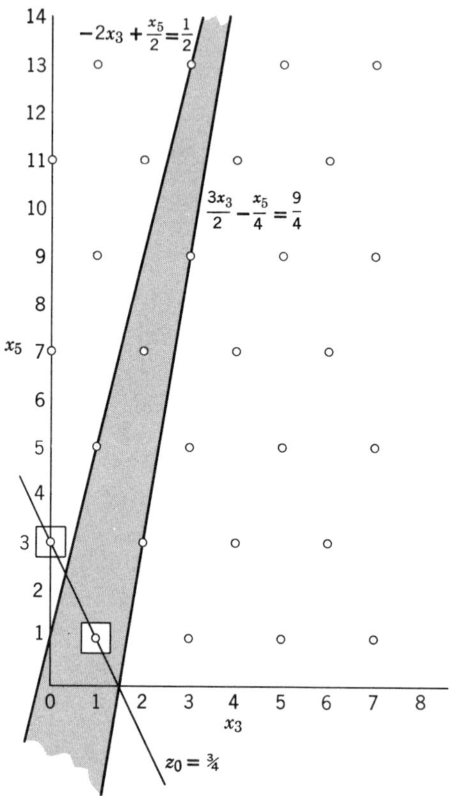

Figure 3

In the relaxation of the ILP to the ILPC the omitted constraints in terms of (x_3, x_5), are

$$\begin{aligned} \frac{3x_3}{2} - \frac{x_5}{4} &\leq \frac{9}{4} \\ -2x_3 + \frac{x_5}{2} &\leq \frac{1}{2} \\ -\frac{x_3}{2} + \frac{x_5}{4} &\leq \frac{11}{4} \end{aligned} \tag{17}$$

The constraints (17) are represented by the shaded polyhedron in Figure 3. In terms of (x_3, x_5), the feasible solutions to the ILP are represented by the circles within the shaded polyhedron. The point $x_N^* = (x_3^*, x_5^*) = (1, 1)$ is an optimal solution to the ILPC; it satisfies the constraints (17) and therefore yields an optimal solution to the ILP. To find the corresponding optimal values of (x_1, x_2, x_4), substitute x_N^* into (14), which yields

$$x_2^* = \frac{9}{4} + \frac{1}{4} - \frac{3}{2} = 1$$

$$x_4^* = \frac{1}{2} - \frac{1}{2} + 2 = 2$$

$$x_1^* = \frac{11}{4} - \frac{1}{4} + \frac{1}{2} = 3$$

7.2. EQUIVALENT ILPC REPRESENTATIONS

Our objective is to transform the ILPC constraints of (11), by changing variables, into a form more suitable for analysis. Some classical results on the solution of simultaneous linear equations in integers provide the background. Let

$$S = \{x | Bx = b, x \text{ integer}\}, T = \{y | \bar{B}y = \bar{b}, y \text{ integer}\}$$

where B and $\bar{B}$ are mth-order integer matrices, and b and $\bar{b}$ are m-dimensional integer vectors. If there is a one-to-one correspondence between the elements of S and T given by $y = Px$, where P is an mth-order integer matrix, then $Bx = b$, x integer, and $\bar{B}y = \bar{b}$, y integer, are said to be *equivalent representations.*

To obtain a representation equivalent to $Bx = b$, x integer, Theorems 1 and 2 are used.

Theorem 1: *Let E be an mth-order unimodular integer matrix. For every integer y, there exists a unique integer x such that $y = Ex$.*

PROOF: From the unimodularity of E (see Section 3.3), it follows that E^{-1} exists and is an integer matrix. Choose an arbitrary integer vector $\hat{y}$. Let $\hat{x} = E^{-1}\hat{y}$ so that $\hat{y} = E\hat{x}$. The uniqueness of $\hat{x}$ follows from the uniqueness of E^{-1}. ■

Theorem 2: *Let R and C be mth-order unimodular integer matrices and let $RBC = \hat{B}$. Then $Bx = b$, x integer, and $\hat{B}y = Rb$, y integer, are equivalent representations.*

PROOF: Multiplying $Bx = b$ on the left by R yields

$$RBx = Rb \tag{18}$$

Since C is unimodular, C^{-1} exists, and thus

$$RB = \hat{B}C^{-1} \tag{19}$$

Substituting (19) into (18) yields

$$\hat{B}C^{-1}x = Rb \tag{20}$$

Now let

$$y = C^{-1}x \tag{21}$$

Note that C unimodular and integer implies that C^{-1} is unimodular and integer. Therefore, from Theorem 1, there is a one-to-one correspondence between the integer values of x and y in (21). Substituting (21) into (20) yields

$$\hat{B}y = Rb \tag{22}$$

and the result that an integer vector $\hat{y}$ solves (22) if and only if $\hat{x} = C\hat{y}$ is such that $RB\hat{x} = Rb$. But $Rb\hat{x} = Rb$ if and only if $B\hat{x} = b$, since R is non-singular. ■

Theorem 2 can be applied to the ILPC constraints of (11). Let $x_N = x'_N$ and consider

$$Bx_B = b - Nx'_N, \qquad x_B \text{ integer} \tag{23}$$

Letting R and C be mth-order integer unimodular matrices and $RBC = \hat{B}$, an equivalent representation of (23) is

$$\hat{B}y = R(b - Nx'_N), \qquad y \text{ integer} \tag{24}$$

Problem (11) can be made easier to analyze by obtaining a simple form for $\hat{B}$.

Example

Using the example of Section 7.1 and the optimal LP basis yields

$$\begin{aligned} x_1 + x_2 \phantom{{}+x_4} &= 5 - x_3 \\ -x_1 + x_2 + x_4 &= 0 \\ 6x_1 + 2x_2 \phantom{{}+x_4} &= 21 \qquad - x_5 \end{aligned}$$

Thus (23) is

$$\begin{bmatrix} 1 & 1 & 0 \\ -1 & 1 & 1 \\ 6 & 2 & 0 \end{bmatrix} \begin{bmatrix} x_1 \\ x_2 \\ x_4 \end{bmatrix} = \begin{bmatrix} 5 \\ 0 \\ 21 \end{bmatrix} - \begin{bmatrix} 1 & 0 \\ 0 & 0 \\ 0 & 1 \end{bmatrix} \begin{bmatrix} x_3 \\ x_5 \end{bmatrix} \tag{25}$$

$$x_1, x_2, x_4 \text{ integer}$$

Let

$$R = \begin{bmatrix} 1 & 0 & 0 \\ 1 & 1 & 0 \\ -6 & 0 & 1 \end{bmatrix} \quad \text{and } C = \begin{bmatrix} 1 & 0 & 1 \\ 0 & 0 & -1 \\ 0 & 1 & 2 \end{bmatrix}$$

By direct calculation, it is easy to see that R and C are unimodular. This particular choice of R and C will be justified shortly.

To obtain (24), note that

$$RBC = \begin{bmatrix} 1 & 0 & 0 \\ 0 & 1 & 0 \\ 0 & 0 & 4 \end{bmatrix} \qquad Rb = \begin{bmatrix} 5 \\ 5 \\ -9 \end{bmatrix}$$

$$RNx_N = \begin{bmatrix} x_3 \\ x_3 \\ -6x_3 + x_5 \end{bmatrix}$$

Let $y = (y_1, y_2, y_3)$, then (24) is

$$\begin{aligned} y_1 &= 5 - x_3 \\ y_2 &= 5 - x_3 \\ 4y_3 &= -9 + 6x_3 - x_5 \end{aligned} \tag{26}$$

$$y_1, y_2, y_3 \text{ integer}$$

Now (26) is a simpler representation than (25), in the sense that it immediately provides necessary and sufficient conditions on (x_3, x_5) for y to be integer, and equivalently for x_B to be integer. In particular, any (x_3, x_5) integer yields (y_1, y_2) integer, and y_3 is integer if and only if

$$-9 + 6x_3 - x_5 \equiv 0 \pmod 4$$

or

$$3 + 2x_3 + 3x_5 \equiv 0 \pmod 4 \tag{27}$$

or

$$2x_3 + 3x_5 \equiv 1 \pmod 4$$

Thus $(x_3, x_5) \geq 0$ and integer yields a feasible solution to the ILPC if and only if (27) holds. The representation of the ILPC example given by (15) of Section 7.1 has been reduced to

$$\begin{aligned} \min z_0 = \frac{x_3}{2} &+ \frac{x_5}{4} \\ 2x_3 + 3x_5 &\equiv 1 \pmod 4 \\ x_3, x_5 &\geq 0 \text{ integer} \end{aligned} \tag{28}$$

It can be seen from Section 6.10 that (28) is a group knapsack problem over $G(4)$.

Canonical Representations

The notion of equivalent problems will be useful as a general tool, only if it is possible to represent all ILPC's in a suitable canonical form. The existence of two such canonical representations is established by Theorems 3 and 4, both of which have been known since the middle of the nineteenth century.

Theorem 3 (Hermite Normal Form): *Given an mth-order nonsingular integer matrix B, there exists an mth-order, unimodular integer matrix K such that* $(f_{ij}) = F = BK$ *with:*

1. $f_{ij} = 0$, *for all i and* $j > i$.
2. $f_{ii} > 0$, *for all i.*
3. $f_{ij} < 0$ *and* $|f_{ij}| < f_{ii}$, *for all i and* $j < i$.

The matrix F is called the Hermite Normal Form of B and is unique for a given B. The calculation and application of the Hermite Normal Form to the ILPC are discussed in Exercises 10–14. The canonical representation that will be pursued is given by

Theorem 4 (Smith Normal Form): *Given an mth-order nonsingular integer matrix B, there exist mth-order, unimodular integer matrices R and C such that* $(\delta_{ij}) = \Delta = RBC$, *where:*

1. Δ *is a diagonal matrix* $(\delta_{ij} = 0,\ i \neq j)$.
2. *The diagonal elements, denoted by* $\delta_1, \ldots, \delta_m$, *are positive integers.*
3. δ_i *is a divisor of* δ_{i+1}, $i = 1, \ldots, m - 1$.

The matrix Δ is called the Smith Normal Form of B. For a given B, Δ is unique, but R and C are not necessarily unique. Since

$$\det \Delta = |\det RBC| = |\det R|\,|\det B|\,|\det C| = |\det B| = D$$

it follows that

$$\prod_{i=1}^{m} \delta_i = D$$

In Section 7.3, it will be shown that an ILPC can be stated as a group knapsack problem over a direct sum group of order D. By using Smith Normal Form, the unique canonical representation of the direct sum group given in Section 6.6 will be obtained. In Section 7.4 an algorithm for computing Smith Normal Form will be given.

7.3. GROUP KNAPSACK REPRESENTATION OF AN ILPC

Letting $\hat{B} = \Delta$ (Smith Normal Form of B) in (24) yields

$$\Delta y = R(b - Nx_N), \qquad y \text{ integer} \tag{29}$$

From Section 7.2, it follows that (29) is equivalent to

$$Bx_B + Nx_N = b, \qquad x_B \text{ integer}$$

where $\Delta = RBC$ and $y = C^{-1}x_B$. Therefore, the ILPC (13) can be stated as

$$\begin{aligned} \min z_0 &= (c_B B^{-1} N - c_N)x_N \\ \Delta y &= R(b - Nx_N) \\ &y \text{ integer} \\ &x_N \geq 0 \text{ integer} \end{aligned} \tag{30}$$

Denote the ith row of R by R_i, $i = 1, \ldots, m$. Then the ith row of

$$\Delta y = R(b - Nx_N)$$

is

$$\delta_i y_i = R_i(b - Nx_N) \tag{31}$$

Since for x_N integer, the right-hand side of (31) is an integer, there exists an integer y_i satisfying (31) if and only if

$$R_i(b - Nx_N) \equiv 0 \pmod{\delta_i}$$

or equivalently

$$R_i N x_N \equiv R_i b \pmod{\delta_i}, \qquad i = 1, \ldots, m \tag{32}$$

Note that if $\delta_i = 1$, (32) is superfluous, in the sense that it is satisfied by any integer vector x_N.

If $\delta_1 = \cdots = \delta_{k-1} = 1$, $\delta_k > 1$, then from (32), (30) can be stated as

$$\begin{aligned} \min z_0 &= (c_B B^{-1} N - c_N) x_N \\ R_i N x_N &\equiv R_i b \pmod{\delta_i}, \qquad i = k, \ldots, m \\ x_N &\geq 0 \text{ integer} \end{aligned} \tag{33}$$

Note that $D > 1$ implies that $\delta_m > 1$.

Suppose that $k = m$ so that there is exactly one constraint in (33). Let $x_N = (x_1, \ldots, x_r)$, $c_B B^{-1} a_j - c_j = y_{0j}$, and

$$\alpha_j = R_m a_j \pmod{\delta_m}, \; \alpha_0 = R_m b \pmod{\delta_m}$$

Then (33) is

$$\begin{aligned} \min z_0 &= \sum_{j=1}^{r} y_{0j} x_j \\ &\sum_{j=1}^{r} \alpha_j x_j \equiv \alpha_0 \pmod{\delta_m} \\ x_j &\geq 0 \text{ integer}, \quad j = 1, \ldots, r \end{aligned} \tag{34}$$

As shown in Section 6.10 [see (39)–(41)],

$$\sum_{j=1}^{r} \alpha_j x_j \equiv \alpha_0 \pmod{\delta_m}$$

is a group equation over $G(\delta_m)$ and (34) is a group knapsack problem over $G(\delta_m)$. The objective coefficients y_{0j} can be transformed into integers by multiplying the objective function by D.

In the general case, $\alpha_{ij} = R_i a_j \pmod{\delta_i}$ and $\alpha_{i0} = R_i b \pmod{\delta_i}$. Then (33) can be stated as

$$\begin{aligned} \min z_0 &= \sum_{j=1}^{r} y_{0j} x_j \\ &\sum_{j=1}^{r} \alpha_{ij} x_j \equiv \alpha_{i0} \pmod{\delta_i}, \qquad i = k, \ldots, m \\ x_j &\geq 0 \text{ integer}, \quad j = 1, \ldots, r \end{aligned} \tag{35}$$

The congruences of (35) are equivalent to a group equation over $G(\delta_k, \ldots, \delta_m)$ and (35) is a group knapsack problem. In particular, represent α_{ij} by the group element $g_{\alpha_{ij}}$ in $G(\delta_i)$. Denote the elements of the group $G(\delta_k, \ldots, \delta_m)$ by $g_{i_k, \ldots, i_m}$, where $0 \leq i_l < \delta_l$, $l = k, \ldots, m$. Therefore, $g_{\alpha_{kj}, \ldots, \alpha_{mj}}$ is an element of the group $G(\delta_k, \ldots, \delta_m)$.

Example

For the example of Section 7.2, the representation (34) is

$$\begin{aligned} \min z_0 &= \frac{x_3}{2} + \frac{x_5}{4} \\ 2x_3 + 3x_5 &\equiv 1 \pmod 4 \\ x_3, x_5 &\geq 0 \text{ integer} \end{aligned} \tag{36}$$

as can be seen from (28). The group knapsack problem corresponding to (36) is

$$\begin{aligned} &\min 2t_2 + t_3 \\ &t_2 g_2 \oplus t_3 g_3 = g_1 \\ &t_2, t_3 \geq 0 \text{ integer} \end{aligned} \tag{37}$$

over the group $G(4)$, where t_2 corresponds to x_3 and t_3 corresponds to x_5. By inspection, (37) has two optimal solutions, $t_2 = t_3 = 1$, and $t_2 = 0$, $t_3 = 3$. Thus (36) has two corresponding optimal solutions, $x_3 = x_5 = 1$, and $x_3 = 0$, $x_5 = 3$. Now substituting the optimal solutions of (36) into

$$x_B = B^{-1}(b - Nx_N)$$

yields the two corresponding optimal ILPC solutions

$$(x_1, \ldots, x_5) = (3, 1, 1, 2, 1) \text{ and } (2, 3, 0, -1, 3)$$

The former is feasible and thus is optimal to the ILP, but the latter is infeasible.

Example

To give an example for which the group is not cyclic, consider

$$\begin{aligned} \max x_0 = \qquad\qquad & - x_3 - 2x_4 \\ 2x_1 + 4x_2 + x_3 \qquad &= 12 \\ 12x_1 + 8x_2 \qquad + x_4 &= 60 \\ x_1, \ldots, x_4 \geq 0 & \text{ integer} \end{aligned} \tag{38}$$

The optimal LP solution is $(x_1, \ldots, x_4) = (\frac{9}{2}, \frac{3}{4}, 0, 0)$. Therefore, the optimal LP basis is

$$B = \begin{bmatrix} 2 & 4 \\ 12 & 8 \end{bmatrix}$$

To diagonalize B into Smith Normal Form, let

$$R = \begin{bmatrix} 1 & 0 \\ 6 & -1 \end{bmatrix}, C = \begin{bmatrix} 1 & -2 \\ 0 & 1 \end{bmatrix}$$

with the result that

$$RBC = \begin{bmatrix} 2 & 0 \\ 0 & 16 \end{bmatrix}$$

Thus $\delta_1 = 2$, $\delta_2 = 16$, and $D = 32$. Note that

$$N = \begin{bmatrix} 1 & 0 \\ 0 & 1 \end{bmatrix}, RN = R, Rb = \begin{bmatrix} 12 \\ 12 \end{bmatrix}$$

Therefore the ILPC associated with the optimal LP basis of (38) is

$$\begin{aligned} \min z_0 &= x_3 + 2x_4 \\ x_3 &\equiv 12 \pmod{2} \\ 6x_3 - x_4 &\equiv 12 \pmod{16} \\ x_3, x_4 &\geq 0 \text{ integer} \end{aligned}$$

or

$$\begin{aligned} \min z_0 &= x_3 + 2x_4 \\ x_3 &\equiv 0 \pmod{2} \\ 6x_3 + 15x_4 &\equiv 12 \pmod{16} \\ x_3, x_4 &\geq 0 \text{ integer} \end{aligned} \tag{39}$$

We can represent (39) as a group knapsack problem over the group $G(2, 16)$. In particular the coefficients of x_3 and x_4 correspond to $g_{1,6}$ and $g_{0,15}$ respectively. Thus with t_1 and t_2 corresponding to x_3 and x_4 respectively, the group knapsack problem is

$$\begin{aligned} &\min t_1 + 2t_2 \\ &t_1 g_{1,6} \oplus t_2 g_{0,15} = g_{0,12} \\ &t_1, t_2 \geq 0 \text{ integer} \end{aligned} \tag{40}$$

The group equation of (40) implies that t_1 is even and since $2g_{1,6} = g_{0,12}$, it is obvious that an optimal solution to (40) is $t_1 = 2$, $t_2 = 0$. This yields an optimal solution $x_3 = 2$, $x_4 = 0$ in (39), and an optimal ILPC solution $x_1 = 5$, $x_2 = 0$, $x_3 = 2$, $x_4 = 0$, which is feasible to the ILP and therefore optimal.

To obtain the group knapsack problem, the basis matrix B has been diagonalized into Smith Normal Form. However, it is clear that any unimodular R and C such that $RBC = \Delta'$, where Δ' is a diagonal matrix with positive integer diagonal elements $(\delta'_1, \ldots, \delta'_m)$ will yield a group knapsack problem. Smith Normal Form is preferred for computation because it yields the simplest representation of the group. In particular, from Smith Normal Form we obtain the canonical representation $G(\delta_k, \ldots, \delta_m)$, where $\delta_1 = \cdots =$

$\delta_{k-1} = 1$ and δ_p divides δ_{p+1}, $p = k, \ldots, m - 1$. Any other isomorphic representation $G(\delta_t', \ldots, \delta_m')$, where $\delta_1' = \cdots = \delta_{t-1}' = 1$, must have $t \leq k$ (see Section 6.6). Since the addition of two elements in a direct sum group that is the sum of q cyclic groups involves q modulo additions, the amount of computation required for the addition of group elements is minimized by using the canonical representation determined from Smith Normal Form.

7.4. CALCULATION OF SMITH NORMAL FORM

The calculation of R and C such that $RBC = \Delta$ involves applying the following elementary row and column operations to B:

1. Interchanging rows (columns) p and q.
2. Adding an integer multiple, say θ, of row (column) p to row (column) q.
3. Multiplying a row or column p by -1.

The first elementary row (column) operation corresponds to multiplying B on the left (right) by the mth-order matrix $R\,(C)$, where $R = C$ and $r_{pq} = r_{qp} = 1$, $r_{ii} = 1$, $i \neq p, q$, $r_{ij} = 0$ otherwise. This operation does not change the value of $|\det B| = D$, and since

$$|\det (RB)| = |\det R| \cdot D$$

$|\det R| = 1$ and $R\,(C)$ is unimodular.

The second elementary row (column) operation corresponds to multiplying B on the left (right) by the mth-order matrix $R\,(C)$, where $C = R^T$ and $r_{ii} = 1$, $i = 1, \ldots, m$, $r_{qp} = \theta$, and $r_{ij} = 0$ otherwise. Again, since the operation does not change D, $R\,(C)$ is unimodular.

The third elementary row (column) operation corresponds to multiplying B on the left (right) by the mth-order unimodular matrix $R\,(C)$ where $r_{ij} = 0$, $i \neq j$, $r_{ii} = 1$, $i \neq p$, $r_{pp} = -1$, and $C = R$.

Suppose that B is transformed into Δ using s elementary row operations and t elementary column operations. Thus

$$\Delta = R_s \cdots R_1 B C_1 \cdots C_t = RBC \tag{41}$$

where $R = R_s \cdots R_1$ and $C = C_1 \cdots C_t$. Because

$$|\det R| = \prod_{i=1}^{s} |\det R_i| = 1, \qquad |\det C| = \prod_{i=1}^{t} |\det C_i| = 1$$

R and C are unimodular. Since $R = R_s \cdots R_1 I$ and $C = IC_1 \cdots C_t$, applying the respective elementary row (column) operations to I yields R (C). Although C is not needed in the formulation of (33), in manual calculations it can be used to check (41).

Outline of an Algorithm for Smith Normal Form

The algorithm involves a systematic application of the operations (1)–(3) and is closely related to the Euclidean algorithm for finding the greatest common divisor of a set of integers (see Exercise 7). Beginning with the northwest corner of the matrix, successively larger square matrices are put into Smith Normal Form.

Considering the general step, suppose that B has been reduced to

$$\bar{B} = \left[\begin{array}{ccc|ccc} \delta_1 & & 0 & & & \\ & \ddots & & & 0 & \\ 0 & & \delta_{m-k} & & & \\ \hline & & & \alpha_{11} & \cdots & \alpha_{1k} \\ & 0 & & \vdots & & \vdots \\ & & & \alpha_{k1} & \cdots & \alpha_{kk} \end{array}\right] = \begin{bmatrix} \bar{B}_{11} & \bar{B}_{12} \\ \bar{B}_{21} & \bar{B}_{22} \end{bmatrix}$$

where δ_i divides δ_{i+1}, $i = 1, \ldots, m - k - 1$, and δ_{m-k} divides all of the elements of $\bar{B}_{22}$. At the beginning of the algorithm, $m - k = 0$ and $\bar{B}_{22} = B$.

In the basic step of the algorithm, the objective is to operate on $\bar{B}_{22}$ to obtain

$$\hat{B}_{22} = \left[\begin{array}{c|ccc} \delta_{m-k+1} & & 0 & \\ \hline & \hat{\alpha}_{22} & \cdots & \hat{\alpha}_{2k} \\ 0 & \vdots & & \vdots \\ & \hat{\alpha}_{k2} & \cdots & \hat{\alpha}_{kk} \end{array}\right] = \begin{bmatrix} \delta_{m-k+1} & 0 \\ 0 & \underline{B}_{22} \end{bmatrix}$$

with the property that δ_{m-k} divides δ_{m-k+1} and δ_{m-k+1} divides every element of $\underline{B}_{22}$.

Note that row or column operations on $\bar{B}$, involving only rows and/or columns $m - k + 1$ to m, cannot change $\bar{B}_{11}$, $\bar{B}_{12}$, or $\bar{B}_{21}$. Furthermore, when adding integer multiples of one row or column to another, it will still be the case that δ_{m-k} divides all of the elements of the transformed $\bar{B}_{22}$. This is true because if δ divides a and b, then δ divides $ma + nb$, where m and n are arbitrary integers.

Thus in performing row and column operations on $\bar{B}$, it is only necessary to consider $\bar{B}_{22}$. Let the rows and columns of $\bar{B}_{22}$ be denoted by r_i and c_i, respectively, $i = 1, \ldots, k$. To avoid cumbersome notation, we will continue to refer to the elements of the transformed matrix as α_{ij}.

Let p and q be such that

$$\alpha_{pq} = \min_{i,j} |\alpha_{ij}|, \qquad \alpha_{ij} \neq 0$$

Since a necessary condition for α_{11} to divide all other elements of $\bar{B}_{22}$ is $p = q = 1$, if $p \neq 1$ interchange r_p and r_1; if $q \neq 1$ interchange c_q and c_1. Now if $\alpha_{11} < 0$ (the element just put in the northwest corner), multiply r_1 by -1.

For $i, j = 1, \ldots k$, let

$$\gamma_{ij} = \alpha_{ij} \,(\mathrm{mod}\ \alpha_{11})$$

and

$$\alpha_{ij} = n_{ij}\alpha_{11} + \gamma_{ij}$$

If $\gamma_{i1} = 0$, $i = 2, \ldots, k$, and $\gamma_{1j} = 0$, $j = 2, \ldots, k$, let

$$\hat{r}_i = r_i - n_{i1}r_1, \qquad i = 2, \ldots, k$$
$$\hat{c}_j = c_j - n_{1j}c_1, \qquad j = 2, \ldots, k$$

to obtain the desired zeros in row 1 and column 1.

On the other hand, suppose that there exists i such that $\gamma_{i1} \neq 0$ or j such that $\gamma_{1j} \neq 0$. In particular, suppose that $\gamma_{p1} \neq 0$. Let

$$\hat{r}_p = r_p - n_{p1}r_1$$

so that $\alpha_{p1} < \alpha_{11}$. There is now at least one nonzero element in row p which is smaller in absolute value than α_{11}. Interchange rows and columns again to put the smallest element of absolute value in the northwest corner and then repeat the process just given. In a finite number of steps $\gamma_{i1} = 0$ and $\gamma_{1j} = 0$, $i, j = 2, \ldots, k$ is obtained, since in each step α_{11} is reduced by an integer. Note that $\alpha_{11} = 1$ is a sufficient condition for $\gamma_{i1} = 0$ and $\gamma_{1j} = 0$, $i, j = 2, \ldots, k$.

Before continuing with the general description, consider the example

$$B = \bar{B}_{22} = \begin{bmatrix} 5 & 10 & 5 \\ 8 & 5 & 9 \\ 5 & 0 & 5 \end{bmatrix} \tag{42}$$

Note that $\alpha_{11} = \min |\alpha_{ij}|$, $\alpha_{ij} \neq 0$. Since $\gamma_{21} = 3$ and $n_{21} = 1$, let $\hat{r}_2 = r_2 - r_1$ yielding

$$\bar{B}_{22} = \begin{bmatrix} 5 & 10 & 5 \\ 3 & -5 & 4 \\ 5 & 0 & 5 \end{bmatrix}$$

The nonzero element of smallest absolute value is $\alpha_{21} = 3$, so that r_1 and r_2 are interchanged with the result that

$$\bar{B}_{22} = \begin{bmatrix} 3 & -5 & 4 \\ 5 & 10 & 5 \\ 5 & 0 & 5 \end{bmatrix}$$

Now $\gamma_{21} = 2$ and $n_{21} = 1$. Thus let $\hat{r}_2 = r_2 - r_1$, which yields

$$\bar{B}_{22} = \begin{bmatrix} 3 & -5 & 4 \\ 2 & 15 & 1 \\ 5 & 0 & 5 \end{bmatrix}$$

Since $\alpha_{23} = \min |\alpha_{ij}|$, $\alpha_{ij} \neq 0$, interchange r_2 with r_1 and then c_3 with c_1 to obtain

$$\bar{B}_{22} = \begin{bmatrix} 1 & 15 & 2 \\ 4 & -5 & 3 \\ 5 & 0 & 5 \end{bmatrix}$$

Since $\alpha_{11} = 1$, $\gamma_{1j} = \gamma_{i1} = 0$, $i, j = 2, 3$. Note that $n_{21} = 4$ and $n_{31} = 5$ so that

$$\hat{r}_2 = r_2 - 4r_1$$
$$\hat{r}_3 = r_3 - 5r_1$$

yielding

$$\bar{B}_{22} = \begin{bmatrix} 1 & 15 & 2 \\ 0 & -65 & -5 \\ 0 & -75 & -5 \end{bmatrix}$$

Since $n_{12} = 15$ and $n_{13} = 2$, let

$$\hat{c}_2 = c_2 - 15c_1$$
$$\hat{c}_3 = c_3 - 2c_1$$

yielding the desired result

$$\bar{B}_{22} = \begin{bmatrix} 1 & 0 & 0 \\ 0 & -65 & -5 \\ 0 & -75 & -5 \end{bmatrix} = \hat{B}_{22} \tag{43}$$

Assume that $\gamma_{i1} = \gamma_{1j} = 0$, $i, j = 2, \ldots, k$. Now if $\gamma_{ij} = 0$, $i, j = 2, \ldots, k$, $\hat{B}_{22}$ has been obtained, which is the case in (43). Suppose, however, that $\gamma_{ij} \neq 0$ exists. Let

$$\alpha_{pq} = \min |\alpha_{ij}|, \qquad \alpha_{ij} \neq 0$$

If $p \neq 1$, again interchange r_p and r_1, and if $q \neq 1$, interchange c_q and c_1. Repeat the process given above to obtain $\gamma_{i1} = \gamma_{1j} = 0, i, j = 2, \ldots, k$.

If $\alpha_{pq} = \alpha_{11}$ and $\gamma_{st} \neq 0$ for $s, t > 1$, let

$$\hat{r}_s = r_s + n_{st} r_1$$

and

$$\hat{c}_t = c_t - c_1$$

yielding $\alpha_{st} = \gamma_{st} < \alpha_{11}$. There is now at least one nonzero element in the matrix smaller in absolute value than α_{11}. Put the smallest element in the northwest corner and then repeat the whole process. Each time it is repeated, α_{11} is reduced by at least one and $\alpha_{11} = 1$ is a sufficient condition for $\gamma_{ij} = 0$, for all i and j. Thus in a finite number of steps $\hat{B}_{22}$ is obtained.

In (43), $\bar{B}_{22}$ has the property that $\gamma_{ij} = 0, i, j = 2, 3$, so that the whole process would now be repeated on the submatrix

$$\begin{bmatrix} -65 & -5 \\ -75 & -5 \end{bmatrix}$$

To illustrate the case, $\alpha_{pq} = \alpha_{11}$ and $\gamma_{st} > 0$ for some $s, t > 1$, consider

$$\bar{B}_{22} = \begin{bmatrix} 2 & 0 & 0 \\ 0 & 3 & 8 \\ 0 & 0 & 8 \end{bmatrix}$$

Now $\gamma_{22} = 1$ and $n_{22} = 1$. Let $\hat{r}_2 = r_2 + r_1$, yielding

$$\bar{B}_{22} = \begin{bmatrix} 2 & 0 & 0 \\ 2 & 3 & 8 \\ 0 & 0 & 8 \end{bmatrix}$$

and then $\hat{c}_2 = c_2 - c_1$ gives

$$\bar{B}_{22} = \begin{bmatrix} 2 & -2 & 0 \\ 2 & 1 & 8 \\ 0 & 0 & 8 \end{bmatrix}$$

Interchanging rows and columns, we obtain

$$\bar{B}_{22} = \begin{bmatrix} 1 & 2 & 8 \\ -2 & 2 & 0 \\ 0 & 0 & 8 \end{bmatrix}$$

Now let $\hat{r}_2 = r_2 + 2r_1$, yielding

$$\bar{B}_{22} = \begin{bmatrix} 1 & 2 & 8 \\ 0 & 6 & 16 \\ 0 & 0 & 8 \end{bmatrix}$$

Then $\hat{c}_2 = c_2 - 2c_1$, $\hat{c}_3 = c_3 - 8c_1$ gives the desired result,

$$\bar{B}_{22} = \begin{bmatrix} 1 & 0 & 0 \\ 0 & 6 & 16 \\ 0 & 0 & 8 \end{bmatrix} = \hat{B}_{22}$$

Table I

R			B			C			Comments
1	0	0	5	10	5	1	0	0	Step 1: No work.
0	1	0	8	5	9	0	1	0	Step 2: $\gamma_{21} \neq 0$.
0	0	1	5	0	5	0	0	1	Step 3: $\hat{r}_2 = r_2 - r_1$.
1	0	0	5	10	5				Step 1: Interchange r_1
−1	1	0	3	−5	4				and r_2.
0	0	1	5	0	5				
−1	1	0	3	−5	4				Step 2: $\gamma_{21} \neq 0$.
1	0	0	5	10	5				Step 3: $\hat{r}_2 = r_2 - r_1$.
0	0	1	5	0	5				
−1	1	0	3	−5	4				Step 1: Interchange r_2 and
2	−1	0	2	15	1				r_1, and then c_3 and c_1.
0	0	1	5	0	5				
2	−1	0	1	15	2	0	0	1	Step 2:
−1	1	0	4	−5	3	0	1	0	$\hat{r}_2 = r_2 - 4r_1$
0	0	1	5	0	5	1	0	0	$\hat{r}_3 = r_3 - 5r_1$
									$\hat{c}_2 = c_2 - 15c_1$
									$\hat{c}_3 = c_3 - 2c_1$.
2	−1	0	1	0	0	0	0	1	Step 1: Interchange c_2 and c_3
−9	5	0	0	−65	−5	0	1	0	and multiply c_2 by −1.
−10	5	1	0	−75	−5	1	−15	−2	
			1	0	0	0	−1	0	Step 2:
			0	5	−65	0	0	1	$\hat{r}_3 = r_3 - r_2$
			0	5	−75	1	2	−15	$\hat{c}_3 = c_3 + 13c_2$.
2	−1	0	1	0	0	0	−1	−13	Step 1: $\hat{c}_3 = -c_3$.
−9	5	0	0	5	0	0	0	1	
−1	0	1	0	0	−10	1	2	11	
			1	0	0	0	−1	13	Complete.
			0	5	0	0	0	−1	
			0	0	10	1	2	−11	

The Algorithm

The stepwise procedure for transforming $\bar{B}_{22}$ into $\hat{B}_{22}$ is summarized below. The procedure must be applied m times to obtain the complete diagonalization.

STEP 1: Let $\alpha_{pq} = \min_{i,j} |\alpha_{ij}|$, $\alpha_{ij} \neq 0$. If $p \neq 1$, interchange r_p and r_1, and if $q \neq 1$, interchange c_q with c_1. If $\alpha_{11} < 0$, let $\hat{r}_1 = -r_1$ (or $\hat{c}_1 = -c_1$). Go to Step 2.

STEP 2: Let $\alpha_{ij} = n_{ij}\alpha_{11} + \gamma_{ij}$, $0 \leq \gamma_{ij} < \alpha_{11}$. If $\gamma_{i1} = \gamma_{1j} = 0$, $i, j = 2, \ldots, k$, let

$$\hat{r}_i = r_i - n_{i1}r_1, \qquad i = 2, \ldots, k$$

and then

$$\hat{c}_j = c_j - n_{1j}c_1, \qquad j = 2, \ldots, k$$

and go to Step 4. Otherwise go to Step 3.

STEP 3: Choose any p (or q) such that $\gamma_{p1} \neq 0$ (or $\gamma_{1q} \neq 0$). Let $\hat{r}_p = r_p - n_{p1}r_1$ (or $\hat{c}_q = c_q - n_{1q}c_1$). Return to Step 1.

STEP 4: Is $\alpha_{11} = \min_{i,j} |\alpha_{ij}|$, $\alpha_{ij} \neq 0$? If so, go to Step 5; if not, return to Step 1.

STEP 5: Is $\gamma_{ij} = 0$, $i, j = 2, \ldots, k$? If so, terminate; if not, go to Step 6.

STEP 6: Choose any $\gamma_{st} \neq 0$. Let $\hat{r}_s = r_s + n_{st}r_1$ and then $\hat{c}_t = c_t - c_1$. Return to Step 1.

Example

The details of putting the matrix given by (42) into Smith Normal Form are given in Table 1.

7.5. A SUFFICIENT CONDITION FOR AN ILPC TO SOLVE AN ILP

Our objective is to get an upper bound on the magnitude of Nx_N^*, where x_N^* is an optimal solution to the ILPC (13). Then, given an optimal basis B, we obtain a sufficient condition for

$$x_B^* = B^{-1}(b - Nx_N^*) \geq 0 \tag{44}$$

If (44) holds, (x_B^*, x_N^*) solves the ILP (8).

Geometric Interpretation

Before deriving the upper bound, some geometric interpretations are given. If B is an optimal basis, then b is a nonnegative linear combination of the columns of B and is therefore contained in the cone K_B generated by the columns of B (see Figure 4).

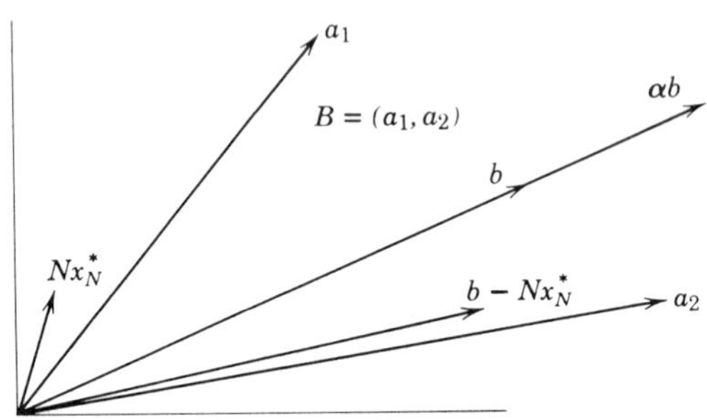

Figure 4

Thus (44) is satisfied if and only if $b - Nx_N^*$ is also contained in K_B, as shown in Figure 4. Assume that every component of $B^{-1}b$ is positive so that b is not on the boundary of K_B. Consider αb for $\alpha > 1$. As α increases, αb moves farther away from the boundary. Thus, for fixed Nx_N^*, there exists sufficiently large α such that $B^{-1}(\alpha b - Nx_N^*) \geq 0$. This explains the asymptotic relationship between an ILPC and an ILP.

The sufficient condition will be based on an upper bound on the Euclidean length of Nx_N^*. Let

$$\|y\| = \left(\sum_{j=1}^{m} y_j^2\right)^{1/2}$$

be the *Euclidean length* of the vector y and

$$K_B(d) = \{y \mid y \in K_B \text{ and for every } z \text{ on the boundary of } K_B,\ \|y - z\| \geq d\}$$

The cone K_B and the translated cone $K_B(d)$ are shown in Figure 5.

Suppose that $b \in K_B(d)$ and $\|Nx_N^*\| \leq d$, then (44) follows. Furthermore, if b is not on the boundary of K_B, for any d there exists sufficiently large α such that $\alpha b \in K_B(d)$.

An Upper Bound on $\|Nx_N^*\|$

An upper bound on $\|Nx_N^*\|$, depending only on N and D, will be obtained. The bound is mainly of theoretical interest, since it is frequently very loose.

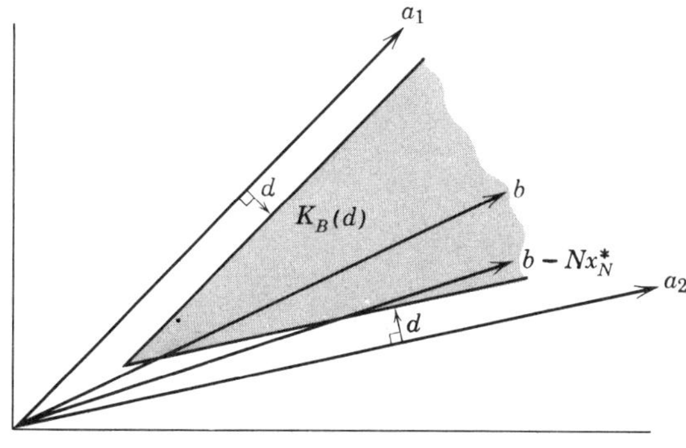

Figure 5

The upper bound is derived from a bound on the variables in the corresponding group knapsack problem. Consider

$$\begin{aligned} &\min \sum_{j \in Q} d_j t_j \\ &\bigoplus_{j \in Q} t_j g_j = g^* \\ &t_j \geq 0 \text{ integer}, \qquad j \in Q \end{aligned} \tag{45}$$

over the group G. Recall from Section 6.7 that if (45) has a feasible solution, it has an optimal solution t^* with $t_j^* \leq |G| - 1$, for all $j \in Q$.

A better upper bound on the t_j's is given in

Theorem 5: *If* (45) *has a feasible solution, it has an optimal solution* t^* *such that* $\sum_{j \in Q} t_j^* \leq |G| - 1$.

PROOF: Let

$$U = \{t \mid t \text{ a feasible solution to (45), with } t_j \leq |G| - 1, j \in Q\}$$

and

$$V = \{t \mid t \text{ an extreme point of the convex hull of } U\}$$

Note that $V \subseteq U$ and from the linearity of the objective function, if (45) has an optimal solution, there exists an optimal solution $t^* \in V$. The theorem is proved by showing that

$$t \in V \Rightarrow \sum_{j \in Q} t_j \leq |G| - 1 \tag{46}$$

We will prove (46) by contradiction. Suppose that there exists $t' \in V$ such that

$$\sum_{j \in Q} t'_j \geq |G| \tag{47}$$

We will show that if t' satisfies (47), then it satisfies

$$\prod_{j \in Q} (1 + t'_j) \geq |G| + 1 \tag{48}$$

Consider

$$\begin{aligned} \min \prod_{j \in Q} (1 + t_j) & \\ \sum_{j \in Q} t_j \geq |G| & \\ t_j \geq 0 \text{ integer}, \qquad j \in Q & \end{aligned} \tag{49}$$

It can easily be demonstrated (see Exercise 17) that an optimal solution to (49) is given by $t_k^o = |G|$, $t_j^o = 0, j \in Q - \{k\}$, k arbitrary. This yields

$$\min \prod_{j \in Q} (1 + t_j) = |G| + 1$$

which establishes that (47) implies (48).

It will now be shown that (48) implies $t' \notin V$. Let

$$R(t') = \{s | 0 \leq s_j \leq t'_j, s_j \text{ integer}, j \in Q\}$$

Note that

$$|R(t')| = \prod_{j \in Q} (1 + t'_j) \geq |G| + 1$$

For any $s \in R(t')$, define

$$f(s) = \bigoplus_{j \in Q} s_j g_j$$

Now, since $|R(t')| > |G|$, there must be a group element g' and $s^1, s^2, s^1 \neq s^2$, such that

$$f(s^1) = f(s^2) = g' \tag{50}$$

Since t' is a feasible solution to (45)

$$f(t') = g^* \tag{51}$$

From (50) and (51) it follows that

$$f(t' + s^1 - s^2) = f(t' - s^1 + s^2) = g^*$$

Therefore $(t' + s^1 - s^2)$ and $(t' - s^1 + s^2)$ are feasible solutions to (45). But

$$t' = \frac{(t' + s^1 - s^2)}{2} + \frac{(t' - s^1 + s^2)}{2}$$

contradicting the assumption that $t' \in V$. ■

Since (45) corresponds to an ILPC, it follows from Theorem 5 that there exists an optimal ILPC solution $x_N^* = (x_1^*, \ldots, x_r^*)$ such that

$$\sum_{j=1}^{r} x_j^* \le |G| - 1 = D - 1 \tag{52}$$

Let $N = (a_1, \ldots, a_r)$ and thus

$$\|Nx_N^*\| = \left\| \sum_{j=1}^{r} a_j x_j^* \right\|$$

Let a_p be a column of N with greatest Euclidean length and $L_p = \|a_p\|$. Then

$$\|Nx_N^*\| \le \left\| a_p \sum_{j=1}^{n} x_j^* \right\| = L_p \sum_{j=1}^{n} x_j^* \tag{53}$$

Combining (52) and (53) yields the desired bound

$$\|Nx_N^*\| \le L_p(D - 1) \tag{54}$$

Thus if

$$b \in K_B(L_p(D - 1))$$

(44) holds. (See Exercise 3 for a numerical example.)

7.6. SOLVING AN ILP USING AN ILPC AND BRANCH AND BOUND

From (10) and (13) of Section 7.1, an ILP can be written as

$$\max z(x_N), \qquad x_N \in S_0 \tag{55}$$

where

$$z(x_N) = c_B B^{-1} b - (c_B B^{-1} N - c_N) x_N$$

and

$$S_0 = \{x_N | B^{-1} N x_N \equiv B^{-1} b \ (\text{mod } 1),\ B^{-1}(b - N x_N) \ge 0,\ x_N \ge 0 \text{ integer}\}$$

Let $x_N = (x_1, \ldots, x_r)$ and assume that B is an optimal LP basis.

The ILP (55) will be solved using the branch and bound concept of Section 4.2. The problem at vertex k has constraints that bound x_N from below by an integer vector $p^k = (p_1^k, \ldots, p_r^k) \ge 0$ and is defined as

$$\max z(x_N), \qquad x_N \in S_k \tag{56}$$

where

$$S_k = \{x_N | B^{-1} N x_N \equiv B^{-1} b \ (\text{mod } 1),\ B^{-1}(b - N x_N) \ge 0,\ x_N \ge p^k \text{ integer}\}$$

Bounding

To obtain an upper bound on the value of an optimal solution to (56), solve

$$z(x_N^o(k)) = \max z(x_N), \qquad x_N \in T_k \tag{57}$$

where

$$T_k = \{x_N | B^{-1}Nx_N \equiv B^{-1}b \pmod 1),\ x_N \geq p^k \text{ integer}\}$$

For $k = 0$, $p^0 = 0$, and (57) is equivalent to the ILPC (13). It was shown in Section 7.3 that (13) could be solved as a group knapsack problem.

It will now be shown that (57) is a group knapsack problem for all k. In fact, for different k, the only difference among the group knapsack problems is the element on the right-hand side of the group equation.

Suppose that for $k = 0$, the group knapsack problem is

$$\begin{aligned} \min \sum_{j=1}^{r} d_j t_j \\ \bigoplus_{j=1}^{r} t_j g_j = g^* \\ t_j \geq 0 \text{ integer}, \qquad j = 1, \ldots, r \end{aligned} \tag{58}$$

In (58), it is assumed that t_j corresponds to x_j, $j = 1, \ldots, r$. Thus it may be the case that for $j \neq j'$, $g_j = g_{j'}$. Although such duplicate group elements would be omitted when computing an optimal solution to (58), in deriving the group equation for $k \neq 0$, it is necessary to keep them.

For arbitrary k in (57), $x_1^o(k) = t_1^o, \ldots, x_r^o(k) = t_r^o$ is an optimal solution to (57) if and only if $(t_1^o, \ldots, t_r^o)$ is an optimal solution to

$$\begin{aligned} \min \sum_{j=1}^{r} d_j t_j \\ \bigoplus_{j=1}^{r} t_j g_j = g^* \\ t_j \geq p_j^k \text{ integer}, \qquad j = 1, \ldots, r \end{aligned} \tag{59}$$

Let $v_j = t_j - p_j^k$. Then (59) can be written as

$$\begin{aligned} \min \sum_{j=1}^{r} d_j v_j + \sum_{j=1}^{r} d_j p_j^k \\ \bigoplus_{j=1}^{r} (v_j + p_j^k) g_j = g^* \\ v_j \geq 0 \text{ integer}, \qquad j = 1, \ldots, r \end{aligned} \tag{60}$$

Let

$$\bigoplus_{j=1}^{r} p_j^k g_j = \bar{g}_k$$

and

$$g^* \oplus -\bar{g}_k = g_k^*$$

Then, dropping the constant in the objective function, we can write (60) as the group knapsack problem

$$\min \sum_{j=1}^{r} d_j v_j$$

$$\bigoplus_{j=1}^{r} v_j g_j = g_k^* \tag{61}$$

$$v_j \geq 0 \text{ integer}, \qquad j = 1, \ldots, r$$

When computing an optimal solution to (61), duplicate group elements are dropped, as explained in Section 6.10. It follows that $x_1^o(k) = v_1^o + p_1^k, \ldots, x_r^o(k) = v_r^o + p_r^k$ is an optimal solution to (57) if and only if $(v_1^o, \ldots, v_r^o)$ is an optimal solution to (61).

The only variation in (61), for different k, is g_k^*. Recall that the algorithms given in Sections 6.8 and 6.9 for the group knapsack problem yield a parametric optimal solution as a function of the right-hand side of the group equation. Thus solving one group knapsack problem yields the information needed to find an optimal solution for all right-hand sides, and consequently for all k in (57).

Separation

If vertex k is not fathomed, consider the separation

$$S_k = \bigcup_{j=0}^{r} S_{k(j)}$$

where

$$S_{k(0)} = S_k \cap \{x_N | x_N = p^k\}$$

and

$$S_{k(j)} = S_k \cap \{x_N | x_j \geq p_j^k + 1\}, \qquad j = 1, \ldots, r$$

This separation is clearly not a partition. However, we will show that it is not necessary to consider the problems at vertices $k(j), j = 0, \ldots, j(k) - 1$, where

$$j(k) = \begin{cases} \max\limits_{j \in R_k} j & R_k \neq \varnothing \\ 1 & R_k = \varnothing \end{cases}$$

and $R_k = \{j | p_j^k > 0\}$. There are two cases

($j = 0$): Either $S_{k(0)} = \varnothing$ or $S_{k(0)} = \{p^k\}$. If $p_k \in S_{k(0)}$, then $p_k \in S_k \subseteq T_k$. Since, by virtue of dual feasibility, the objective coefficients in (57) are non-positive, $p_k \in T_k$ implies that $x_N = p_k$ is an optimal solution to (57) at vertex k. Thus $S_{k(0)} = \{p_k\}$ implies that vertex k would have been fathomed. Clearly, there is no point in branching to vertex $k(0)$ if $S_{k(0)} = \varnothing$.

($j = 1, \ldots, j(k) - 1$): Consider j^*, $1 \leq j^* \leq j(k) - 1$. Let k' be such that

$$p_j^{k'} = \begin{cases} p_j^k & \text{for } j \neq j^*, j(k) \\ p_j^k + 1 & \text{for } j = j^* \\ p_j^k - 1 & \text{for } j = j(k) \end{cases}$$

Then $j(k') \leq j(k)$ and $S_{k'(j(k))} = S_{k(j^*)}$. In other words, the problem at vertex $k(j^*)$ does not need to be considered, since the identical problem will, if necessary, be considered at vertex $k'(j(k))$. Thus, in the absence of fathoming, there is a unique path from vertex 0 to vertex k' with $p^{k'} = (p_1^{k'}, p_2^{k'}, \ldots, p_r^{k'})$. In particular, there is a unique path from vertex 0 to vertex k_1 with $p^{k_1} = (p_1^{k'}, 0, \ldots, 0)$; if $p_1^{k'} = 0$, $k_1 = 0$. Then there is a unique path from vertex k_1 to vertex k_2 with $p^{k_2} = (p_1^{k'}, p_2^{k'}. 0, \ldots, 0)$, and so forth.

A possible tree for a three-variable problem is illustrated in Figure 6. In separating S_3,

$$S_3 \cap \{x_N | x_1 \geq 1\} \qquad \text{and} \qquad S_3 \cap \{x_N | x_2 \geq 1\}$$

are not considered. However,

$$S_6 = S_3 \cap \{x_N | x_1 \geq 1\} = S_1 \cap \{x_N | x_3 \geq 1\}$$

and, since vertex 2 has been fathomed, $S_3 \cap \{x_N | x_2 \geq 1\}$ has been considered implicitly.

Branching

The rule given in Section 4.1 can be used. Note that if the optimal solution to (57) at vertex k, $x_N^0(k)$, is such that $x_N^0(k) \geq p^{k_1} > p^k$, it is not necessary to solve (57) at vertex k_1, since $x_N^0(k)$ is also an optimal solution to (57) at vertex k_1. For computer implementation it would be desirable to have a branching rule that does not generate vertices such as k_1 (see Exercise 15).

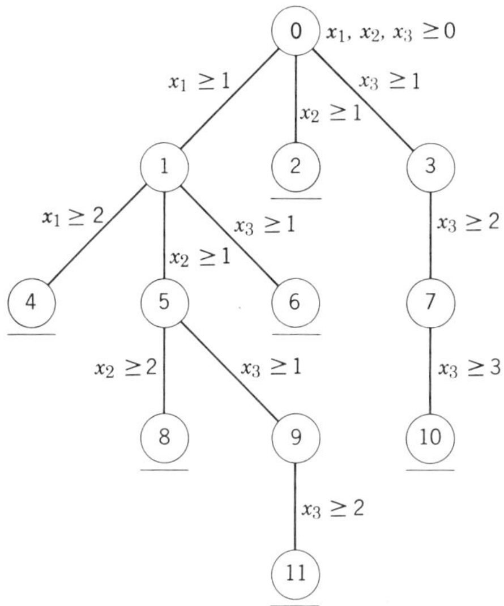

Figure 6

Termination

Suppose that the ILP (8) is bounded, which implies the existence of u_j such that $x_j \leq u_j, j = 1, \ldots, r$. Let

$$H_k = \{x_N | 0 \leq p_j^k \leq x_j \leq u_j, x_j \text{ integer}\}$$

Then $|H_0| = \prod_{j=1}^r (1 + u_j)$. If vertex k' is a successor of vertex k, there exists j^* such that $p_{j^*}^{k'} = p_{j^*}^k + 1$ and $p_j^{k'} = p_j^k$ for $j \neq j^*$. Thus

$$|H_{k'}| < |H_k|$$

Bounded enumeration can then be assured, if upper bounds are known, by keeping track of the lower bounds p^k and never creating a vertex k such that $p_j^k > u_j$ for any j. This implies that a vertex k having $|H_k| = 1$ has no successors.

Example

Consider the ILP

$$\begin{aligned} \max x_0 = 3x_1 &+ 2x_2 \\ x_1 + x_2 + x_3 \qquad\qquad &= 5 \\ -x_1 + x_2 \qquad + x_4 \qquad &= 0 \qquad (62) \\ 6x_1 + 2x_2 \qquad\qquad + x_5 &= 21 \\ x_1, \ldots, x_5 \geq 0 &\text{ integer} \end{aligned}$$

This problem is the same as the one given in Section 7.1, except that $c_1 = 3$ and $c_2 = 2$. The optimal solution to the LP corresponding to (62) is $x_0 = \frac{51}{4}$, $x_2 = \frac{9}{4}$, $x_4 = \frac{1}{2}$, $x_1 = \frac{11}{4}$, and $x_3 = x_5 = 0$.

From the solution to the LP, we obtain the ILPC

$$\max z_0 = \frac{51}{4} - \frac{x_5}{4} - \frac{3x_3}{2}$$

$$\frac{9}{4} + \frac{x_5}{4} - \frac{3x_3}{2} \equiv 0 \pmod 1$$

$$\frac{1}{2} - \frac{x_5}{2} + 2x_3 \equiv 0 \pmod 1$$

$$\frac{11}{4} - \frac{x_5}{4} + \frac{x_3}{2} \equiv 0 \pmod 1$$

$$x_3, x_5 \geq 0 \text{ integer}$$

Corresponding to the ILPC, the group problem (58) at vertex 0 is

$$\begin{aligned} &\min 6t_2 + t_3 \\ &t_2 g_2 \oplus t_3 g_3 = g_1 \\ &t_2, t_3 \geq 0 \text{ integer} \end{aligned} \tag{63}$$

over the group $G(4)$, where t_2 corresponds to x_3 and t_3 to x_5 [see (36) and (37)]. The optimal solution to (63) is $t_3 = 3$, $t_2 = 0$. The solution $x_5 = 3$, $x_3 = 0$, yields $x_2 = 3$, $x_4 = -1$, $x_1 = 2$, $x_0 = 12$, which is infeasible to (62).

Now branch to vertex 1 as in Figure 7. At vertex 1, the group problem is

$$\begin{aligned} &\min 6v_2 + v_3 \\ &v_2 g_2 \oplus v_3 g_3 = g_3 \\ &v_2, v_3 \geq 0 \text{ integer} \end{aligned} \tag{64}$$

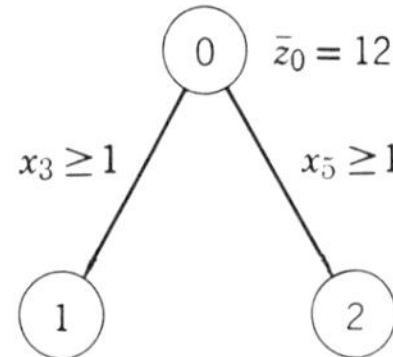

Figure 7

where $v_2 = t_2 - 1$, and $v_3 = t_3$. The optimal solution to (64) is $v_2 = 0$, $v_3 = 1$, yielding $x_3 = x_5 = 1$ and, in turn, the feasible solution to (62),

$$(x_0, x_1, \ldots, x_5) = (11, 3, 1, 1, 2, 1)$$

Thus vertex 1 is fathomed and $z_0 = 11$.

The optimal solution $(x_3, x_5) = (0, 3)$ at vertex 0 implies that the optimal solution at vertex 2 is also $(x_3, x_5) = (0, 3)$. Note that vertex 3 is the only successor of vertex 2, and vertex 4 is the only successor of vertex 3 (see Figure 8). For the reason just given, $(x_3, x_5) = (0, 3)$ is an optimal solution at vertices 3 and 4.

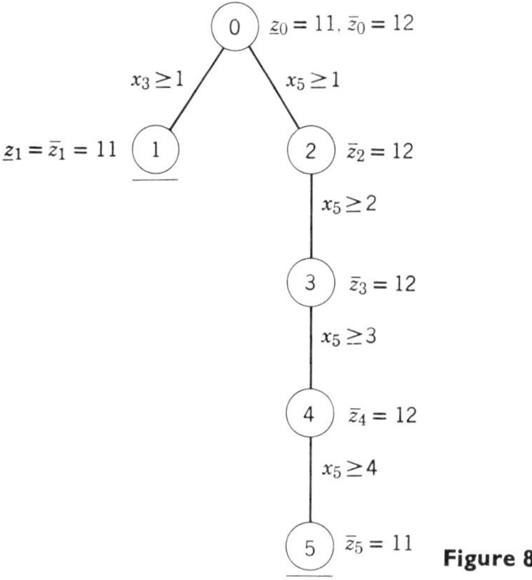

Figure 8

At vertex 5 the group problem is

$$\begin{aligned} \min\ & 6v_2 + v_3 \\ & v_2 g_2 \oplus v_3 g_3 = g_1 \\ & v_2, v_3 \geq 0 \text{ integer} \end{aligned} \tag{65}$$

where $v_2 = t_2$ and $v_3 = t_3 - 4$. The optimal solution to (65) is $v_2 = 0$, $v_3 = 3$, yielding $x_3 = 0$, $x_5 = 7$, and $x_0 = 11$. Thus $\bar{z}_5 = 11 = \underline{z}_0$ and vertex 5 is fathomed. This yields $(x_0, x_1, \ldots, x_5) = (11, 3, 1, 1, 2, 1)$, determined at vertex 1, as an optimal solution to (62).

7.7. PROPERTIES OF CORNER POLYHEDRA

We have seen that a bounded ILP can, in principle, be represented as an LP with constraints chosen as the inequalities that define the convex hull of feasible integer solutions. If these inequalities (sometimes called faces) could be easily obtained, all of the power of linear programming could be brought to bear on the solutions of ILP's. Unfortunately, such results are available only in special cases. For example, in Chapter 3, the faces for certain matching problems were given.

In this section, characterizations, in terms of extreme points and faces are given for the corner polyhedra introduced in Section 7.1. Since ILPC's generally are not models of real problems, one might think that results for corner polyhedra are only of theoretical interest. However a corner polyhedron contains the convex hull of feasible solutions to a corresponding ILP. Thus a face for a corner polyhedron is a valid cut for a corresponding ILP. In fact, from the definition of corner polyhedra, such cuts are the deepest that can be obtained by requiring integrality on all variables and nonnegativity on the nonbasic variables.

Faces and Extreme Points of Corner Polyhedra

Suppose that a polyhedron P in n-space is given by m' inequalities as

$$P = \{x | a_i x \leq b_i, i = 1, \ldots, m'\} \tag{66}$$

The inequality $a_k x \leq b_k$ in (66) is said to be *superfluous* if

$$\{x | a_i x \leq b_i, i = 1, \ldots, m', i \neq k\} = P$$

Suppose that all superfluous inequalities (if any) have been removed from (66) and the remaining ones are renumbered such that

$$P = \{x | a_i x \leq b_i, i = 1, \ldots, m \leq m'\} \tag{67}$$

The sets of points

$$F_i = \{x | a_i x = b_i\} \cap P, \qquad i = 1, \ldots, m$$

are generally called the *faces* (facets) of P. The term facet is a descriptive one if one thinks about a diamond. For convenience, and to follow the literature upon which this section is based, we will refer to the inequalities $a_i x \leq b_i$, $i = 1, \ldots, m$, as the faces of P. The connection between faces and extreme points is familiar from linear programming. In particular x^* is an extreme point of P given by (67) if and only if (x^*, s^*) is a basic solution to

$$a_i x + s_i = b_i, \qquad i = 1, \ldots, m \tag{68}$$

and satisfies $s^* \geq 0$.

Given B, N, and b in a dual feasible solution to an LP, the ILPC constraints are

$$\begin{aligned} B^{-1}Nx_N &\equiv B^{-1}b \,(\text{mod } 1) \\ x_N &\geq 0 \text{ integer} \end{aligned} \tag{69}$$

where $x_N = (x_1, \ldots, x_r)$. The corner polyhedron, denoted by $P(B, N, b)$, is the unbounded polyhedron that is the convex hull of feasible solutions to (69). Our objective is to obtain characterizations of the faces and extreme points of $P(B, N, b)$ for arbitrary B, N, and b.

Recall from Section 7.3 that (69) can be represented by a group knapsack constraint

$$\begin{aligned} \bigoplus_{i \in Q} t_i g_i &= g^* = g_\rho \\ t_i &\geq 0 \text{ integer}, i \in Q \end{aligned} \tag{70}$$

over a given finite abelian group G, where $|Q| \leq r$. Denote by $P(G, Q, \rho)$ the polyhedron that is the convex hull of feasible solutions to (70). Let

$$T_i = \{j | B^{-1}a_j \text{ correspond to } g_i\}, \qquad i \in Q$$

There is a simple relationship between $P(B, N, b)$ and $P(G, Q, \rho)$ which follows from the result that t^0 is a feasible solution to (70) if and only if x_N^0, satisfying

$$\sum_{j \in T_i} x_j^0 = t_i^0, \qquad i \in Q \tag{71}$$

is a feasible solution to (69).

Now, $P(G, Q, \rho)$ is a polyhedron in $|Q|$-dimensional space and $P(B, N, b)$ is in r-dimensional space. If $|T_i| = 1$ for all $i \in Q$, (71) reduces to $x_j^0 = t_i^0$ and the two polyhedra are identical. In general, it is more convenient to work with the lower dimensional polyhedron $P(G, Q, \rho)$ and to compute $P(B, N, b)$ from it. This computation involves nothing more than the change of variables given by (71). Thus if

$$\sum_{i \in Q} \alpha_{ki} t_i \geq \alpha_{k0}, \qquad k = 1, \ldots, m \tag{72}$$

are the faces of $P(G, Q, \rho)$, then

$$\sum_{i \in Q} \alpha_{ki} \left(\sum_{j \in T_i} x_j \right) \geq \alpha_{k0}, \qquad k = 1, \ldots, m \tag{73}$$

are the faces of $P(B, N, b)$. For all $j \in T_i$, every x_j in (73) has the same coefficient. Thus, geometrically, $P(B, N, b)$ is symmetric with respect to all x_j,

$j \in T_i$, and $P(G, Q, \rho)$ can be considered as a compact representation of $P(B, N, b)$ in which all redundant information has been deleted.

Because the coefficients for the x_j's, $j \in T_i$, are identical in (73), an extreme point of $P(B, N, b)$ must satisfy

$$x_j x_{j'} = 0, \qquad j, j' \in T_i, j \neq j' \tag{74}$$

Thus t° is an extreme point of $P(G, Q, \rho)$ if and only if x_N° satisfying (71) and (74) is an extreme point of $P(B, N, b)$.

Example

Consider the group knapsack constraints

$$\begin{aligned} t_1 g_1 \oplus t_2 g_2 &= g_3 \\ t_1, t_2 &\geq 0 \text{ integer} \end{aligned} \tag{75}$$

over the group $G(4)$. The corresponding polyhedron is $P(G(4), \{1, 2\}, 3)$. It will be shown later in this section that the faces of this polyhedron are given by

$$\begin{aligned} t_1 \qquad &\geq 1 \\ t_2 &\geq 0 \\ \frac{t_1}{3} + \frac{2t_2}{3} &\geq 1 \end{aligned} \tag{76}$$

The constraints (76) are shown in Figure 9. It is easily seen that the extreme points of $P(G(4), \{1, 2\}, 3)$ are $(t_1, t_2) = (3, 0)$ and $(1, 1)$.

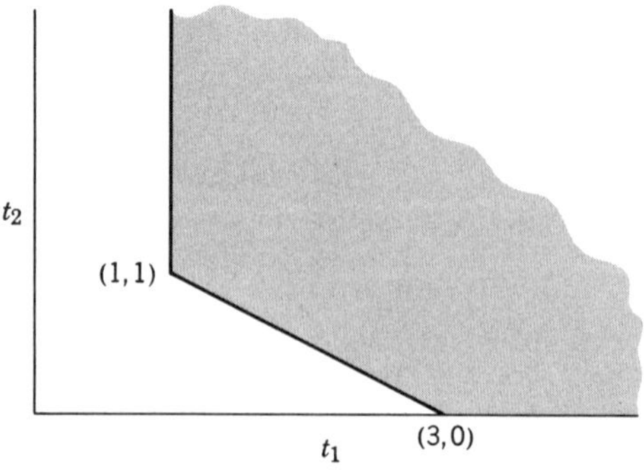

Figure 9

Suppose that the corresponding $P(B, N, b)$ is three-dimensional with $T_1 = \{1, 2\}$ and $T_2 = \{3\}$. Then, from (73) and (76), the faces of $P(B, N, b)$ are given by

$$\begin{aligned} x_1 + x_2 \qquad &\geq 1 \\ x_3 &\geq 0 \\ \frac{x_1}{3} + \frac{x_2}{3} + \frac{2x_3}{3} &\geq 1 \end{aligned} \tag{77}$$

The constraints (77) are shown in Figure 10. The extreme points are $(x_1, x_2, x_3) = (3, 0, 0), (0, 3, 0), (1, 0, 1)$, and $(0, 1, 1)$. Note that the trapezoid

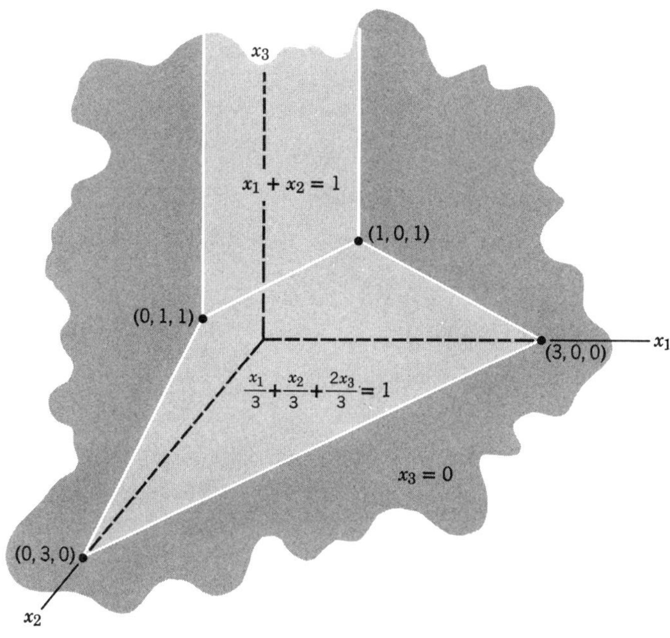

Figure 10

determined by the extreme points in Figure 10 corresponds to the line from (1, 1) to (3, 0) in Figure 9 and the plane parallel to the x_3-axis in Figure 10 corresponds to the line parallel to the t_2-axis in Figure 9.

Master Polyhedra

Although each (G, Q, ρ) combination generates a different polyhedron, for a given G and ρ, the polyhedra for all Q can be obtained from one of them called the *master polyhedron*. The master polyhedron for given G and ρ is the corner polyhedron $P(G, Q^*, \rho) = P(G, \rho)$, where Q^* is the index set corresponding to all elements of the group except the identity element.

First, the simple manipulations involved in obtaining $P(G, Q, \rho)$ from $P(G, \rho)$ for any $Q \subseteq Q^*$ will be given. Then, the main result of this section, the characterization of the faces of $P(G, \rho)$ for arbitrary G and ρ, will be presented.

Note that $P(G, \rho)$ is the convex hull of feasible solutions to

$$\begin{aligned} \bigoplus_{i \in Q^*} t_i g_i &= g_\rho \\ t_i \geq 0 \text{ integer}, &\ i \in Q^* \end{aligned} \tag{78}$$

Extended to $|G| - 1 = |Q^*|$-dimensional space, $P(G, Q, \rho)$ is the convex hull of feasible solutions to

$$\begin{aligned} \bigoplus_{i \in Q^*} t_i g_i &= g_\rho \\ t_i \geq 0 \text{ integer}, &\quad i \in Q \\ t_i = 0, &\quad i \in Q^* - Q \end{aligned} \tag{79}$$

From (78) and (79), it follows that in $|Q^*|$-dimensional space,

$$P(G, Q, \rho) = P(G, \rho) \cap \{t \,|\, t_i = 0,\ i \in Q^* - Q\} \tag{80}$$

Assume that the faces of $P(G, \rho)$ are

$$\sum_{i \in Q^*} \alpha_{ki} t_i \geq \alpha_{k0}, \qquad k = 1, \ldots, M \tag{81}$$

Then from (80), $P(G,Q,\rho)$ in $|Q^*|$-dimensional space is given by the constraints (81) and

$$t_i = 0, \qquad i \in Q^* - Q$$

The projecting of $P(G, Q, \rho)$ back into Q-space is algebraically equivalent to eliminating t_i, $i \in Q^* - Q$, from (81), yielding

$$\sum_{i \in Q} \alpha_{ki} t_i \geq \alpha_{k0}, \qquad k = 1, \ldots, M \tag{82}$$

Now, it may be the case that some constraints (say $m + 1, \ldots, M$) of (82) are superfluous. After eliminating them, the faces of $P(G, Q, \rho)$ are obtained as

$$\sum_{i \in Q} \alpha_{ki} t_i \geq \alpha_{k0}, \qquad k = 1, \ldots, m \tag{83}$$

The relationship between the extreme points of $P(G, Q, \rho)$ and $P(G, \rho)$ is given in Theorem 6. For notational convenience, index the elements of G so that the $|Q^*|$-dimensional vector $t = (u, v)$ is such that u is a $|Q|$-dimensional vector corresponding to the elements of the index set Q.

Theorem 6: *u^o is an extreme point of* (83) *if and only if t^o is an extreme point of* (81), *where $u_i^o = t_i^o, i = 1, \ldots, |Q|$, and $t_i^o = 0, i = |Q| + 1, \ldots, |Q^*|$.*

PROOF: Suppose that t^o is an extreme point of (81) satisfying the conditions of the theorem and u^o as given in the theorem is not an extreme point of (83). Since $t_i^o = 0$, $i = |Q^*| + 1, \ldots, |Q|$, and the constraints of (83) are a subset of the constraints of (82), u^o satisfies (83). Since u^o is not an extreme point of (83), let

$$u^o = \lambda \bar{u}^o + (1 - \lambda)\hat{u}^o, \qquad 0 < \lambda < 1$$

where $\bar{u}^o$ and $\hat{u}^o$ satisfy the constraints of (83). Since (83) is obtained from (82) by deleting superfluous constraints, $\bar{u}^o$ and $\hat{u}^o$ satisfy (82). Comparing (82) with (81), it is clear that $\bar{t}^o = (\bar{u}^o, 0)$ and $\hat{t}^o = (\hat{u}^o, 0)$ are solutions to (81). However,

$$t^o = \lambda \bar{t}^o + (1 - \lambda)\hat{t}^o$$

which contradicts the assumption that t^o is an extreme point of (81). The proof of the "only if" clause proceeds similarly by contradiction and is asked for in Exercise 22. ■

Example, Continued

It will be shown that the five faces of $P(G(4), 3)$ are

$$\begin{aligned} \frac{t_1}{3} + \frac{2t_2}{3} + t_3 &\geq 1 \\ t_1 \qquad\quad + t_3 &\geq 1 \\ t_1, t_2, t_3 &\geq 0 \end{aligned} \tag{84}$$

Thus the constraints (82) of $P(G(4), \{1, 2\}, 3)$ obtained by putting $t_3 = 0$ are

$$\begin{aligned} \frac{t_1}{3} + \frac{2t_2}{3} &\geq 1 \\ t_1 \qquad\quad &\geq 1 \\ t_1, t_2 &\geq 0 \end{aligned} \tag{85}$$

However, $t_1 \geq 0$ is superfluous in (85), so that the faces of $P(G(4), \{1, 2\}, 3)$ are

$$\begin{aligned} \frac{t_1}{3} + \frac{2t_2}{3} &\geq 1 \\ t_1 \qquad\quad &\geq 1 \\ t_2 &\geq 0 \end{aligned} \tag{86}$$

as in (76).

The extreme points of $P(G(4), 3)$ are $(t_1, t_2, t_3) = (3, 0, 0)$, $(1, 1, 0)$ and $(0, 0, 1)$. Thus the extreme points of $P(G(4), \{1, 2\}, 3)$ are, from Theorem 6, $(t_1, t_2) = (3, 0)$, $(1, 1)$, as shown in Figure 9. $P(G(4), 3)$ is shown in Figure 11. By comparing Figure 9 with the $t_3 = 0$ plane in Figure 11, the projection of the master polyhedron into lower dimensions can be seen geometrically.

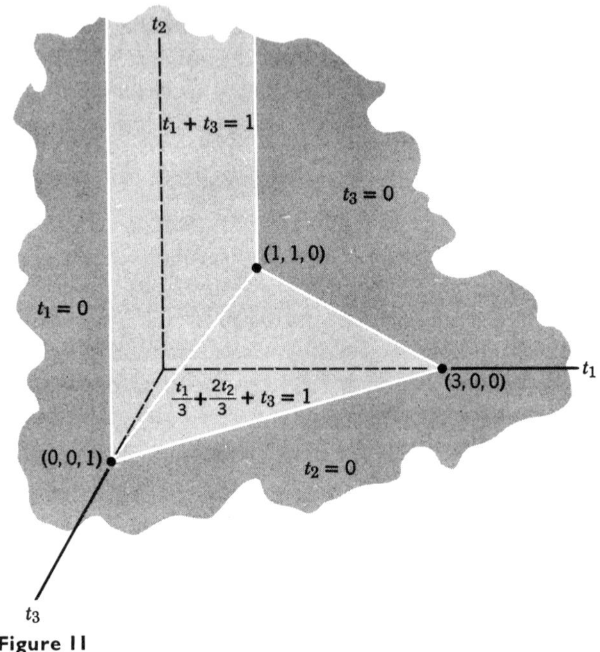

Figure 11

Faces of Master Polyhedra

The problem of finding extreme points and faces of corner polyhedra has now been simplified to finding faces of $P(G, \rho)$. The faces of $P(G, \rho)$ are characterized in Theorem 7. This theorem has an involved proof beyond the scope of our treatment.

Theorem 7: *All faces of $P(G, \rho)$ are of the form $t_i \geq 0$ or $\sum_{i \in Q^*} \alpha_i t_i \geq 1$ and*

1. *$t_i \geq 0$, $i = 1, \ldots, |Q^*|$ are faces if and only if $G \neq G(2)$.*
2. *$\sum_{i \in Q^*} \alpha_i t_i \geq 1$ is a face if and only if $(\alpha_1, \ldots, \alpha_{|Q^*|})$ corresponds to a basic feasible solution to the following system of equalities and inequalities:*

(a) $\alpha_\rho = 1$, *if* $g_\rho \neq g_0$ (*identity element of the group*).
(b) $\alpha_i + \alpha_{\rho(i)} = 1$, $i = 1, \ldots, |Q^*|$, $i \neq \rho$, *where* $g_{\rho(i)} = g_\rho \oplus - g_i$.
(c) $\alpha_i + \alpha_j - \alpha_k \geq 0$, $i, j = 1, \ldots, |Q^*|$, *where* $g_i \oplus g_j = g_k \neq g_0$.
(d) $\alpha_i \geq 0$, $i = 1, \ldots, |Q^*|$.

Case 2 of Theorem 7 involves computing all basic feasible solutions to a set of approximately $|Q^*|^2/2$ equations. However, the constraints are of a simple form and the calculations only have to be done once. The results can be saved in tables (see Section 7.9).

Example, Continued

For $P(G(4), 3)$, case 1 yields the faces $t_1, t_2, t_3 \geq 0$. To determine the remaining faces, find all basic feasible solutions to

$$\begin{array}{rll}
\alpha_3 = 1 & & \text{(case a)} \\
\alpha_1 + \alpha_2 \qquad = 1 & & \text{(case b, } i = 1;\ i = 2 \text{ yields the same constraint)} \\
2\alpha_1 - \alpha_2 \qquad \geq 0 & & [\text{case c, } (i, j) = (1, 1)] \\
\alpha_1 + \alpha_2 - \alpha_3 \geq 0 & & [\text{case c, } (i, j) = (1, 2)] \\
-\alpha_1 + \alpha_2 + \alpha_3 \geq 0 & & [\text{case c, } (i, j) = (2, 3)] \\
-\alpha_2 + 2\alpha_3 \geq 0 & & [\text{case c, } (i, j) = (3, 3)] \\
\alpha_1, \alpha_2, \alpha_3 \geq 0 & & \text{(case d)}
\end{array}$$

By substitution and elimination of redundancies, the above system simplifies to

$$\begin{aligned}
\alpha_1 + \alpha_2 \quad &= 1 \\
2\alpha_1 - \alpha_2 \quad &\geq 0 \\
-\alpha_1 + \alpha_2 \quad &\geq -1 \\
\alpha_3 &= 1 \\
\alpha_1, \alpha_2 &\geq 0
\end{aligned} \tag{87}$$

The basic feasible solutions to (87) are $(\alpha_1, \alpha_2, \alpha_3) = (1, 0, 1)$ and $(\frac{1}{3}, \frac{2}{3}, 1)$. Thus the faces of $P(G(4), 3)$ are as given by (84).

Group Automorphisms in Face Calculations

In certain instances, for $\rho' \neq \rho$, it is possible to obtain $P(G, \rho')$ directly from $P(G, \rho)$. Suppose there exists a permutation $\phi(i)$, other than $\phi(i) = i$ for all i, that renumbers the group elements and preserves group addition. That is, if

$$g_i \oplus g_j = g_k$$

then

$$g_{\phi(i)} \oplus g_{\phi(j)} = g_{\phi(k)}$$

for all i, j, and k. Such a permutation ϕ is called an *automorphism* of the group. Note that $G(4)$ has the automorphism, $\phi(0) = 0$, $\phi(1) = 3$, $\phi(2) = 2$, $\phi(3) = 1$. Let ϕ^{-1} be such that if $\phi(i) = j$, then $\phi^{-1}(j) = i$. In $G(4)$, $\phi^{-1}(0) = 0$, $\phi^{-1}(3) = 1$, $\phi^{-1}(2) = 2$, and $\phi^{-1}(1) = 3$.

If G has a nontrivial automorphism with $\phi(\rho) = \rho'$, Theorem 8 provides the faces and extreme points of $P(G, \rho')$ from the corresponding ones of $P(G, \rho)$. A proof of Theorem 8 is asked for in Exercise 23.

Theorem 8: *Let G have an automorphism ϕ such that $\phi(\rho) = \rho'$. Then*

1. $\sum \alpha_i t_i \geq \alpha_0$ *is a face of* $P(G, \rho)$ *if and only if* $\sum \alpha_{\phi^{-1}(i)}\, t_i \geq \alpha_0$ *is a face of* $P(G, \rho')$.
2. $(t_1, \ldots, t_{|G|-1})$ *is an extreme point of* $P(G, \rho)$ *if and only if*

   $(t_{\phi^{-1}(1)}, \ldots, t_{\phi^{-1}(|G|-1)})$ *is an extreme point of* $P(G, \rho')$.

Example, Continued

Since $\phi(3) = 1$ in $G(4)$, the faces and extreme points of $P(G(4), 1)$ can be obtained from the faces and extreme points of $P(G(4), 3)$. The results are shown in Table 2.

Table 2

	$P(G(4), 3)$	$P(G(4), 1)$
Faces	$t_1 + t_3 \geq 1$ $t_1/3 + 2t_2/3 + t_3 \geq 1$ $t_1 \geq 0, t_2 \geq 0, t_3 \geq 0$	$t_1 + t_3 \geq 1$ $t_1 + 2t_2/3 + t_3/3 \geq 1$ $t_1 \geq 0, t_2 \geq 0, t_3 \geq 0$
Extreme Points (t_1, t_2, t_3)	(3, 0, 0) (1, 1, 0) (0, 0, 1)	(0, 0, 3) (0, 1, 1) (1, 0, 0)

7.8. EXERCISES

1. Diagonalize

$$B = \begin{bmatrix} 3 & 2 & 0 \\ 1 & 4 & 0 \\ 3 & 3 & 1 \end{bmatrix}$$

into Smith Normal Form.

2. Solve the ILPC corresponding to the optimal LP basis for the problem

$$\begin{aligned} \max x_0 = 4x_1 + 5x_2 + x_3 & \\ 3x_1 + 2x_2 \qquad & \leq 10 \\ x_1 + 4x_2 \qquad & \leq 11 \\ 3x_1 + 3x_2 + x_3 & \leq 13 \\ x_1, x_2, x_3 & \geq 0 \text{ integer} \end{aligned}$$

Show that this solution yields an optimal solution to the ILP.

3. For the problem of Exercise 2, compute the bound $L_p(D - 1)$ given by (54). Is it true that $b = (10, 11, 13) \in K_B(d)$? Give the linear inequalities that define $K_B(d)$.

4. Change b of Exercise 2 to (10, 19, 15). Solve the ILPC corresponding to the optimal LP basis. Show that the optimal solution to the ILPC does not yield a feasible solution to the ILP. Now solve the ILP using the branch and bound algorithm of Section 7.6.

5. Solve, using the methodology of this chapter,

$$\begin{aligned} \max x_0 = 2x_1 + x_2 & \\ x_1 + x_2 & \geq 5 \\ 2x_1 + 3x_2 & \leq 20 \\ 4x_1 + 3x_2 & \leq 25 \\ x_1, x_2 & \geq 0 \text{ integer} \end{aligned}$$

6. Solve, using the methodology of this chapter,

$$\begin{aligned} \min x_0 = 10x_1 + 10x_2 + 10x_3 + 16x_4 & \\ x_1 + x_2 \qquad + x_4 & \geq 1 \\ x_2 + x_3 + x_4 & \geq 1 \\ x_1 \qquad + x_3 + x_4 & \geq 1 \\ x_1, \ldots, x_4 & = 0, 1 \end{aligned}$$

7. Prove that if the Smith Normal Form of B is Δ, then δ_1 is the greatest common divisor of the elements of B.

8. Devise an algorithm for computing Smith Normal Form that proceeds as follows:

(a) Specialize the procedure of Section 7.4 to diagonalize B to Λ where the diagonal elements of Λ are $\lambda_1, \lambda_2, \ldots, \lambda_m$. The λ_i are integers but do not necessarily satisfy the divisibility property of Smith Normal Form.

(b) Specialize the procedure of Section 7.4 to put Λ into Smith Normal Form.

How does this procedure compare with the algorithm given in Section 7.4?

9. Use the method of Exercise 8 to solve Exercise 1.

10. Give an algorithm for transforming a nonsingular matrix into Hermite Normal Form (Theorem 3 of Section 7.2). Note that only elementary column operations are to be used.

11. Transform B of Exercise 1 into Hermite Normal Form.

12. Show that an ILPC can be stated as

$$\begin{aligned} \min z_0 &= (c_B B^{-1} N - c_N) x_N \\ Fy &= b - N x_N \\ &y \text{ integer} \\ x_N &\geq 0 \text{ integer} \end{aligned}$$

where F is in Hermite Normal Form and B is a dual feasible LP basis matrix.

13. Devise a Fourier-Motzkin elimination scheme for solving ILPC's in Hermite Normal Form. (See Exercises 14–17 of Chapter 6). Establish that when applying Fourier-Motzkin elimination to matrices in Hermite Normal Form, the number of inequalities does not increase.

14. Use the procedure of Exercise 13 to solve the ILP of Exercise 2.

15. Devise a branching rule for the branch-and-bound algorithm of Section 7.6 that will avoid generating some implicitly fathomed vertices. In other words, if $p^k < p^{k_1}$ and $x_N^o(k) \geq p^{k_1}$, the branching rule should not generate vertex k_1.

16. Let B be a dual feasible basis matrix for the LP corresponding to a graph matching problem (see Section 3.6). Show that the Smith Normal Form for B is $\delta_1 = \cdots = \delta_k = 1$, $\delta_{k+1} = \cdots = \delta_m = 2$, where $1 \leq k \leq m$. Thus prove that an ILPC for a graph matching problem corresponds to a group knapsack problem over $G(2, 2, \ldots, 2)$.

17. Prove that an optimal solution to (49) is given by $t_k^o = |G|$, $t_j^o = 0$, $j \in Q - \{k\}$, where k is arbitrary.

18. Find all faces and vertices of the master polyhedron $P(G(8), 6)$.

19. Use the results of Exercise 18 to find all faces and vertices of $P(G(8), 2)$.

20. Use the results of Exercise 18 to find all faces and vertices of $P(G(8), \{2, 3, 5\}, 6)$.

21. Find all faces and vertices of the master polyhedron $P(G(2, 4), 0, 2)$.

22. Prove the "only if" clause of Theorem 6.

23. Prove Theorem 8.

24. Let S be the set of cuts obtainable from (69) of Section 5.15 from a given tableau and all integer combinations of rows. Suppose that two such cuts are $\sum_{j \in R} d_j x_j \geq d_0$ and $\sum_{j \in R} \hat{d}_j x_j \geq \hat{d}_0$. Let $d_j^1 = (\hat{d}_j + d_j) \pmod 1$ for $j \in R \cup \{0\}$.

Prove that the cut $\sum_{j \in R} d_j^1 x_j \geq d_0^1$ is valid and, in fact, is an element of S. Prove that the set S forms a finite abelian group under addition modulo 1 and that the order of the group is no larger than D.

7.9. NOTES

7.1. The ILPC was introduced in Gomory (1965), which has been reprinted in Dantzig and Veinott (1968).

7.2. This section was, in part, motivated by the concept of equivalent integer programs discussed in Bradley (1969), (1970c), and (1971a); and Faaland (1970). The canonical representation given in Theorems 3 and 4 are discussed in several algebra books, for example, MacDuffee (1940).

7.3 The transformation of an ILPC to a group knapsack problem is presented in Gomory (1965) and Shapiro (1968a). Relationships between more general ILPC's (nonnegativity is dropped only on a subset of the basic variables) and group theory are explored in Shapiro (1970) and Wolsey (1971a).

Group theoretic aspects of mixed integer linear programming also have been investigated. Wolsey (1970) and (1971b) shows that an MILPC can be represented as a group knapsack problem. The results, however, do not appear promising regarding computation.

Wolsey (1970) has a nice exposition of the group knapsack representation that motivated our presentation and also contains some interesting ideas for relaxation which reduce the order of the resulting group. These ideas are developed further in Gorry, Shapiro, and Wolsey (1972). Bradley and Wahi (1969) give an algorithm for solving the ILPC based upon the Hermitian Normal Form of Theorem 3 (see Exercises 10–14).

7.4. The calculation of Smith Normal Form essentially involves repeated application of an algorithm for finding the greatest common divisor of n integers. The classical procedure for doing this is the Euclidean algorithm, which is described in a variety of algebra texts, for example, Birkhoff and Maclane (1941). Improved versions of the Euclidean algorithm are given in Blankenship (1963) and Bradley (1970a).

The algorithm given in this section and the variation discussed in Exercise 8 follow Hu (1970). An algorithm for calculating the Hermite and Smith Normal forms based on Bradley's version of the Euclidean algorithm is given in Bradley (1970b).

7.5. The bound (54) is from Gomory (1965). Another bound is given in Shapiro (1968a). In Shapiro (1970), these results are generalized to ILP's for which nonnegativity is dropped only for a subset of the basic variables. Balas (1971d) has shown that (54) can never hold for the class of binary ILP's.

7.6. This section is based on Shapiro (1968b) and Gorry et al. (1970). See Exercise 15 for an improved branching rule. The method of this section is incorporated into a more general algorithm that can use cuts and surrogate constraints in Gorry and Shapiro (1971). Greenberg (1969b), and Glover and Litzler (1969) give different enumerative algorithm for achieving nonnegativity on the basic variables.

7.7. This section summarizes the results of Gomory (1969). Tables of faces and extreme points for all master polyhedra corresponding to groups of order up to 11 are given in Gomory (1969) and Hu (1969). Preliminary results appeared in Gomory (1967), which has been reprinted in Dantzig and Veinott (1968). Glover (1971a) and (1968d) and Devine and Glover (1969) also give algorithms for computing the faces of corner polyhedra. More recent results on the structure of corner polyhedra are given in Gomory and Johnson (1971). These results yield strong cuts for ILP's and MILP's.

Further Notes

A different approach, using an ILPC to try to solve an ILP, is given in Shapiro (1971). If no optimal solution to the ILPC is feasible to the ILP, the family of ILPC's

$$\begin{aligned} \min_{x_N} \; & ((c_B + \lambda)B^{-1}N - c_N)x_N \\ & B^{-1}Nx_N \equiv B^{-1}b \pmod 1 \\ & x_N \geq 0 \text{ integer} \end{aligned} \tag{88}$$

is considered, for nonnegative values of the m-vector λ. The components of λ are called *generalized lagrange multipliers*. Note that (88) with $\lambda = 0$ is the ILPC (13).

The generalized lagrange multiplier (GLM) approach for general mathematical programming problems is developed in Everett (1963). Using Everett's results, it can be shown that if

(a) x_N^* is an optimal solution to (88) with $\lambda = \lambda^* \geq 0$,

(b) $x_B^* = B^{-1}(b - Nx_N^*) \geq 0$,

(c) $\lambda^* x_B^* = 0$,

then (x_B^*, x_N^*) is an optimal solution to the ILP.

Methods for computing λ^*, if it exists, are given in Brooks and Geoffrion (1966) and Nemhauser and Widhelm (1971). Unfortunately, for λ^* to exist such that (a), (b), and (c) can be met, it is necessary for an optimal ILP solution (x_B^0, x_N^0) to exist which satisfies

(d) x_N^o is an extreme point of the corner polyhedron [otherwise (88) could not generate x_N^o].

(e) If (88) with $\lambda = 0$ does not yield a feasible solution to the ILP, then there must exist i^* such that $x_{B_{i^*}}^o = 0$ [otherwise (c) could not be satisfied].

Thus, if the standard ILPC does not solve the ILP, it is unlikely that the appropriate multipliers will exist to satisfy conditions (a), (b), and (c). Nevertheless, this approach may be useful because it also yields cuts that are closely related to the faces of the corner polyhedron.

A different GLM approach to ILP's is given in Lorie and Savage (1955) and Kaplan (1966). Unfortunately, this approach is not practical for finding optimal solutions to an ILP, since an optimal solution to an ILP can be verified by this method only if it is identical to some optimal solution to the corresponding LP. This result is given in Nemhauser and Ullmann (1968).

The existence of a dual problem to an ILP is closely related to the existence of generalized lagrange multipliers. Duality in integer programming has been considered in Gomory and Baumol (1960); Alcaly and Klevorick (1966); and Balas (1969a), (1970a), (1970c), and (1971b). Duality theory for ILP's is considerably weaker than the corresponding theory for LP's. The theoretical significance and computational benefits of duality in integer programming require further investigation.

8 The Set Covering and Partitioning Problems

8.1. INTRODUCTION

Consider a set $I = \{1, \ldots, m\}$, and a set $P = \{P_1, \ldots, P_n\}$, where $P_j \subseteq I$, $j \in J = \{1, \ldots, n\}$. A subset $J^* \subseteq J$ defines a *cover* of I if

$$\bigcup_{j \in J^*} P_j = I \tag{1}$$

If, in addition,

$$j, k \in J^*, j \neq k \Rightarrow P_j \cap P_k = \varnothing \tag{2}$$

J^* defines a *partition* of I. For convenience, we sometimes refer to the index set J^* as the cover or partition.

Let a cost $c_j > 0$ be associated with every $j \in J$. The total cost of the cover J^* is $\sum_{j \in J^*} c_j$. The *set covering problem* (CP) is to find a cover of minimum cost and can be written as the ILP

$$\min x_0 = \sum_{j=1}^{n} c_j x_j \tag{3}$$

$$\sum_{j=1}^{n} a_{ij} x_j \geq 1, \qquad i = 1, \ldots, m \tag{4}$$

$$x_j = 0, 1, \qquad j = 1, \ldots, n \tag{5}$$

where

$$x_j = \begin{cases} 1 & \text{if } j \text{ is in the cover} \\ 0 & \text{otherwise} \end{cases}$$

$$a_{ij} = \begin{cases} 1 & \text{if } i \in P_j \\ 0 & \text{otherwise} \end{cases}$$

Similarly, the *set partitioning problem* (PP) is obtained by replacing (4) with

$$\sum_{j=1}^{n} a_{ij}x_j = 1, \qquad i = 1, \ldots, m \tag{4a}$$

Any x satisfying (4) and (5) [or (4a) and (5)] is called a *cover* (*partition*) *solution*.

Converting a PP to a CP

The CP is the more general of the two problems, in the sense that any PP having a feasible solution can be transformed into a CP by changing the cost vector. In particular, consider any PP defined by $A = (a_{ij})$ and cost vector c. Let $t_j = \sum_{i=1}^{m} a_{ij}$ and choose any $L > \sum_{j=1}^{n} c_j$. Then

Theorem 1: *If the* PP *defined by A and c has a partition, the* CP *defined by A and costs $c'_j = c_j + Lt_j, j = 1, \ldots, n$, has the same set of optimal solutions as the* PP.

PROOF: Let S_c and S_p be the sets of cover and partition solutions, respectively. Observe that $S_c \supseteq S_p$. Assume that $S_c - S_p \neq \varnothing$; otherwise the theorem is trivial. Select any $x^* \in S_p$ and $\hat{x} \in S_c - S_p$. Then

$$c'x^* = cx^* + Lm < L(m + 1)$$

Let

$$\hat{I} = \left\{ i \,\middle|\, \sum_{j=1}^{n} a_{ij}\hat{x}_j > 1 \right\}$$

and note that $|\hat{I}| \geq 1$. Thus

$$c'\hat{x} \geq c\hat{x} + L(m + |\hat{I}|) \geq L(m + 1) > c'x^*$$

It follows that no element of $S_c - S_p$ solves the CP.

Example

Consider the PP having

$$A = \begin{bmatrix} 1 & 0 & 0 & 1 \\ 0 & 1 & 0 & 1 \\ 1 & 0 & 1 & 0 \\ 1 & 0 & 1 & 0 \end{bmatrix} \text{ and } c = (2, 3, 6, 1)$$

The optimal partition solution is $x' = (1, 1, 0, 0)$. The corresponding CP with the same costs has optimal cover solution $x'' = (1, 0, 0, 1)$ which is not a partition solution. However, the transformed CP with $L = 13$ has $c' = (41, 16, 32, 27)$ and optimal cover solution $x^* = (1, 1, 0, 0)$.

The conditions of Theorem 1 assume that a partition exists. For given A, it is not generally obvious whether a partition exists (see Exercise 3). However, a necessary and sufficient condition for a cover to exist is that no row of A be a null vector. If this condition is satisfied, then J is always a cover.

In the next section, it will be seen that both the CP and the PP have many applications. Fortunately, they also have many properties that make them among the most tractable ILP's.

8.2. SOME APPLICATIONS

Information Retrieval

Consider the problem of retrieving information from n files, where the jth file is of length c_j, $j = 1, \ldots, n$. Suppose that m requests for information are received. Each unit of information i is stored in at least one file j indicated by $a_{ij} = 1$. An optimal cover yields a subset of files that minimizes the maximum total length which needs to be searched in order to guarantee retrieval of all of the information.

Disconnecting Paths in a Graph

Consider a graph $G = (V, E)$. Let I correspond to a set of paths of the graph, and let $J = (1, \ldots, |E|\}$. If $a_{ij} = 1$ means that edge j is in path i, and c_j is the cost of removing edge j from the graph, then an optimal cover yields a minimum cost set of edges that disconnects all paths of I.

Truck Deliveries

Consider the problem of making deliveries to m locations by truck. There are n feasible routes to choose from and $a_{ij} = 1$ if location i is on route j. A cost c_j (possibly its length) is assigned to route j. Then an optimal partition gives a minimal cost routing that makes each delivery exactly once.

Political Districting

Here I corresponds to a set of basic population units (counties, census tracts, etc.). A subset P_j of I is in P if the population units form a district that meets requirements on population, contiguity, compactness, and so forth. In addition, it is required to have exactly K districts. If c_j is some measure of the unacceptability of district j, then an optimal solution to the PP, together with the additional constraint $\sum x_j = K$, yields an optimal districting plan.

Airline Crew Scheduling

An airline has a set of m flight "legs," each of which requires a crew. Let $P_j \in P$ if it corresponds to a set that can be handled by a single crew. Let c_j be some measure of the cost of P_j. Depending on whether or not crew members are allowed to be passengers on certain flights, optimal covers or optimal partitions yield optimal schedules.

Coloring Problems

Consider the problem of coloring a map so that no two adjacent areas have the same color. Let I represent the set of areas. A subset P_j is in P if no two elements of P_j correspond to areas having a common boundary. If all costs are unity, an optimal partition indicates the minimum number of colors needed.

There are many other applications for the CP and PP models including problems such as the design of switching circuits and assembly line balancing. The graph covering problem discussed in Section 3.6 is also a CP, where A is the vertex-edge incidence matrix of the graph. A relationship between the general and graph covering problems is discussed in Exercise 20.

8.3. REDUCTIONS

It is often possible, a priori, to eliminate certain rows and columns from A. Some of these reductions are valid for the CP, some for the PP, and some for both. The notation r_i is used for row i, and a_j for column j of A.

REDUCTION 1: (CP, PP) If r_i is a null vector for some i, there is no feasible solution since the ith constraint cannot be satisfied.

REDUCTION 2: (CP, PP) If $r_i = e_k$ (the kth unit vector) for some i, k, then $x_k = 1$ in every feasible solution, and a_k may be deleted. Also, every row t, such that $t \in P_k$ may be deleted.

Since r_i cannot be covered without a_k, and every element of P_k is covered by a_k, this reduction is obvious.

REDUCTION 2A: (PP) In addition to the deletions of Reduction 2, every column $q \neq k$ such that $a_{tq} = a_{tk} = 1$ for some t must be deleted.

Since $x_k = 1$, setting $x_q = 1$ for any such column would result in $\sum_{j=1}^{n} a_{tj}x_j \geq 2$ and row t would be "overcovered."

REDUCTION 3: (CP, PP) If $r_t \geq r_p$ for some t and p, then r_t may be deleted.

This reduction follows because every cover of row p covers row t.

REDUCTION 3A: (PP) In addition to deleting row t in Reduction 3, every column k such that $a_{tk} = 1$ and $a_{pk} = 0$ must be deleted.

This follows because some column q with $a_{pq} = a_{tq} = 1$ must have $x_q = 1$ in order to cover row p. Thus, if $x_k = 1$, row t would be "over-covered."

REDUCTION 4: (PP, CP) If for some set of columns S and some column k, $\sum_{j \in S} a_j = a_k$ and $\sum_{j \in S} c_j \leq c_k$, column k may be deleted.

REDUCTION 4A: (CP) If for some set of columns S and some column k, $\sum_{j \in S} a_j > a_k$ and $\sum_{j \in S} c_j \leq c_k$, column k may be deleted.

The proofs of Reductions 4 and 4A are left as Exercise 4.

Example

In the following example, the data are considered for both the PP and the CP case. Reductions are performed sequentially, and it should be noted that one reduction may make another possible.

	1	2	3	4	5	6	7	8	9
$c =$	10	5	8	6	9	13	11	4	6
$A =$ 1	1	1	1	0	1	0	1	1	0
2	0	1	1	0	0	0	0	1	0
3	0	1	0	0	1	1	0	1	1
4	0	0	0	1	0	0	0	0	0
5	1	0	1	0	1	1	0	0	1
6	0	1	1	0	0	0	1	0	1
7	1	0	0	1	1	0	0	1	1

PP

(3) $r_7 > r_4$; delete r_7.
(3A) Delete a_1, a_5, a_8 and a_9.
(2) $x_4 = 1$; delete a_4 and r_4.
(3) $r_1 > r_2$; delete r_1.
(3A) Delete a_7, resulting in

$$c = \begin{bmatrix} 5 & 8 & 13 \end{bmatrix}$$

$$A = \begin{array}{c|ccc|} 2 & 1 & 1 & 0 \\ 3 & 1 & 0 & 1 \\ 5 & 0 & 1 & 1 \\ 6 & 1 & 1 & 0 \\ \hline & 2 & 3 & 6 \end{array}$$

Clearly, there is no feasible solution.

CP

(3) $r_7 > r_4$; delete r_7.
(2) $x_4 = 1$; delete a_4 and r_4.
(3) $r_1 > r_2$; delete r_1.
(4A) $S = \{2\}$; delete a_7.
(4A) $S = \{3\}$; delete a_1.
(4A) $S = \{5\}$; delete a_6.
(4A) $S = \{9\}$; delete a_5.
(3) $r_6 > r_5$; delete r_6.
(4) $S = \{8\}$; delete a_2, resulting in

$$c = \begin{bmatrix} 8 & 4 & 6 \end{bmatrix}$$

$$A = \begin{array}{c|ccc|} 2 & 1 & 1 & 0 \\ 3 & 0 & 1 & 1 \\ 5 & 1 & 0 & 1 \\ \hline & 3 & 8 & 9 \end{array}$$

Clearly, the optimal solution is $x_8 = x_9 = x_4 = 1$.

8.4. HANDLING BINARY DATA

In the CP and PP all of the data except the costs are binary. Since virtually all scientific computers store their information in binary, one piece of data can be stored as one "bit." For instance, the IBM 7094 has a 36-bit word, so that any column or row of A with 36 bits or fewer can be stored as one word. In general, if the word size of the computer is k bits and a column (row) has q entries, it takes $\langle q/k \rangle$ words to store the column (row). Thus much larger matrices can be saved than would be the case in a general ILP.

Another handy feature of binary storage is that certain calculations can be done using the logical AND and logical OR statements of computer assembly languages. These two operations are performed on vectors as follows:

AND

$$a \circ b = c$$

$$c_j = \begin{cases} 1 & \text{if } a_j = b_j = 1 \\ 0 & \text{otherwise} \end{cases}$$

OR

$$a \diamond b = c$$

$$c_j = \begin{cases} 0 & \text{if } a_j = b_j = 0 \\ 1 & \text{otherwise} \end{cases}$$

The reductions of the last section are handled quite easily by these operations. For example, in Reduction 3, $r_t \geq r_p$ is equivalent to $r_t \circ r_p = r_p$. Also, some of the algorithms described later perform the operations $\cup$ and $\cap$ on sets, where the sets are represented by binary vectors. These set operations correspond to the operations $\diamond$ and $\circ$, respectively, on binary vectors.

8.5. SOME EXTREME POINT PROPERTIES

In this section some relationships between covering and partitioning problems and some related LP's are examined. The usefulness of some of these results will be made clear in later sections. Others have not been exploited, but indicate that covering problems could well be attacked by algorithms based on linear programming.

The LP obtained by replacing (5) with

$$x_j \geq 0, \qquad j = 1, \ldots, n \tag{5a}$$

in the CP is denoted by CP$'$, and in the PP by PP$'$.

Redundant Covers

For any cover J', a column $j^* \in J'$ is said to be *redundant*, if $J' - \{j^*\}$ is also a cover. If a cover contains one or more redundant columns, it is also termed redundant. A cover that is not redundant is called *prime*. Column j^* is redundant with respect to the cover J' if and only if

$$\sum_{j \in J'} a_{ij} \geq 2, \qquad \text{for all } i \in P_{j^*} \tag{6}$$

Equivalently, $\hat{j} \in J'$ is not redundant if and only if

$$I(\hat{j}) = \{i \mid \sum_{j \in J'} a_{ij} = 1, i \in P_{\hat{j}}\} \neq \varnothing \tag{7}$$

Since $c_j > 0$ for all j, every optimal cover is prime.

Now let x^* be any *optimal* solution to CP$'$. Clearly, $0 \leq x_j^* \leq 1$, $j = 1, \ldots, n$, since $c_j > 0$, and for any feasible solution x' to CP$'$, $x_j'' = \min\{1, x_j'\}, j = 1, \ldots, n$ is also feasible to CP$'$. From x^*, a feasible cover solution $\hat{x}$ is easily obtained as

$$\hat{x}_j = \langle x_j^* \rangle, \qquad j = 1, \ldots, n$$

Of course, $\hat{x}$ is not necessarily an optimal cover solution. In fact, $\hat{J} = \{j \mid \hat{x}_j = 1\}$ may be redundant. To obtain a prime cover from $\hat{J}$ (or from any other redundant cover), we drop indices corresponding to redundant columns from $\hat{J}$ one at a time until a prime cover is found. The resulting cover will be a function of the set of redundant columns dropped which is not necessarily unique.

Associating a Basis Matrix with a Prime Cover

Theorem 2: *Let $J' = \{j \mid x'_j = 1\}$ be any prime cover. Then x' is an extreme point of the set of feasible solutions to the CP'.*

PROOF: From (7), $|J'| \leq m$, since each column specified by J' satisfies at least one constraint as an equality. It also follows from (7) that the columns specified by J' are linearly independent. ■

A basis matrix can be associated with J', by adding $m - |J'|$ columns corresponding to surplus variables s_i of the form

$$s_i = 1 - \sum_{j=1}^{n} a_{ij} x_j$$

The rules for selecting the basic surplus variables are:

(i) Select s_i if $\sum_{j \in J'} a_{ij} \geq 2$.

(ii) For every $j \in J'$ such that $|I(j)|$ is greater than one, arbitrarily select surplus variables corresponding to any $|I(j)| - 1$ elements of $I(j)$.

These rules provide exactly $m - |J'|$ surplus variables, since $m =$ number of overcovered rows $+ \sum_{j \in J'} |I(j)|$. After illustrating them with an example, it will be shown that a basis matrix is obtained and that it has a special property.

Example

For the example of Section 8.3, a redundant cover is $J^* = \{1, 2, 3, 4\}$. The redundant columns are 1 and 3; the two prime covers that can be derived from J^* are $\{2, 3, 4\}$ and $\{1, 2, 4\}$. Letting $J' = \{2, 3, 4\}$, by rule (i), s_1, s_2, and s_6 become basic. Also, $I(2) = \{3\}$, $I(3) = \{5\}$, $I(4) = \{4, 7\}$. Thus rule (ii) allows a choice of s_4 or s_7 to be basic. Arbitrarily choosing s_7, we obtain the basis matrix

$$B = \begin{bmatrix} 1 & 1 & 0 & -1 & 0 & 0 & 0 \\ 1 & 1 & 0 & 0 & -1 & 0 & 0 \\ 1 & 0 & 0 & 0 & 0 & 0 & 0 \\ 0 & 0 & 1 & 0 & 0 & 0 & 0 \\ 0 & 1 & 0 & 0 & 0 & 0 & 0 \\ 1 & 1 & 0 & 0 & 0 & -1 & 0 \\ 0 & 0 & 1 & 0 & 0 & 0 & -1 \end{bmatrix}$$

Involutory Basis Matrices

To show that rules (i) and (ii) yield a basis matrix B, we show that B^{-1} exists. In fact, we can say even more. Namely,

Theorem 3: *A matrix $\hat{B}$ can be derived from B by a suitable permutation of rows such that $\hat{B}^{-1} = \hat{B}$.*

PROOF: Let row i of $\hat{B}$, $i = 1, \ldots, |J'|$, be the row of B that is the unit vector e_i. Let row i of $\hat{B}$, $i = |J'| + 1, \ldots, m$ be the row of B having a -1 in column i. The resulting matrix $\hat{B}$ is of the form

$$\hat{B} = \left[\begin{array}{c|c} I_1 & 0 \\ \hline P & -I_2 \end{array}\right]$$

where I_1 is of order $|J'|$ and I_2 is of order $m - |J'|$. Thus

$$\hat{B}\hat{B} = \left[\begin{array}{c|c} I_1 & 0 \\ \hline P & -I_2 \end{array}\right]\left[\begin{array}{c|c} I_1 & 0 \\ \hline P & -I_2 \end{array}\right] = \left[\begin{array}{c|c} I_1 & 0 \\ \hline 0 & I_2 \end{array}\right] \quad \blacksquare$$

Clearly the existence of $\hat{B}^{-1}$ implies the existence of B^{-1}. In our example

$$\hat{B} = \begin{bmatrix} 1 & 0 & 0 & 0 & 0 & 0 & 0 \\ 0 & 1 & 0 & 0 & 0 & 0 & 0 \\ 0 & 0 & 1 & 0 & 0 & 0 & 0 \\ 1 & 1 & 0 & -1 & 0 & 0 & 0 \\ 1 & 1 & 0 & 0 & -1 & 0 & 0 \\ 1 & 1 & 0 & 0 & 0 & -1 & 0 \\ 0 & 0 & 1 & 0 & 0 & 0 & -1 \end{bmatrix} = \hat{B}^{-1}$$

A matrix which is its own inverse is said to be *involutory*.

Extreme Points of PP′

Theorem 4: *Let $J' = \{j \mid x_j' = 1\}$ be any partition. Then x' is an extreme point of the set of feasible solutions to the* PP′.

PROOF: Certainly $|J'| \leq m$, since for all $j \in J'$

$$|I(j)| = |P_j| \geq 1$$

Also, the columns corresponding to the elements of J' have no positive entries in common, so that they are obviously independent. $\blacksquare$

Two further theorems will be given without proof. The proofs are rather long, and the theorems themselves, although interesting, have not yet proved to be applicable.

A basis B for the PP′ will be termed integer if $B^{-1}b$ is a binary vector. Two bases are *adjacent* if they differ in exactly one column.

Theorem 5: *For every feasible integer basis to* PP′, *there are at least as many adjacent feasible integer bases as there are nonbasic columns.*

It should be pointed out, however, that since integer solutions will invariably be degenerate, a large number of these adjacent bases may correspond to the same extreme point.

Theorem 6: *Let x^1 be a partition solution associated with a basis matrix B_1 to PP^1, where R_1 is the index set of the nonbasic columns. Let x^2 be another partition solution at least as good (in objective function value) as x^1. Let J_1 and J_2 be the corresponding partitions. Then a basis matrix B_2 associated with x^2 can be obtained from B_1 by a sequence of $|R_1 \cap J_2|$ pivots, such that each of the solutions is feasible, integer, and not worse than its predecessor.*

One would hope that this theorem would lead directly to a simple algorithm. Unfortunately, some of the pivots defined in the sequence may be on negative elements, and there is no obvious way to protect against cycling.

8.6. A CUTTING PLANE ALGORITHM FOR THE SET COVERING PROBLEM

An algorithm for the CP has been developed that uses cutting planes in a different manner from the approaches of Chapter 5.

Let x' be an integer basic feasible solution to the LP represented by

$$\min x_0 = y_{00} - \sum_{j \in R} y_{0j} x_j$$

$$x_{B_i} = y_{i0} - \sum_{j \in R} y_{ij} x_j, \qquad i = 1, \ldots, m$$

obtained by setting $x_j = 0, j \in R$. If x^* is any other feasible integer solution to the LP, then

$$x_0^* - x_0' = -\sum_{j \in R} y_{0j} x_j^*$$

It follows that a necessary condition for $x_0^* < x_0'$, or equivalently $x_0^* \leq x_0' - 1$, is

$$\sum_{j \in Q} x_j^* \geq 1 \tag{8}$$

where $Q = \{j \mid y_{0j} > 0\}$. Furthermore, $Q = \varnothing$ implies that x' is optimal to the LP and consequently to the corresponding ILP as well.

In order for the cut (8) to be useful in solving the CP, the set Q and consequently B^{-1} must be known, since $y_{0j} = c_B B^{-1} a_j - c_j, j \in R$. However, in Section 8.5 it was shown that from a prime cover, by a suitable permutation of rows it is possible to find a corresponding basis $\hat{B}$ such that $\hat{B} = \hat{B}^{-1}$. Thus, for prime covers, Q can be determined without any matrix inversion.

In Section 8.5, $\hat{B}$ was shown to be of the form

$$\hat{B} = \left[\begin{array}{c|c} I_1 & 0 \\ \hline P & -I_2 \end{array}\right] = \left[\begin{array}{c|c} \hat{B}_1 & \hat{B}_2 \end{array}\right]$$

where $\hat{B}_1$ is associated with original variables and $\hat{B}_2$ with surplus variables. Also

$$c_B \hat{B}^{-1} = (c_{B_1}, 0)\left[\begin{array}{c|c} I_1 & 0 \\ \hline P & -I_2 \end{array}\right] = c_B \tag{9}$$

Now, let $A' = (a_1, \ldots, a_{n+m})$, where a_{n+i} is associated with s_i, $i = 1, \ldots, m$. Also, let $\hat{A}$ be derived from A' by taking the same row permutation as was used in deriving $\hat{B}$ from B. It follows that

$$B^{-1} a_j = \hat{B}^{-1} \hat{a}_j, \qquad j = 1, \ldots, n + m \tag{10}$$

since permuting rows in a matrix has the effect of permuting the corresponding columns of the matrix inverse. Thus from (9) and (10)

$$c_B B^{-1} a_j = c_B \hat{B}^{-1} \hat{a}_j = c_B \hat{a}_j, \qquad j = 1, \ldots, n + m \tag{11}$$

It also follows from (11), and $c_B \geq 0$, $a_j \geq 0$, $j = 1, \ldots, n$, and $a_j \leq 0$, $j = n + 1, \ldots, n + m$, that

$$c_B \hat{a}_j \begin{cases} \geq 0, & j = 1, \ldots, \mathrm{n} \\ \leq 0, & j = n + 1, \ldots, n + m \end{cases}$$

Since $y_{0j} = c_B \hat{a}_j - c_j$ and $c_j = 0$ for $j = n + 1, \ldots, n + m$, Q can only contain indices corresponding to original variables. Thus (8) is of the same form as the constraints (4), and a CP with (8) added is still a CP.

Algorithm

STEP 1: Let the original CP be denoted by CP_1. If any row of A is a null vector, terminate, since there is no cover. Otherwise let $\bar{z} = \sum_{j=1}^{n} c_j$, and let $k = 1$, where k is an iteration counter. Go to Step 2.

STEP 2: Let x^k be any prime cover solution for CP_k, and let $\hat{B}$ be a corresponding involutory basis. If $cx^k < \bar{z}$, let $\bar{z} = cx^k$ and record x^k. Go to Step 3.

STEP 3: Let $Q = \{j | c_B \hat{a}_j - c_j > 0\}$. If $Q = \varnothing$, terminate. The last recorded solution is optimal. If $Q \neq \varnothing$, go to Step 4.

STEP 4: Add the constraint $\sum_{j \in Q} x_j \geq 1$ to CP_k. Let $k = k + 1$ and go to Step 2.

Finiteness of the algorithm is easily shown, since at least one cover is eliminated at every iteration. An upper bound on the number of covers, and consequently on the number of iterations, is 2^n.

Step 2 of the algorithm is vague. However, there are a number of simple ways of finding covers. Some methods are:

METHOD 1: Set $x_j = 1, j = 1, \ldots, n$, to obtain a cover solution for any CP_k.

METHOD 2: If x is a cover solution for CP_k, then $x + e_t$ is a cover solution for CP_{k+1}, for any $t \in Q$, where e_t is the tth unit vector.

METHOD 3: If x is an optimal solution to CP'_k, then $x' = (\langle x_1 \rangle, \ldots, \langle x_n \rangle)$ is a cover solution for CP_k.

The covers obtained by Methods 1–3 may be redundant. These can be reduced to prime covers, as explained in Section 8.5. Also see Section 8.7 for a method of obtaining "good" prime covers.

Use of Linear Programming

Method 3 has considerable appeal. The optimal solution to the LP serves not only to provide a cover but in the usual way provides a lower bound on the value of the optimal cover. Of course, if an optimal CP′ solution is integer, the CP is solved.

Example

$$
\begin{aligned}
\min x_0 = 7x_1 + 3x_2 + 7x_3 + 12x_4 + 6x_5 & \\
x_1 + x_3 + x_4 - s_1 &= 1 \\
x_2 + x_4 - s_2 &= 1 \\
x_1 + x_5 - s_3 &= 1 \\
x_3 + x_5 - s_4 &= 1 \\
x_1, \ldots, x_5 &= 0, 1 \\
s_1, \ldots, s_4 &\geq 0
\end{aligned}
$$

Method 3 is used for Step 3. Let $\bar{z} = 35$ and $k = 1$.

An optimal solution to CP_1' is $x = (x_1, \ldots, x_5) = (\frac{1}{2}, 1, \frac{1}{2}, 0, \frac{1}{2})$. Rounding up yields $x' = (1, 1, 1, 0, 1)$. There are three prime covers available. Arbitrarily choose $x^1 = (1, 1, 1, 0, 0)$ and let $\bar{z} = 17$. Also

$$B = \begin{bmatrix} 1 & 0 & 1 & -1 \\ 0 & 1 & 0 & 0 \\ 1 & 0 & 0 & 0 \\ 0 & 0 & 1 & 0 \end{bmatrix}, \hat{B} = \begin{bmatrix} 1 & 0 & 0 & 0 \\ 0 & 1 & 0 & 0 \\ 0 & 0 & 1 & 0 \\ 1 & 0 & 1 & -1 \end{bmatrix}$$

$$c_B = (7, 3, 7, 0), \hat{A} = \begin{bmatrix} 1 & 0 & 0 & 0 & 1 & 0 & 0 & -1 & 0 \\ 0 & 1 & 0 & 1 & 0 & 0 & -1 & 0 & 0 \\ 0 & 0 & 1 & 0 & 1 & 0 & 0 & 0 & -1 \\ 1 & 0 & 1 & 1 & 0 & -1 & 0 & 0 & 0 \end{bmatrix}$$

and $c_B\hat{A} = (7, 3, 7, 3, 14, 0, -3, -7, -7)$. Thus $Q = \{5\}$ and the cut (8) is $x_5 \geq 1$.

An optimal solution to CP_2' is $x = (\frac{1}{2}, 1, \frac{1}{2}, 0, 1)$. Again, $x' = (1, 1, 1, 0, 1)$, but there are only two prime covers. Arbitrarily choose $x^2 = (1, 1, 0, 0, 1)$. The basis matrices associated with x^2 are

$$\begin{bmatrix} 1 & 0 & 0 & 0 & 0 \\ 0 & 1 & 0 & 0 & 0 \\ 1 & 0 & 1 & -1 & 0 \\ 0 & 0 & 1 & 0 & -1 \\ 0 & 0 & 1 & 0 & 0 \end{bmatrix} \quad \text{and} \quad \begin{bmatrix} 1 & 0 & 0 & 0 & 0 \\ 0 & 1 & 0 & 0 & 0 \\ 1 & 0 & 1 & -1 & 0 \\ 0 & 0 & 1 & 0 & 0 \\ 0 & 0 & 1 & 0 & -1 \end{bmatrix}$$

Choosing the first yields

$$\hat{B} = \begin{bmatrix} 1 & 0 & 0 & 0 & 0 \\ 0 & 1 & 0 & 0 & 0 \\ 0 & 0 & 1 & 0 & 0 \\ 1 & 0 & 1 & -1 & 0 \\ 0 & 0 & 1 & 0 & -1 \end{bmatrix}, c_B\hat{A} = (7, 3, 7, 10, 6, -7, -3, 0, 0, -6)$$

Thus $Q = \varnothing$, and x^2 is an optimal cover solution. Note that this implies that x^2 is also an alternative optimal solution to CP_2'.

8.7. AN ENUMERATION ALGORITHM FOR THE SET COVERING PROBLEM

Good computational results for the CP have been achieved using a variation of the basic implicit enumeration of Section 4.5. (Note that the CP is of the form min cx, $Ax \geq b$, x binary, which is different from the standard form used in Section 4.5)

Consider solving the CP by implicit enumeration. Let $\bar{z}_0 = \infty$ and $W_0 = \varnothing$. At vertex k, with partial solution W_k, we have

$$S_k^+ = \{j \mid j \in W_k \text{ and } x_j = 1\}$$
$$S_k^- = \{j \mid j \in W_k \text{ and } x_j = 0\}$$
$$F_k = \{j \mid j \notin W_k\}$$

Thus the problem at vertex k is

$$\begin{aligned} \min z_k = \sum_{j \in F_k} c_j x_j &+ \sum_{j \in S_k^+} c_j \\ \sum_{j \in F_k} a_{ij} x_j &\geq 1 - \sum_{j \in S_k^+} a_{ij}, \qquad i = 1, \ldots, m \\ x_j &= 0, 1, \qquad j \in F_k \end{aligned} \tag{12}$$

Letting

$$Q_k = \{i \mid a_{ij} = 0 \text{ for all } j \in S_k^+\}$$

and $\underline{z}_k = \sum_{j \in S_k^+} c_j$, (12) can be written as

$$\begin{aligned} \min z_k = \sum_{j \in F_k} c_j x_j &+ \underline{z}_k \\ \sum_{j \in F_k} a_{ij} x_j &\geq 1, \qquad i \in Q_k \\ x_j &= 0, 1, \qquad j \in F_k \end{aligned} \tag{13}$$

Now, (13) is also a CP. Consider the CP′ associated with (13) and the nature of its solution. There are three possibilities:

(i) The CP′ has an optimal, noninteger solution given by (z_k^*, x^*).
(ii) The CP′ has an optimal, integer solution given by (z_k^*, x^*). Note that in the trivial case $Q_k = \varnothing$, the appropriate solution is $(\underline{z}_k, 0)$.
(iii) The CP′ has no feasible solution. This can only occur if $\sum_{j \in F_k} a_{ij} = 0$ for some $i \in Q_k$. In this case, we let $z_k^* = \infty$.

In cases (ii) and (iii), backtrack and replace $\bar{z}_0$ by $\min\{\bar{z}_0, z_k^*\}$. In case (i), if $\langle z_k^* \rangle \geq \bar{z}_0$, backtrack. If $\langle z_k^* \rangle < \bar{z}_0$, vertex k is not fathomed. However, a cover solution x' for the original CP is obtained by letting $x_j' = \langle x_j^* \rangle$, $j \in F_k$, $x_j' = 1, j \in S_k^+$, $x_j' = 0, j \in S_k^-$. From x' a prime cover solution x^k can be obtained. Since the algorithm is enumerative, it is important to achieve tight upper bounds at every partial solution. To this end a heuristic is used that attempts to produce a prime cover having a low objective value.

A Method for Producing a Prime Cover from a Redundant Cover

STEP 1: Let $J^k = \{j | x'_j = 1, j \in F_k\}$ and let R^k be the set of redundant columns with respect to the cover J^k. Go to Step 2.

STEP 2: Define

$$f(j) = c_j / \textstyle\sum_{i \in Q^k} a_{ij}, \qquad j \in R^k$$

and choose j^* such that

$$f(j^*) = \min_{j \in R^k} f(j).$$

Let $J^k = J^k - \{j^*\}$. Go to Step 3.

STEP 3: If $R^k \neq \varnothing$, go to Step 2. If $R^k = \varnothing$, terminate. The cover solution x^k for (13) is given by $x_j^k = 1, j \in J^k$, $x_j^k = 0$ otherwise.

Intuitively, the heuristic attempts to exclude those variables which have the maximum cost per constraint satisfied. Once x^k is obtained, replace $\bar{z}_0$ by $\min \{\bar{z}_0, z^k\}$, where $z^k = \sum_{j \in J^k} c_j + \underline{z}_k$. Branch to the partial solution W_{k+1} given by

$$S_{k+1}^+ = S_k^+ \cup J^k$$
$$S_{k+1}^- = S_k^-$$

Since W_{k+1} corresponds to the cover $J^k \cup S_k^+$, backtracking occurs immediately.

Example

We repeat the example of Section 8.5 (see Figure 1 for the enumeration tree).

$$\begin{aligned} \min x_0 = 7x_1 + 3x_2 + 7x_3 + 12x_4 + 6x_5 & \\ x_1 + x_3 + x_4 & \geq 1 \\ x_2 + x_4 & \geq 1 \\ x_1 + x_5 & \geq 1 \\ x_3 + x_5 & \geq 1 \\ x_1, \ldots, x_5 = 0, 1 & \end{aligned} \tag{14}$$

Initially, $W_0 = \varnothing$. The optimal solution to CP'_0 is $x_0^* = (\frac{1}{2}, 1, \frac{1}{2}, 0, \frac{1}{2})$, $z_0^* = 13$. This yields $x' = (1, 1, 1, 0, 1)$. Entering the heuristic, we have

STEP 1: $J^0 = \{1, 2, 3, 5\}$, $R^k = \{1, 3, 5\}$.

STEP 2: $f(1) = 7/2$, $f(3) = 7/2$, $f(5) = 3$. Arbitrarily choose $j^* = 3$, so that $J^0 = \{1, 2, 5\}$. Go to Step 3.

STEP 3: $R^0 = \varnothing$. Terminate with cover J^0 having cost $cx^0 = \bar{z}_0 = 16$.

Branch to the partial solution $W_1 = (1, 2, 5)$ and backtrack to $W_2 = (1, 2, \underline{5})$. This yields CP'_2

$$\begin{aligned} \min z_2 = 7x_3 + 12x_4 + 10 & \\ x_3 \qquad\qquad \geq 1 & \\ x_3, x_4 \geq 0 & \end{aligned}$$

with optimal solution $x_3^* = 1$, $x_4^* = 0$, $z_2^* = 17$. Thus we backtrack to $W_3 = (1, \underline{2})$ and CP'_3

$$\begin{aligned} \min z_3 = 7x_3 + 12x_4 + 6x_5 + 7 & \\ x_4 \qquad \geq 1 & \\ x_3 \qquad\quad + x_5 \geq 1 & \\ x_3, x_4, x_5 \geq 0 & \end{aligned}$$

with optimal solution $x_3^* = 0$, $x_4^* = 1$, $x_5^* = 1$, $z_3^* = 25$. Backtrack to $W_4 = (\underline{1})$, and CP'_4

$$\begin{aligned} \min z_4 = 3x_2 + 7x_3 + 12x_4 + 6x_5 & \\ x_3 + x_4 \qquad \geq 1 & \\ x_2 \qquad + x_4 \qquad \geq 1 & \\ x_5 \geq 1 & \\ x_3 \qquad\quad + x_5 \geq 1 & \\ x_2, \ldots, x_5 \geq 0 & \end{aligned}$$

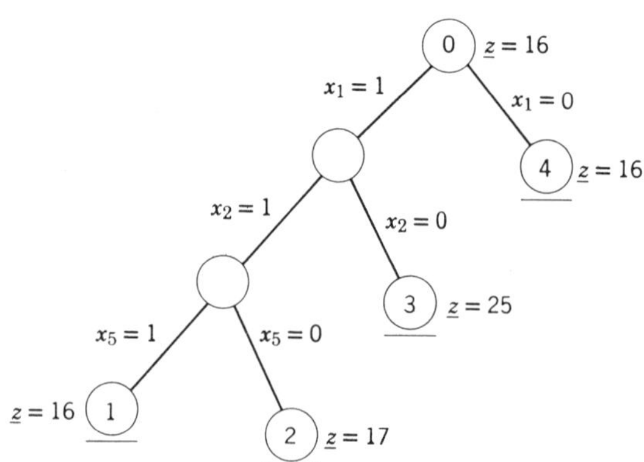

Figure 1

The optimal solution to CP_4' is $x_2^* = 1$, $x_3^* = 1$, $x_4^* = 0$, $x_5^* = 1$, $z_4^* = 16$. Thus enumeration is complete and two optimal solutions, (1, 1, 0, 0, 1) and (0, 1, 1, 0, 1) to (14), have been found.

8.8. AN ENUMERATION ALGORITHM FOR THE SET PARTITIONING PROBLEM

For the PP, a very specialized form of the standard implicit enumeration algorithm is considered. Extensive use is made of the results of Section 8.4 on handling binary data.

Placing Columns in Lists

Initially, every column of A is placed in one of m lists. Column j is placed in list k if $k = \min\{i \mid a_{ij} = 1\}$. Note that some lists may be empty. However, if list 1 is empty, there is no feasible solution. Within each list, the columns are ordered by the cost coefficients. In particular,

$$k, t \in \text{list } i \quad \text{and} \quad k \text{ precedes } t \Rightarrow c_k \leq c_t$$

Example

$$\begin{array}{lllllllll}
 & \overbrace{}^{\text{List 1}} & \overbrace{}^{\text{List 2}} & \overbrace{}^{\text{List 3}} & \overbrace{}^{\text{List 4}} & \\
\min & 3x_1 + 7x_2 & + 5x_3 + 8x_4 + 10x_5 & + 4x_6 + 6x_7 & + 9x_8 & \\
 & x_1 + x_2 & & & & = 1 \\
 & & x_3 + x_4 + x_5 & & & = 1 \\
 & & x_5 & + x_6 + x_7 & & = 1 \\
 & & & x_7 & + x_8 & = 1 \\
 & x_2 & + x_4 & + x_6 & & = 1 \\
\end{array}$$

$$x_1, \ldots, x_8 = 0, 1$$

It is convenient to introduce notation slightly different from that of Section 8.7. Let S denote a partial solution, $S^+ = \{j \mid x_j = 1, j \in S\}$ and $z(S) = \sum_{j \in S^+} c_j$. Let the constraints satisfied by S be $Q(S) = \bigcup_{j \in S^+} P_j$.

Algorithm

STEP 1: (Initialization.) Let $S = \varnothing$, $\bar{z} = \infty$. Go to Step 2.

STEP 2: (Choose next list.) Let $i^* = \min \{i \mid i \notin Q(S)\}$. Set an indicator at the top (lowest cost element) of list i^*. Go to Step 3.

STEP 3: (Test for an augmenting variable.) Begin at the indicated position in list i^* and examine the columns of the list in order. If a column j is found such that $Q(S) \cap P_j = \varnothing$ and $z(S) + c_j < \bar{z}$, go to Step 4. If a column j is reached such that $z(S) + c_j \geq \bar{z}$, or if list i^* is exhausted, go to Step 5.

STEP 4: (Test for solution.) Let $S^+ = S^+ \cup \{j\}$. If $Q(S) = I$, a better solution has been found, let $\bar{z} = z(S)$, and go to Step 5. If $Q(S) \neq I$, go to Step 2.

STEP 5: (Backtrack.) If $S^+ = \varnothing$, go to Step 6. If $S^+ \neq \varnothing$, let $\{k\}$ be the last element included in S^+. Let $S^+ = S^+ - \{k\}$. Let i^* be the list in which $\{k\}$ is found, and set an indicator directly below $\{k\}$ in list i^*. Go to Step 3.

STEP 6: (Termination). If $\bar{z} = \infty$, no partition solution exists. If $\bar{z} < \infty$, the solution that gave $\bar{z}$ is optimal.

It should be clear that this algorithm is actually a special case of the basic enumeration scheme of Section 4.5. The important difference is in Step 3, where the search for an augmenting variable can be limited to list i^*. This search limitation is possible because if $x_j = 1$, every $i \in P_j$ can be skipped.

Since every partition solution contains an element of list 1, the search actually terminates when list 1 is exhausted. For this reason, it seems logical to have list 1 be as small as possible. This may be achieved by permuting the rows such that

$$\sum_j a_{1j} = \min_i \sum_j a_{ij}$$

It is extremely important for efficient computation that the basic test in Step 3, $Q(S) \cap P_j \stackrel{?}{=} \varnothing$, be implemented by using the logical AND operation described in Section 8.4.

Example

The example introduced earlier in this section is now solved.

Step
1: $S = \varnothing, \bar{z} = \infty$.
2: $Q(S) = \varnothing, i^* = 1$.
3: $Q(S) \cap P_1 = \varnothing, 0 + 3 < \infty, j = 1$.
4: $S^+ = \{1\}$, $Q(S) = \{1\}$, $z(S) = 3$.

2: $i^* = 2$.
3: $Q(S) \cap P_3 = \varnothing$, $5 + 3 < \infty$, $j = 3$.
4: $S^+ = \{1, 3\}$, $Q(S) = \{1, 2\}$, $z(S) = 8$.
2: $i^* = 3$.
3: $j = 6$.
4: $S^+ = \{1, 3, 6\}$, $Q(S) = \{1, 2, 3, 5\}$, $z(S) = 12$.
2: $i^* = 4$.
3: $j = 8$.
4: $S^+ = \{1, 3, 6, 8\}$, $Q(S) = I$, $\bar{z} = z(S) = 21$.
5: $S^+ = \{1, 3, 6\}$, $i^* = 4$.
3: List 4 is exhausted.
5: $S^+ = \{1, 3\}$, $i^* = 3$.
3: $j = 7$.
4: $S^+ = \{1, 3, 7\}$, $Q(S) = \{1, 2, 3, 4\}$, $z(S) = 14$.
2: $i^* = 5$.
3: List 5 is empty.
5: $S^+ = \{1, 3\}$, $i^* = 3$.
3: List 3 is exhausted.
5: $S^+ = \{1\}$, $i^* = 2$.
3: $j = 4$.
4: $S^+ = \{1, 4\}$, $Q(S) = \{1, 2, 5\}$, $z(S) = 11$.
2: $i^* = 3$.
3: $j = 7$.
4: $S^+ = \{1, 4, 7\}$, $Q(S) = I$, $\bar{z} = z(S) = 17$.
5: $S^+ = \{1, 4\}$, $i^* = 3$.
3: List 3 is exhausted.
5: $S^+ = \{1\}$, $i^* = 2$.
3: $j = 5$.
4: $S^+ = \{1, 5\}$, $Q(S) = \{1, 2, 3\}$, $z(S) = 13$.
2: $i^* = 4$.
3: No j found.
5: $S^+ = \{1\}$, $i^* = 2$.
3: List 2 is exhausted.
5: $S^+ = \varnothing$, $i^* = 1$.
3: $j = 2$.
4: $S^+ = \{2\}$, $Q(S) = \{1, 5\}$, $z(S) = 7$.
2: $i^* = 2$.
3: $j = 3$.
4: $S^+ = \{2, 3\}$, $Q(S) = \{1, 2, 5\}$, $z(S) = 12$.
2: $i^* = 3$.
3: No j found.
5: $S^+ = \{2\}$, $i^* = 2$.

3: No j found.
5: $S^+ = \varnothing$, $i^* = 1$.
3: List 1 is exhausted.
6: Terminate. An optimal solution is $x_1 = x_4 = x_7 = 1$, $x_j = 0$ otherwise.

A Simple Surrogate Constraint

A simple surrogate constraint (see Section 4.7) can also be employed. At any partial solution S, the resulting PP is

$$\begin{aligned} \min z_k &= \sum_{j \notin S} c_j x_j + z(S) \\ \sum_{j \notin S} a_{ij} x_j &= 1, \qquad i \notin Q(S) \\ x_j &= 0, 1, \qquad j \notin S \end{aligned} \tag{15}$$

Adding the constraints of (15) yields the problem

$$\begin{aligned} \min \sum_{j \notin S} c_j x_j &+ z(S) \\ \sum_{j \notin S} v_j x_j &= t \\ x_j &= 0, 1, \qquad j \notin S \end{aligned} \tag{16}$$

where $v_j = \sum_{i \notin Q(S)} a_{ij}$ and $t = m - |Q(S)|$. Problem (16) is a version of the knapsack problem, discussed in Chapter 6, and could be solved to yield a lower bound on z_k. A weaker bound that is more easily obtained is

$$\underline{z}(S) = \frac{tc_{j^*}}{v_{j^*}} + z(S)$$

where $c_{j^*}/v_{j^*} = \min_{j \notin S} c_j/v_j$. Backtracking occurs if $\langle \underline{z}(S) \rangle \geq \bar{z}$.

8.9. EXERCISES

1. In Theorem 1 it was shown that every PP with a feasible solution has an equivalent CP. Is the converse true?

2. Show that $c_j > 0$ for all $j \in J$ involves no loss of generality:

(a) In the CP, since $c_j \leq 0$ implies that there exist optimal covers having $x_j = 1$.

(b) In the PP, since the PP with transformed costs $c'_j = c_j + d_j$, where $d_j = \sum_{i=1}^{m} k_i a_{ij}$ and k_i are arbitrary real numbers, is equivalent.

3. In Section 8.1 a necessary and sufficient condition for a CP to have an optimal solution was given. Is such a condition available for the PP?

4. Prove that Reductions 4 and 4A are valid.

5. Prove the following reduction theorem for the PP.

Theorem: *Let r_k and r_t be any two rows of A such that it is not true that $r_k \geq r_t$ or $r_t \geq r_k$. Let $K = \{j \mid a_{kj} > a_{tj}\}$ and $T = \{j \mid a_{tj} > a_{kj}\}$. If there exists a row s with $a_{sj} = 1$ for all $j \in K$ and $a_{sj} = 1$ for some $j^* \in T$, then column j^* may be deleted.*

6. Use the result of Exercise 5 to eliminate columns 2 and 5 from

$$A = \begin{bmatrix} 0 & 0 & 1 & 1 & 0 & 0 \\ 1 & 1 & 0 & 1 & 1 & 0 \\ 0 & 1 & 1 & 0 & 1 & 0 \\ 1 & 0 & 0 & 1 & 1 & 1 \end{bmatrix}$$

7. Using the reductions of Section 8.3 and of Exercise 5, eliminate as many rows and columns as you can for the following problem:

(a) As a CP
(b) As a PP

$$A = \begin{bmatrix} 1 & 0 & 1 & 0 & 0 & 0 & 1 \\ 0 & 1 & 0 & 1 & 0 & 1 & 0 \\ 1 & 0 & 0 & 0 & 1 & 0 & 1 \\ 0 & 1 & 0 & 1 & 0 & 1 & 0 \\ 1 & 1 & 0 & 1 & 0 & 0 & 0 \\ 1 & 0 & 1 & 1 & 0 & 0 & 1 \end{bmatrix} \qquad c = (11, 6, 3, 12, 4, 3, 8)$$

8. How can the reduction of Exercise 5 be implemented using the logical AND and OR instructions?

9. Does the concept of a "redundant partition" make sense?

10. In Section 8.5 it was shown that any optimal solution to the CP′ can be rounded up to yield a cover.

(a) Is the statement true with the word "optimal" replaced by "basic feasible"?
(b) Can partitions be derived from optimal solutions to the PP′?

11. In Section 8.5 a technique was given for associating a basis matrix with a prime cover. Is it true that every basis matrix corresponding to the prime cover can be generated by that technique?

12. In Step 2 of the algorithm of Section 8.6, there may be more than one $\hat{B}$ available.

(a) Show, with an example, that the number of iterations can be affected by the choice of $\hat{B}$.

(b) Suppose that method 3 is used in Step 2. Note that the cut derived from CP_k may not be violated by the optimal solution to CP'_k. In that case, CP'_{k+1} will already be solved, and no new LP information will be gained. Thus a reasonable objective in selecting $\hat{B}$ is to ensure that a new LP solution will be needed. Derive a technique that is likely to achieve this objective.

13. Solve the following problem using the algorithm of Section 8.6.

(a) Use Methods 1 and 2 in Step 2.

(b) Use Method 3 in Step 2.

$$\begin{aligned}
\min x_0 = 6x_1 + 8x_2 + 4x_3 + 3x_4 + 5x_5 & \\
x_1 + x_2 \qquad\quad + x_4 \qquad\quad & \geq 1 \\
x_1 \qquad\quad + x_3 \qquad\quad + x_5 & \geq 1 \\
x_2 \qquad\quad\qquad\quad + x_5 & \geq 1 \\
x_1 + x_2 + x_3 \qquad\quad\qquad\quad & \geq 1 \\
x_1, \ldots, x_5 = 0, 1 &
\end{aligned}$$

14. Note that the cut of Section 8.6 has the form of the CP constraints. Would this be true for any of the cuts introduced in Chapter 5?

15. Solve the problem of Exercise 13 by the algorithm of Section 8.7.

16. For the following example

(a) Solve by the algorithm of Section 8.8 without the surrogate constraint.

(b) Solve by the same algorithm with the surrogate constraint.

$$\begin{aligned}
\min 18x_1 + 22x_2 + 14x_3 + 36x_4 + 17x_5 + 14x_6 + 8x_7 + 24x_8 + 14x_9 + 7x_{10} & \\
x_3 + x_5 & = 1 \\
x_4 + x_5 + x_6 & = 1 \\
x_2 + x_4 + x_8 + x_9 & = 1 \\
x_3 + x_4 + x_8 + x_{10} & = 1 \\
x_1 + x_2 + x_7 + x_9 & = 1 \\
x_1 + x_2 + x_4 + x_6 + x_8 & = 1 \\
x_1, \ldots, x_{10} = 0, 1 &
\end{aligned}$$

17. In Exercise 2b it was shown that a PP could be converted into an equivalent PP with costs $c'_j = c_j + \sum_{i=1}^m k_i a_{ij}$. Suppose one were interested in determining k_i such that $\sum_j (c_j - c'_j)x_j$ is maximized, subject to $c_j \geq c'_j \geq 0$ for all j and x a partition solution. Show that the problem is equivalent to the dual of the PP'.

18. What modifications are necessary to the algorithm of Section 8.8 to allow it to solve the CP? Is the resulting algorithm computationally attractive?

19. Consider two sets defined on the matrix A

$$\text{cov}(A) = \{x | Ax \geq 1, x \text{ binary}\}$$
$$\text{cl}(A) = \{x | x \geq r_i, \text{ some } i \in I\}$$

Prove:

(a) $\text{cov}(A) = \text{cl}(\text{cov}(A))$.
(b) $\text{cov}(A) = \text{cov}(\text{cl}(A))$.
(c) $\text{cov}(A) = \text{cov}(B) \Leftrightarrow \text{cl}(A) = \text{cl}(B)$.
(d) $\text{cov}(\text{cov}(A)) = \text{cl}(A)$.
(e) $\text{cov}(\text{cl}(A) \cup \text{cl}(B)) = \text{cov}(\text{cl}(A)) \cap \text{cov}(\text{cl}(B))$.
(f) $\text{cov}(\text{cl}(A) \cap \text{cl}(B)) = \text{cov}(\text{cl}(A)) \cup \text{cov}(\text{cl}(B))$.

20. Consider a CP having $c_j = 1, j = 1, \ldots, n$. Construct a bipartite graph $G = (V, E)$ with vertex set $V = V_1 \cup V_2$, where $V_1 = \{1, \ldots, n\}$ and $V_2 = \{n + 1, \ldots, n + m\}$. There is an edge $(j, n + i)$ if and only if $a_{ij} = 1$.

(a) Prove that $C \subseteq V_1$ is a cover if and only if for all $k \in V_2$, there exists $j \in C$ such that $(j, k) \in E$.
(b) Prove that the cover C is not optimal if and only if G contains as a subgraph a tree $T = (W, F)$ such that
   (i) Each vertex of $V_2 \cap W$ is incident with exactly two edges of F, where one of these edges is incident to a vertex in $C \cap W$ and the other to a vertex in $(V_1 - C) \cap W = Q$.
   (ii) Each vertex in Q is incident to exactly two edges of F.
   (iii) $C' = (C - W) \cup Q$ is a cover.

8.10. NOTES

8.1. A survey of set covering and partitioning is given in Garfinkel and Nemhauser (1972). The conversion of a PP to a CP is from Lemke, Salkin, and Spielberg (1971).

8.2. The information retrieval application is found in Day (1965). A reference to disconnecting graphs is Bellmore, Greenberg, and Jarvis (1970). The truck delivery application is given in Balinski and Quandt (1964). The application of the partitioning problem to political districting is found in Wagner (1968) and Garfinkel and Nemhauser (1970). The airline crew scheduling problem is discussed in Kolner (1966) and Arabeyre et al. (1969). Coloring problems are discussed by Bessiere (1965). The problem of designing optimal switching circuits, or equivalently finding minimal representations for logical functions, is considered by Cobham, Fridshal, and North (1961) and (1962); Cobham and North (1963); Quine (1955); and Pyne and McCluskey (1961). Other applications of the CP are found in Paul and Unger (1959), Bellmore and Ratliff (1971b), and Toregas et al. (1971). Some variations on the PP are given by Dantzig and Ramser (1959), Clarke and Wright (1964), and Jensen (1971).

A particular CP is the graph covering problem discussed in Sections 3.6 and 3.7. Edmonds (1962) establishes some relationships between general CP's and graphs (see Exercise 20).

8.3. The reductions described in this section have been developed in a number of papers, including Balinski and Quandt (1964); Garfinkel and Nemhauser (1969); and Cobham, Fridshal, and North (1961). In many applications, substantial reductions are possible. For problems in which the A matrix is not given, it has been found that the difficult generation process could be shortened significantly by avoiding the generation of columns which would later be deleted by reductions.

8.5. The technique for associating a prime CP solution with a basis matrix and the involutory property is due to Bellmore and Ratliff (1971a). Theorems 2 and 4 are found in Andrew, Hoffman, and Krabek (1968). Theorems 5 and 6 are from Balas and Padberg (1970).

8.6. The reference for this section is Bellmore and Ratliff (1971a). The results are an extension of the work of House, Nelson, and Rado (1966).

8.7. This section is adapted from Lemke, Salkin, and Spielberg (1971). Glover (1971b) gives some extensions.

8.8. The references for this section are Pierce (1968) and Garfinkel and Nemhauser (1969). Some modifications to the basic algorithm, which include using surrogate constraints and LP (see Exercise 17) are given in Pierce and Lasky (1970). A different enumerative algorithm for the PP is given in Marsten (1972).

Further Notes

An approximate method for solving CP's is given in Roth (1969). Thiriez (1969) applies the group theoretic results of Chapter 7 to CP's. Jensen (1971) gives a dynamic programming approach for solving PP's with an underlying graphical structure. Some interesting properties of CP's (see Exercise 19) are given by Lawler (1966).

9 Approximate Methods

9.1. INTRODUCTION

An approximate method (heuristic) for ILP's is one that is designed to provide "good" solutions but cannot guarantee optimality. Methods of this kind are valuable for several reasons. First, computational experience with (exact) algorithms has not yet been completely encouraging. Many real problems of interest are too large to be solved exactly. Second, enumerative algorithms invariably benefit from beginning with a good feasible solution. Finally, a feasible solution provides a lower bound $\underline{z}$ on the optimal objective function value. This bound can be used for fathoming in enumerative algorithms and, in cutting plane algorithms, as a cut $cx \geq \underline{z}$, or as a source row for a cut.

A variety of approximate methods has been proposed. Those discussed in this chapter are representative of the types of techniques available.

Neighborhoods

The concept of a neighborhood of a point x^* is a familiar one from calculus. In Euclidean space it is defined to be an open hypersphere (interval is one dimension) containing x^*. In this chapter the concept of neighborhoods will be used to determine local optima to ILP's. Because the variables are required to be integer, different kinds of neighborhoods are needed. In general, every neighborhood of an integer n-vector x^* will be a set of integer vectors including x^* and, in some sense, near x^*. Some neighborhoods useful in integer programming are given below.

The *unit neighborhood* of x^* is defined as

$$R(x^*) = \{x | x_j = x_j^* - 1, x_j^*, x_j^* + 1, j = 1, \ldots, n\}$$

Thus $|R(x^*)| = 3^n$.

The *m-variable neighborhood* of x^*, $N_m(x^*)$, is the set of integer vectors, each of which differs from x^* in not more than m components.

Local Optima

Consider the ILP

$$\max z = cx, x \in S = \{x | Ax \leq b, x \geq 0 \text{ integer}\} \tag{1}$$

The point $x^o \in S$ is a *global maximum* of (1) if $cx^o \geq cx$, all $x \in S$. In all previous chapters, global optima have been the only ones considered. The point $x' \in S$ is a *local maximum* of (1) with respect to some neighborhood $N(x)$ if

$$cx' \geq cx, \qquad x \in S \cap N(x')$$

A global maximum x^o is a local maximum with respect to any neighborhood containing x^o. Thus enumeration of all local maxima, with respect to any neighborhood, would be sufficient to find a global maximum. The techniques of Sections 9.2 and 9.3 generate local maxima. In general, however, it is not possible to tell whether the generation is complete.

9.2. FINDING LOCAL OPTIMA BY DIRECT SEARCH

In this section two quite simple, but general, approximate methods for solving ILP's are given. The second method is a generalization of the first.

A Basic Approximate Method (AM1)

STEP 1: Select a neighborhood $N(x)$ to be used at every iteration. Choose an $x^* \in S$ and let $\underline{z} = z^* = cx^*$. Go to Step 2.

STEP 2: Solve the problem

$$\max z = cx, \qquad x \in S \cap N(x^*) \tag{2}$$

Go to Step 3.

STEP 3: If (2) is unbounded, the ILP (1) is also unbounded. Suppose that (2) has an optimal solution (z^o, x^o). If $z^o > z^*$, let $x^* = x^o$ and go to Step 2. If $z^o = z^*$, go to Step 4.

STEP 4: The solution x^o is locally optimal with respect to N. Let $\underline{z} = \max\{z^o, \underline{z}\}$. If another iteration is desired, select $x^* \in S$ (hopefully x^* has not previously been encountered) and go to Step 2. If no additional iteration is desired, terminate. The solution that yielded $\underline{z}$ is the best known solution.

AMI—with Neighborhood N_1

If N_1 is used as the neighborhood, Step 2 requires the solution of n trivial ILP's of the form

$$\begin{aligned} z_k^0 = \max z_k &= c_k x_k + \sum_{j \neq k} c_j x_j^* \\ a_{ik} x_k &\leq b_i - \sum_{j \neq k} a_{ij} x_j^*, \qquad i = 1, \ldots, m \\ x_k &\geq 0 \text{ integer} \end{aligned} \tag{3}$$

Let $\Delta x_k = x_k - x_k^*$; then (3) can be written

$$\begin{aligned} z_k^0 = \max z_k &= c_k \Delta x_k + \sum_{j=1}^{n} c_j x_j^* \\ a_{ik} \Delta x_k &\leq b_i - \sum_{j=1}^{n} a_{ij} x_j^* = y_i, \qquad i = 1, \ldots, m \\ \Delta x_k &\geq -x_k^*, \ \Delta x_k \text{ integer} \end{aligned} \tag{4}$$

Problem (4) is equivalent to

$$\begin{aligned} &\max c_k \Delta x_k \\ &\Delta x_k \leq \left[\frac{y_i}{a_{ik}}\right], \qquad i \in S_1 = \{i \mid a_{ik} > 0\} \\ &\Delta x_k \geq \left\langle \frac{y_i}{a_{ik}} \right\rangle, \qquad i \in S_2 = \{i \mid a_{ik} < 0\} \\ &\Delta x_k \geq -x_k^* \end{aligned} \tag{5}$$

Letting

$$\begin{aligned} q_1 &= \begin{cases} \min\limits_{i \in S_1} [y_i / a_{ik}] & \text{if } S_1 \neq \varnothing \\ \infty & \text{if } S_1 = \varnothing \end{cases} \\ q_2 &= \begin{cases} \max\limits_{i \in S_2} \langle y_i / a_{ik} \rangle & \text{if } S_2 \neq \varnothing \\ -\infty & \text{if } S_2 = \varnothing \end{cases} \end{aligned} \tag{6}$$

an optimal solution to (5) is

$$\begin{aligned} c_k \geq 0 &\Rightarrow \Delta x_k^0 = q_1 \\ c_k < 0 &\Rightarrow \Delta x_k^0 = \max\{q_2, -x_k^*\} \end{aligned} \tag{7}$$

From (6) and (7), it follows that (5) is unbounded if and only if $c_k > 0$ and $S_1 = \varnothing$. If (5) is bounded for every k, let

$$z^0 = z_{k'}^0 = \max z_k^0, \qquad k \in \{1, \ldots, n\}$$

and

$$x^o = (x_1^o, \ldots, x_n^o) = (x_1^*, \ldots, x_{k'-1}^*, x_{k'}^* + \Delta x_{k'}^o, x_{k'+1}^*, \ldots, x_n^*)$$

Example

$$\begin{aligned} \max\ & 7x_1 + 2x_2 + 4x_3 + 5x_4 + 8x_5 + 6x_6 + 2x_7 + 10x_8 + 3x_9 \\ & -x_1 + 4x_2 + 3x_4 + x_5 + 2x_6 - 2x_7 + x_9 \le 8 \\ & 2x_1 + 3x_3 - x_4 - 2x_5 + 7x_7 + 2x_8 - x_9 \le 16 \\ & x_1 + x_2 + 2x_3 + 3x_4 + 2x_5 + 4x_6 - 6x_7 + 3x_8 \le 23 \\ & x_1, \ldots, x_9 \ge 0 \text{ integer} \end{aligned} \tag{8}$$

STEP 1: $N(x) = N_1$, $x^* = (0, \ldots, 0)$, $\underline{z} = 0$. The results for Steps 2 and 3 are shown in Table 1.

Table 1

k'	$\Delta x_{k'}^o$	z^o	y
8	+7	70	(8, 2, 2)
9	+8	94	(0, 10, 2)
1	+2	108	(2, 6, 0)*
9	+2	114	(0, 8, 0)
7	+1	116	(2, 1, 6)
5	+2	132	(0, 5, 2)
1	+2	146	(2, 1, 0)
9	+2	152	(0, 3, 0)

STEP 4: $x^o = (4, 0, 0, 0, 2, 0, 1, 7, 12)$ is a local maximum with respect to N_1. Let $\underline{z} = 152$. Select $x^* = (0, 0, 0, 0, 8, 0, 0, 0, 0)$ and go to Step 2. The results for Steps 2 and 3 are given in Table 2.

Table 2

k'	$\Delta x_{k'}^o$	z^o	y
1	+7	113	(7, 18, 0)*
9	+7	134	(0, 25, 0)
7	+3	140	(6, 4, 18)
5	+6	188	(0, 16, 6)
1	+6	230	(6, 4, 0)
9	+6	248	(0, 10, 0)
7	+1	250	(2, 3, 6)
5	+2	266	(0, 7, 2)
1	+2	280	(2, 3, 0)
9	+2	286	(0, 5, 0)

STEP 4: $x^o = (15, 0, 0, 0, 16, 0, 4, 0, 15)$ is a local maximum with respect to N_1. Let $\underline{z} = 286$. Assume that no more iterations are desired and terminate. The solution of iteration 2 is the best known.

The vectors $y^1 = (2, 6, 0)$ in Table 1 and $y^2 = (7, 18, 0)$ in Table 2 are starred. Since $y^2 > y^1$, it follows that the sequence $\{k', \Delta x^o_{k'}\} = \{(9, 2), (7, 1), \ldots, (9, 2)\}$, which produced an increase in z of 44 in Table 1, would do the same in Table 2. However, a sequence $\{k', \Delta x^o_{k'}\}$, which yields a larger increase in z, may be available for y_2 as is the case in Table 2. Clearly, if y^2 is encountered after y^1 within an iteration and $y^2 \geq y^1$, then by the above reasoning the problem is unbounded.

Options for AM1

Step 4 of AM1 is vague. All that can really be said about continuing is that, based on the previous iterations and on the user's knowledge of the real-world problem, two quantities may be estimated and compared. These are: a, the value to the user of the expected increase in $\underline{z}$ from continuing, and b, the expected additional computing cost. One proceeds only if $a > b$.

Another option available is to add the constraint $cx \geq \underline{z} + 1$ to S every time $\underline{z}$ is updated. Implementing this option, however, would make the search for a new feasible solution much more difficult. In any case, the entire matter of finding initial feasible solutions has thus far been ignored. It will be discussed in Section 9.3.

Finally, Step 2 could be simplified by searching for any feasible solution x^o to

$$cx \geq cx^* + 1, \qquad x \in S \cap N(x^*)$$

This option can be useful if the neighborhood definition is such that local maxima are hard to find. It may also reduce the probability of being trapped by a local maximum near a boundary, which is likely to happen for certain neighborhoods (e.g., N_1). If this option is used to solve (8), and the variables are considered in order of their indices, the first iteration might be

STEP 1: $N(x) = N_1$, $x^* = (0, \ldots, 0)$, $\underline{z} = 0$.
The results for Steps 2 and 3 are given in Table 3.

Table 3

k'	$\Delta x^o_{k'}$	z^o	y
1	8	56	(16, 0, 15)
2	4	64	(0, 0, 11)

STEP 4: The point (8, 4, 0, . . ., 0) is a local maximum.

A Generalized Approximate Method (AM2)

When AM1 reaches a local maximum, one must either terminate, or find a new feasible solution. In a sense, one is "trapped" by local maxima. A possible alternative is, at a local maximum x^o, to use a different neighborhood, not contained in $N(x^o)$. In general, one might use an easily calculated neighborhood like N_1 until a local maximum x^o is reached, and then a larger neighborhood like $N_2(x^o)$ in an attempt to "escape." If a point $x' \in N_2$ is found such that $cx' > cx^o$, it is reasonable to switch back to N_1. Of course, if no such point is found, another neighborhood can be tried. In algorithmic form:

STEP 1: Select q neighborhoods $N^1(x), \ldots, N^q(x)$ such that $N^k(x) \not\supseteq N^t(x)$, $t > k$. Choose $x^* \in S$ and let $\underline{z} = z^* = cx^*$. Go to Step 2.

STEP 2: Solve

$$\max z = cx, \qquad x \in S \cap N^1(x^*) \tag{9}$$

Go to Step 3.

STEP 3: If x^o solves (9) and $z^o > z^*$, let $x^* = x^o$ and go to Step 2. If $z^o = z^*$, let $\underline{z} = \max\{z^o, \underline{z}\}$, and $i = 2$. Go to Step 4.

STEP 4: Attempt to find a point x' that satisfies

$$cx' > cx^*, \qquad x \in S \cap N^i(x^*) \tag{10}$$

If x' is found, let $x^* = x'$ and go to Step 2. If no such x' is found and $i < q$, let $i = i + 1$ and repeat Step 4. If $i = q$, go to Step 5.

STEP 5: The solution that yielded $\underline{z}$ is the best known. Either choose another x^* and go to Step 2, or terminate.

Of course, AM1 is a special case of AM2 with $q = 1$.

Example

Return to the example (8).

STEP 1: Let $q = 3$ and $N^1(x) = N_1(x)$, $N^2(x) = N_2(x)$, $N^3(x) = N_3(x)$. Choose $x^* = (0, \ldots, 0)$, $\underline{z} = 0$.

The local maximum with respect to N_1, $x^o = (4, 0, 0, 0, 2, 0, 1, 7, 12)$ is again reached, having $z^o = 152$ and $y^o = (0, 3, 0)$. Using N_2, at most $\binom{9}{2}$ or 36, two variable problems will be considered. They are of the form

$$\begin{aligned} c_p\Delta x_p + c_r\Delta x_r &\geq 1 \\ a_{ip}\Delta x_p + a_{ir}\Delta x_r &\leq y_i^o, \qquad i = 1, \ldots, m \\ \Delta x_p &\geq -x_p^o \\ \Delta x_r &\geq -x_r^o \\ \Delta x_p, \Delta x_r &\text{ integer} \end{aligned} \tag{11}$$

For $p = 1$, $r = 2$, (11) is

$$\begin{aligned} 7\Delta x_1 + 2\Delta x_2 &\geq 1 \\ -\Delta x_1 + 4\Delta x_2 &\leq 0 \\ 2\Delta x_1 \qquad &\leq 3 \\ \Delta x_1 + \Delta x_2 &\leq 0 \\ \Delta x_1 \qquad &\geq -4 \\ \Delta x_2 &\geq 0 \\ \Delta x_1, \Delta x_2 &\text{ integer} \end{aligned}$$

Adding the second and fourth constraints yields $5\Delta x_2 \leq 0$ which, with $\Delta x_2 \geq 0$, yields $\Delta x_2 = 0$. Then the first and fourth constraints result in the contradictory statements $\Delta x_1 \geq 1$, $\Delta x_1 \leq 0$.

A problem that has a feasible solution is obtained with $p = 5$, $r = 7$

$$\begin{aligned} 8\Delta x_5 + 2\Delta x_7 &\geq 1 \\ \Delta x_5 - 2\Delta x_7 &\leq 0 \\ -2\Delta x_5 + 7\Delta x_7 &\leq 3 \\ 2\Delta x_5 - 6\Delta x_7 &\leq 0 \\ \Delta x_5 \qquad &\geq -2 \\ \Delta x_7 &\geq -1 \\ \Delta x_5, \Delta x_7 &\text{ integer} \end{aligned} \tag{12}$$

A feasible solution to (12) is $\Delta x_5 = 2$, $\Delta x_7 = 1$. This solution yields $x^* = (4, 0, 0, 0, 4, 0, 2, 7, 12)$, $z^* = 170$, and $y^* = (0, 0, 2)$, and is also a local maximum with respect to N_1.

Dominated Variables

A variable x_p in (1) is said to be *dominated* by a set $V \subseteq \{1, \ldots, n\}$ if (a) $\sum_{j \in V} a_j \leq a_p$ and (b) $\sum_{j \in V} c_j \geq c_p$. Clearly, if (1) has an optimal solution, one exists with $x_p = 0$. Thus column p may be deleted with no possibility of losing all global maxima. However, depending on the neighborhood used, there may be local maxima with $x_p \geq 1$. It is easy to see that if (b) holds as a strict inequality, and if the neighborhood $N_k(x)$ is used, where $k \geq |V| + 1$, then any local maximum having $x_p \geq 1$ can be avoided.

In the example (8), there are five dominated variables. The variable x_2 is dominated by $V = \{9\}$, x_3 by $V = \{1\}$, x_4 and x_6 by $V = \{5\}$, and x_8 by $V = \{1, 9\}$. Deleting these five dominated variables from (8), we obtain

$$\begin{aligned} \max z = 7x_1 + 8x_5 + 2x_7 + 3x_9 & \\ -x_1 + x_5 - 2x_7 + x_9 &\leq 8 \\ 2x_1 - 2x_5 + 7x_7 - x_9 &\leq 16 \\ x_1 + 2x_5 - 6x_7 &\leq 23 \\ x_1, x_5, x_7, x_9 \geq 0 \text{ integer} & \end{aligned} \tag{13}$$

Solving (13) using AM1, N_1 and $x^* = (0, 0, 0, 0)$, the local maximum obtained is $x_1 = 15$, $x_5 = 16$, $x_7 = 4$, $x_9 = 15$, $z = 286$, an improvement of 89 percent over the solution obtained to (8) using N_1 and $x^* = (0, \ldots, 0)$.

9.3. FINDING FEASIBLE SOLUTIONS

The approximate methods of Section 9.2 require a feasible solution to begin each iteration. If no obvious feasible solution exists, approximate methods for finding them can be used as a phase 1 to precede the search for local maxima.

A measure of the "infeasibility" of a solution is needed. A reasonable one to use is the measure of Section 4.5

$$I(x) = \sum_{i=1}^{m} \max\left\{0, \sum_{j=1}^{n} a_{ij}x_j - b_i\right\}$$

Also, let $T = \{x | x \geq 0 \text{ integer}\}$. Phase 1 begins with an $x^* \in T$ such that $I(x) > 0$ and attempts to reduce $I(x)$ to zero.

Phase I—Method I

STEP 1: Select $N(x)$ and x^*. Go to Step 2.

STEP 2: Solve

$$\min I(x), \qquad x \in N(x^*) \cap T \tag{14}$$

Go to Step 3.

STEP 3: If x^o solves (14) and $I(x^o) = 0$, go to Step 4. If $0 < I(x^o) < I(x^*)$, let $x^* = x^o$ and go to Step 2. If $0 < I(x^o) = I(x^*)$, there are three options:

(a) Choose a new x^* and return to Step 2.
(b) Choose a new $N(x)$ and return to Step 2.
(c) Terminate with no feasible solution. (This does not necessarily mean that $S = \varnothing$.)

STEP 4: Let $x^* = x^o \in S$ and go to Step 1 of AM2.

A reasonable starting point for phase 1–method 1 is any element of

$$Q(x') = \{x | x_j = [x'_j], \langle x'_j \rangle, j = 1, \ldots, n\}$$

where x' is any feasible solution to the LP

$$\max cx, \qquad Ax \leq b, x \geq 0 \tag{15}$$

corresponding to (1). Of course, it may be possible to identify easily an $x^o \in Q(x') \cap S$, in which case phase 1 is unnecessary. However, it is also possible that $Q(x') \cap S = \varnothing$, even though $S \neq \varnothing$.

Phase 1–method 1 is a direct extension of AM1. (The modifications required to develop a phase 1 method corresponding to AM2 are to be provided in Exercise 11.)

Choosing Good Starting Points

When a number of feasible points are available, starting points should be chosen that somehow have a high probability of being near the optimal solution. A good set of points to examine, for instance, would be those near $\hat{x}$, the optimal solution to the LP (15). Thus any point in $Q(\hat{x})$ would seem to be a reasonable starting point for an approximate method.

To indicate the importance of a good starting point, let us return to the example given by (13). The optimal solution to the LP (15) is $\hat{x}_1 = 9.44$, $\hat{x}_5 = 38.78$, $\hat{x}_7 = 10.67$, $\hat{x}_9 = 0$, $\hat{z} = 397.67$. Since $Q(\hat{x}) \cap S = \varnothing$, we choose an element of $Q(\hat{x})$ to start phase 1. Letting $x^* = (x_1^*, x_5^*, x_7^*, x_9^*) = (9, 38, 10, 0)$ to begin phase 1, and using the neighborhood N_1, we obtain the feasible point $x^* = (9, 37, 10, 0)$. Its value $z^* = 379$ is within 5 percent of $\hat{z}$. Switching to AM2 with $q = 3$, $N^1 = N_1$, $N^2 = N_2$ and $N^3 = N_3$, we find that $x^* = (9, 37, 10, 0)$ is optimal with respect to N_1 and N_2. Finally, switching to N_3 yields $x^o = (10, 36, 10, 2)$, $z^o = 384$, which is optimal with respect to N_1, N_2, and N_3, and is the optimal solution to (13).

Phase I—Method 2

Let an optimal solution to the LP (15) be given by $\hat{x} = (\hat{x}_B, \hat{x}_N) = (B^{-1}b, 0)$, where B is an optimal basis matrix. Also, let $E = \{i | \sum_{j=1}^{n} a_{ij}\hat{x}_j = b_i\}$ be the index set of binding constraints and L be the index set of the basic variables excluding slack variables. The approach is to find an integer vector

x^o near $\hat{x}$ that satisfies the constraints of E. Hopefully, x^o will also be feasible to (1), since the remaining constraints are not binding at $\hat{x}$. If not, x^o provides a starting point for phase 1–method 1.

Consider the vector b' given by

$$b'_i = \begin{cases} b_i - 1/2 \sum_{j \in L} |a_{ij}| & \text{if } i \in E \\ b_i & \text{if } i \notin E \end{cases} \tag{16}$$

and the vector $x' = (B^{-1}b', 0)$. Also, let x^* be the integer vector given by $x_j^* = [x'_j + 0.5]$, $j = 1, \ldots, n$. It follows that

$$\sum_{j=1}^{n} a_{ij}x_j^* \leq b_i, \qquad i \in E \tag{17}$$

To prove (17), it is sufficient to prove

$$\sum_{j \in L} a_{ij}x_j^* \leq b_i, \qquad i \in E$$

since $x_j^* = 0$, $j \notin L$. Now $x'_j - \frac{1}{2} < x_j^* \leq x'_j + \frac{1}{2}$ implies that

$$a_{ij}x_j^* \leq a_{ij}x'_j + \tfrac{1}{2}|a_{ij}|, \qquad j \in L \tag{18}$$

From (16) and (18) we obtain

$$\sum_{j \in L} a_{ij}x_j^* \leq \sum_{j \in L} (a_{ij}x'_j + \tfrac{1}{2}|a_{ij}|) = b_i, \qquad i \in E \qquad \blacksquare$$

The integer vector x^* is a feasible solution to (1) if it satisfies

(a) $\sum_{j \in L} a_{ij}x_j^* \leq b_i, \qquad i \notin E$

(b) $x_j^* \geq 0, \qquad j \in L$

If (a) or (b) is not satisfied, a reasonable procedure is to consider points on the line $P = \{x | x = \alpha\hat{x} + (1 - \alpha)x', \alpha \in [0, 1]\}$. Every element of P has $x_j = 0, j \notin L$, and it would not be unreasonable to expect that a feasible solution x^* to (1), given by $x^* = [x_j + 0.5]$, $j = 1, \ldots, n$, can be found for some $x \in P$. If such a point is found, it would have the advantage of being near $\hat{x}$.

Example

Consider the constraint set

$$\begin{aligned} 4x_1 - 3x_2 &\leq 6 \\ 2x_1 + x_2 &\leq 10 \\ x_1 + 3x_2 &\leq 18 \end{aligned}$$

and assume that $E = \{1, 2\}$ and $L = \{1, 2\}$. Then $b' = (\frac{5}{2}, \frac{17}{2}, 18)$. The points $\hat{x}$, x', and x^* are shown in Figure 1. We obtain $x^* = ([x_1' + 0.5], [x_2' + 0.5]) = (3, 3)$. In this example, no other point on P rounds to a different feasible solution.

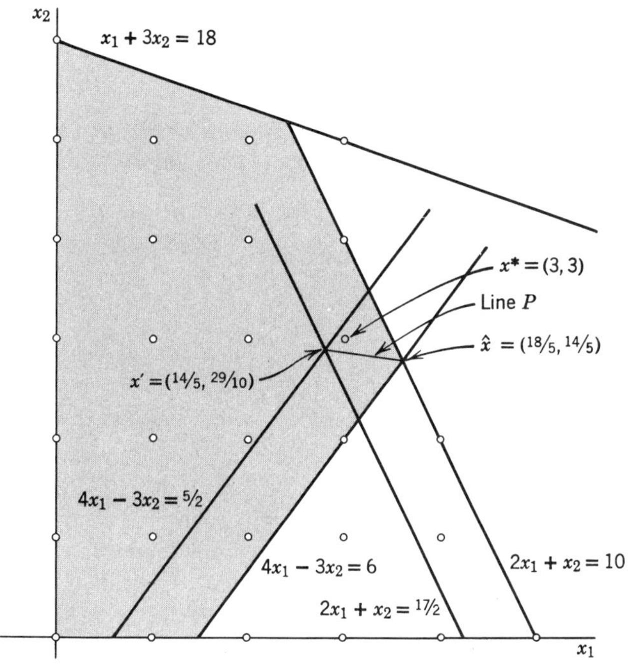

Figure 1

Example

In the example (13), $\hat{x} = (\hat{x}_1, \hat{x}_5, \hat{x}_7, \hat{x}_9) = (9.44, 38.78, 10.67, 0)$. Thus $L = (1, 5, 7)$, $E = (1, 2, 3)$, and

$$B = \begin{bmatrix} -1 & 1 & -2 \\ 2 & -2 & 7 \\ 1 & 2 & -6 \end{bmatrix}$$

Then, $b' = (6, \frac{21}{2}, \frac{37}{2})$ and $x' = (\frac{43}{6}, \frac{169}{6}, \frac{15}{2}, 0)$ so that $x^* = (7, 28, 8, 0)$. Since $x^* \geq 0$ and all constraints are binding, x^* is feasible to (13). It follows that x^* could be used as the starting point for AM2, although it is possible that some other $x \in P$ might round to a better feasible solution.

9.4. EXERCISES

1. What other neighborhoods, besides those given in Section 9.1, might be useful for ILP's?

2. Suppose that x' is a local maximum with respect to N, but there exists another point $x^* \in S \cap N(x')$ such that $cx^* = cx'$. How can AM1 be modified to allow passage from x' to x^*? Is x^* necessarily a local maximum with respect to N?

3. For the following problem,

(a) Do one iteration of AM1 using N_1 and $x^* = (0, \ldots, 0)$.

(b) Continue using AM1 and any other feasible point you can find.

$$\begin{aligned}
\max z = -7x_1 - 3x_2 + 2x_3 + 4x_4 + x_5 & \\
-3x_1 - x_2 \qquad\quad - 4x_4 + 7x_5 &\leq 6 \\
-5x_1 - 2x_2 + 2x_3 + 4x_4 + x_5 &\leq 5 \\
4x_2 + x_3 + 3x_4 - x_5 &\leq 3 \\
x_1, \ldots, x_5 \geq 0 \text{ integer} &
\end{aligned}$$

4. Would a modification of AM2 which stores the y-vectors encountered, in order to test for repetition, be practical?

5. In the subsection "Options for AM1" (p. 329), it was suggested that two quantities a and b could be compared in order to determine whether or not to continue. Under what circumstances are these equivalent to a' the expected cost of the next iteration, and b' the expected improvement in z from the next iteration?

6. Treat the problem of Exercise 3 using AM2, $x^* = (0, \ldots, 0)$, $q = 2$, $N^1 = N_1$, and $N^2 = R$.

7. Construct an example having a dominated variable x_k which when attacked by AM1, N_1 and starting point x^*, yields a local maximum x°; but when attacked by AM1, N_1, starting point x^*, and x_k deleted, yields a local maximum $\bar{x}$, with $cx^\circ > c\bar{x}$.

8. Treat the problem below by phase 1–method 1, $N = N_1$ and $x^* = (0, \ldots, 0)$. When a feasible point is found, continue with AM2, where $q = 2$, $N^1 = N_1$, and $N^2 = N_2$.

$$\begin{aligned}
\max z = 9x_1 - 200x_2 + 7x_3 + 6x_4 - 13x_5 & \\
x_1 - 2x_2 - 11x_3 + x_4 + 2x_5 - 3x_6 &\leq 1 \\
x_1 - x_2 + x_3 + 4x_4 &\leq 3 \\
-5x_1 - x_3 - 20x_4 + x_5 + x_6 &\leq -15 \\
x_1, \ldots, x_6 \geq 0 \text{ integer} &
\end{aligned}$$

An optimal solution to the corresponding LP is $\hat{x}_1 = \frac{1}{3}$, $\hat{x}_4 = \frac{2}{3}$, $\hat{x}_j = 0$ otherwise. Repeat the exercise with $x^* \in Q(\hat{x})$.

9. Attack the problems of Exercises 3 and 8 by phase 1–method 2 and AM2. Optimal basis matrices are

$$\text{Exercise 3:} \quad B = \begin{bmatrix} 1 & 0 & -4 \\ 0 & 2 & 4 \\ 0 & 1 & 3 \end{bmatrix} \qquad B^{-1} = \begin{bmatrix} 1 & -2 & 4 \\ 0 & \frac{3}{2} & -2 \\ 0 & -\frac{1}{2} & 1 \end{bmatrix}$$

$$\text{Exercise 8:} \quad B = \begin{bmatrix} 1 & 1 & -3 \\ 1 & 4 & 0 \\ -5 & -20 & 1 \end{bmatrix} \qquad B^{-1} = \begin{bmatrix} \frac{4}{3} & \frac{59}{3} & 4 \\ -\frac{1}{3} & -\frac{14}{3} & -1 \\ 0 & 5 & 1 \end{bmatrix}$$

10. In the probabilistic approach described in Section 9.5 under "Further Notes," can the assumption of "equally likely maps" be relaxed without destroying the basic results? Discuss the validity of this approach.

11. Extend phase 1–method 1 in a manner analogous to the way AM1 was extended to AM2.

9.5. NOTES

9.2. A general direct search technique for discrete optimization is given in Reiter and Sherman (1965). This paper also contains some statistical tests, based on sampling from the set of feasible starting points, which are used to decide whether to terminate after a given iteration.

Search techniques, including some that are different from the ones we have given are included in Healey (1964), Reiter and Rice (1966), Echols and Cooper (1968), Senju and Toyoda (1968), Trauth and Woolsey (1968), Roth (1970), and Mueller-Merbach (1970). Roth's method is for binary ILP's. He also develops equations for determining the probability of finding an optimal solution as a function of the number of starting points. Mueller-Merbach gives a method for generating cutting planes, in the method of integer forms and primal and dual all-integer algorithms, from approximate solutions.

Rubin (1972) discusses the elimination of dominated variables and superfluous constraints in ILP's.

9.3. Phase 1–method 2, and some associated approximate methods are due to Hillier (1969a).

Further Notes

A quite different approach is taken in Graves and Whinston (1968), where the problem max cx, $Ax \leq b$, x binary is treated. The 2^n possible binary solutions are considered to be maps from $\{1, \ldots, n\}$ to the set of binary n-vectors. The maps are taken to be equally likely, so that in any given map prob $(x_j = 1) =$ prob $(x_j = 0) = \frac{1}{2}$. Using the equally likely assumption, the variables are also shown to be independent. Then the left-hand side of each constraint can be considered to be the sum of independent random variables. When the number of variables is large, this sum has a distribution that approaches the normal distribution. Using an implicit enumeration algorithm (see Section 4.5), at every vertex tests are made to determine the probability of the existence of a feasible completion. Vertices are fathomed if the probability is less than a specified value. Kalymon (1971) comments on this approach.

10 Integer Nonlinear Programming

10.1. EXTENSIONS OF THE LINEAR TECHNIQUES

The general integer nonlinear programming problem (INLP) considered in this chapter is

$$\begin{aligned} \max\ & z(x) \\ & g_i(x) \leq 0, \qquad i = 1, \ldots, m \\ & x \text{ integer} \end{aligned} \tag{1}$$

where z and g_i are real-valued functions. The ILP is a special case of (1) having z and g_i linear.

If the integrality constraints of (1) are dropped, the resulting problem

$$\begin{aligned} &\max z(x) \\ &\quad g_i(x) \leq 0, \qquad i = 1, \ldots, m \end{aligned} \tag{2}$$

is known as a nonlinear programming problem (NLP) of which the LP is a special case. In general, NLP's are much more difficult to solve than LP's, and it is to be expected that the same relationship holds for INLP's and ILP's.

As indicated in Section 1.4, it may be possible to convert an INLP into an ILP by "linearization" techniques. In this chapter, however, the INLP is treated directly. In the remainder of this section, extensions of the techniques developed for ILP's are explored.

Branch and Bound

A general branch and bound algorithm that applies to INLP's was given in Section 4.2. At vertex j, if

$$S_j = \{x \mid g_i(x) \leq 0,\ i = 1, \ldots, m,\ x \text{ integer}\} \cap Q_j$$

where $Q_j = \{x | \alpha^j \leq x \leq \beta^j\}$ then some possible choices for T_j are

$$T_j = \{x | g_i(x) \leq 0, i = 1, \ldots, m\} \cap Q_j \tag{2a}$$

$$T_j = \{x | g_i(x) \leq 0, i \in I, \text{ where } I \text{ (possibly empty) is the index set of the linear constraints in (1)}\} \cap Q_j \tag{2b}$$

$$T_j = \{x | Ax \leq b\} \supseteq S_j \tag{2c}$$

Note that if all of the constraints are linear, (2a), (2b), and (2c) are equivalent. The problem (possibly an LP or NLP)

$$\max z(x), \qquad x \in T_j \tag{3}$$

is solved, and the vertex is either fathomed or separation of S_j is required. If vertex j is not fathomed, partitioning can be accomplished as in Section 4.3. If the optimal solution to (3) is $x^\circ(j)$, not all integer, then for some k, $x_k^\circ(j) > [x_k^\circ(j)]$. The partitioning set can be

$$S_j^* = \{S_j \cap \{x | x_k \leq [x_k^\circ(j)]\}, S_j \cap \{x | x_k \geq \langle x_k^\circ(j) \rangle\}\} \tag{4}$$

If all variables have upper and lower bounds, the resulting algorithm is finite. Of course, the computational feasibility of the branch and bound procedure will depend on the ease of solving (3), and the quality of the bounds obtained from (3).

Example

$$\begin{aligned} \max z(x) = 3x_1^2 &+ 2x_1x_2 \\ x_1 + 2x_2 &\leq 5 \\ 2x_1 + x_2^2 &\leq 7 \\ -x_1 \quad &\leq 0 \\ -x_2 &\leq 0 \\ x_1, x_2 &\text{ integer} \end{aligned} \tag{5}$$

Define T_j as in (2a) and let $\beta^0 = (\infty, \infty)$. At v_0 solve

$$\begin{aligned} \max z(x) = 3x_1^2 &+ 2x_1x_2 \\ x_1 + 2x_2 &\leq 5 \\ 2x_1 + x_2^2 &\leq 7 \\ x_1, x_2 &\geq 0 \end{aligned} \tag{6}$$

The optimal solution is $x_1^\circ(0) = 3.45$, $x_2^\circ(0) = 0.32$, $z_0^\circ = 37.9$. Since the data are integer, take $\bar{z}_0 = 37$ and partition on x_1 or x_2. We choose to partition on x_2 (see Figure 1).

At v_1 we solve (6) with the additional constraint $x_2 \leq 0$ (or equivalently $x_2 = 0$). The optimal solution is $x_1^\circ(1) = 3.5$, $x_2^\circ(1) = 0$, $z_1^\circ = 36.75$. We choose

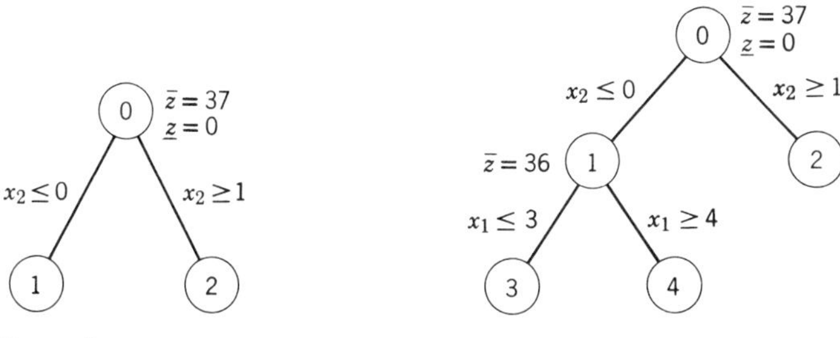

Figure 1

Figure 2

to separate S_1 next and therefore partition on x_1 (see Figure 2). From the second constraint of (5), v_4 can be fathomed immediately. At v_3 the optimal solution is $x_1^o(3) = 3$, $x_2^o(3) = 0$, $z_3^o = 27$ which is integer, and v_3 is fathomed (see Figure 3).

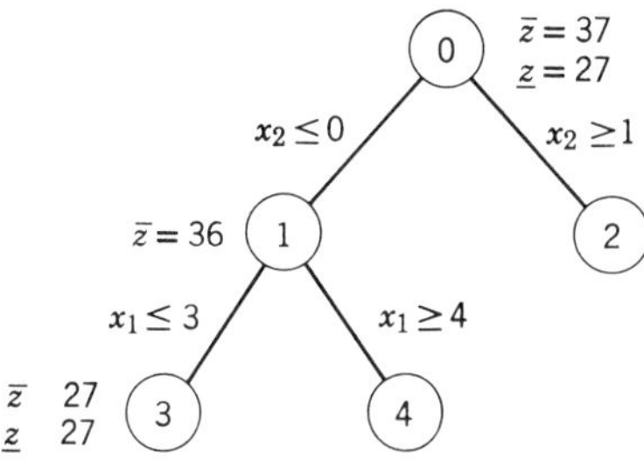

Figure 3

Returning to v_2, the optimal solution is $x_1^o(2) = 3$, $x_2^o(2) = 1$, $z_2^o = 33$. Since $x^o(2)$ is integer, v_2 is fathomed and $x^o(2)$ is the optimal solution to (5).

Implicit Enumeration

The problem

$$\begin{aligned} &\max z(x) \\ &\quad g_i(x) \le 0, \qquad i = 1, \ldots, m \\ &\quad x \text{ binary} \end{aligned} \tag{7}$$

can certainly be attacked by a variant of the implicit enumeration algorithm of Section 4.5, since the fundamental tree structure depends only on the fact that x is binary.

If z is linear, a simple change of variables (see Section 4.5) can be used to make z monotone nonincreasing in each component of x. Thus vertices may be fathomed when a feasible solution is obtained.

If g_i, $i \in I \subseteq \{1, \ldots, m\}$ are linear, then those tests of Section 4.6 which do not depend on the objective function may be employed (see Exercise 5). If all m constraints are linear and $z(x)$ is a negative semidefinite quadratic form, some explicit tests have been derived (see Exercises 2-3).

Cutting Planes

The algorithms of Chapter 5 are all based on deriving valid linear inequalities that are implied by integer solutions to linear equations. Extensions of these methods to general INLP's have not been developed except for certain cases in which z or g_i are quadratic (see Section 10.7).

Approximate Methods

The extensions required to solve INLP's by the approximate methods of Chapter 9 are quite straightforward. Problem (9) of Section 9.2 becomes

$$\max z(x), \qquad x \in S \cap N^1(x^*) \tag{8}$$

where $S = \{x | g_i(x) \leq 0, i = 1, \ldots, m, x \text{ integer}\}$. The computational tractability of the INLP (8) clearly depends on the neighborhood chosen. Similarly, in place of (10) of Section 9.2, one requires a solution x to

$$z(x) > z(x^*), \qquad x \in S \cap N^i(x^*) \tag{9}$$

In Section 10.5 an approximate method is applied to the traveling salesman problem.

10.2. A LEXICOGRAPHIC ENUMERATION ALGORITHM

In this section an algorithm is presented for the binary INLP

$$\begin{aligned} &\max z(x) \\ &\quad g_{i1}(x) - g_{i2}(x) \leq 0, \qquad i = 1, \ldots, m \\ &\quad x \text{ binary} \end{aligned} \tag{10}$$

where the functions $z, g_{i1}, \ldots, g_{m2}$ are monotone nonincreasing in each of the variables $x_1, \ldots, x_n$. A function h is monotone nonincreasing in every component if

$$x^1 \geq x^2 \Rightarrow h(x^1) \leq h(x^2)$$

Binary ILP's can easily be transformed into the form (10). The functions g_{i1} and g_{i2} are determined by grouping the coefficients of like sign in the ith row. The objective function can be made monotone nonincreasing by a simple transformation of variables, as shown in Section 4.5. More generally, any binary INLP can be put in the form of (10) by transforming arbitrary functions into polynomials and grouping coefficients of like sign.

In particular, consider the function g defined on the set of binary n-vectors. Any binary n-vector x is uniquely determined by the equation

$$\prod_{j\in T} x_j \prod_{j\in J-T} (1 - x_j) = 1 \tag{11}$$

where $T \subseteq J = \{1, \ldots, n\}$. The unique binary solution to (11) is $x_j = 1$, $j \in T$, $x_j = 0$, $j \in J - T$. Let J^* be the set of subsets of J; for any $T \in J^*$, let $x(T)$ be the unique solution to (11). It follows that

$$g(x) = \sum_{T\in J^*} g(x(T)) \prod_{j\in T} x_j \prod_{j\in J-T} (1 - x_j) \tag{12}$$

However, the computation involved in achieving (12) may be on the order of total enumeration of the 2^n binary n-vectors.

If $z(x)$ is not monotone nonincreasing, we can add the constraint $-x_0 - z(x) \leq 0$ and maximize $-x_0$.

Base 2 Representations

In the decimal (base 10) number system, an integer k is represented as

$$\begin{gathered} k = k_0 + 10k_1 + 10^2k_2 + \cdots \\ 0 \leq k_i \leq 9, \qquad i = 0, 1, \ldots \end{gathered}$$

and is written as $k = (\ldots k_2k_1k_0)$ with zeros to the left omitted. For a general base q, where $q \geq 2$ and integer, any integer k such that $0 \leq k < q^{n+1}$ may be uniquely written

$$k = k_0 + k_1q + k_2q^2 + \cdots + k_nq^n$$

where $0 \leq k_i \leq q - 1$, $i = 0, \ldots, n$. In vector form $k = (k_n, \ldots, k_0)$.

In this section, integers are expressed base 2. In particular, there is a one-to-one correspondence between the binary n-vectors and the integers $0, \ldots, 2^n - 1$ given by

$$p(x) = x_n + 2x_{n-1} + \cdots + 2^{n-2}x_2 + 2^{n-1}x_1$$

It follows that the 2^n binary n-vectors can be ordered by p and enumerated in that order to solve (10). The order given by p is precisely a *lexicographically increasing* ordering of the binary n-vectors. The method below is a modification of total enumeration in lexicographically increasing order.

Skipping Vectors

Here we develop methods for "skipping" certain of the 2^n binary n-vectors in the enumeration process. If the binary n-vectors are ordered by p, and x has just been considered, skipping means that the next vector enumerated is x^* with $p(x^*) > p(x) + 1$. Such skipping is permissible when it can be shown that every y such that $p(x) < p(y) < p(x^*)$ is not optimal to (10).

Define for an arbitrary binary n-vector x,

$$T(x) = \{y | y > x, y \text{ a binary } n\text{-vector}\}$$
$$Q(x) = \{y | p(y) > p(x), y \text{ a binary } n\text{-vector}\}.$$

Clearly, for any x, $T(x) \subseteq Q(x)$. From a particular vector x, it is often possible to skip to the vector x^*, where

$$p(x^*) = \min p(y), \qquad y \in Q(x) - T(x)$$

Let position k contain the rightmost one in x, and position t the rightmost zero to the left of k. If x^* exists, it is obtained from x by

1. Putting $x_t^* = 1$.
2. Putting $x_j^* = 0, t + 1 \leq j \leq n$.
3. Putting $x_j^* = x_j, 1 \leq j \leq t - 1$.

For example, if $x = (10110)$, then $k = 4$, $t = 2$, and $x^* = (11000)$. To verify the construction for this example, note that $Q(x) = \{(10111), (11000), (11001), (11010), (11011), (11100), (11101), (11110), (11111)\}$, and $T(x) = \{(10111), (11110), (11111)\}$.

PROOF OF CONSTRUCTION: Suppose that the construction cannot be carried out. Then either $x = (0, \ldots, 0)$, or x is one of the vectors $(1, 0, \ldots, 0)$, $(1, 1, 0, \ldots, 0), \ldots, (1, 1, \ldots, 1)$. But in all of these cases $Q(x) - T(x) = \varnothing$, so that x^* does not exist.

If the construction can be carried out, then

$$p(x) = x_1 2^{n-1} + \cdots + x_{t-1} 2^{n-t+1} + 2^{n-t-1} + \cdots + 2^{n-k}$$

and the construction yields

$$p(x^*) = x_1 2^{n-1} + \cdots + x_{t-1} 2^{n-t+1} + 2^{n-t}$$

Thus $p(x^*) - p(x) = 2^{n-t} - 2^{n-t-1} - \cdots - 2^{n-k} > 0$, so that $x^* \in Q(x)$. Also, since $x_k^* = 0$, $x^* \notin T(x)$. Now, suppose that some other $x' \in Q(x) - T(x)$ satisfies $p(x') < p(x^*)$. It follows from $p(x) < p(x') < p(x^*)$ that $x_j' = x_j^* = x_j, j = 1, \ldots, t - 1$, and $x_t' = 0$. But then $p(x') > p(x) \Rightarrow x_j' = 1, j = t + 1, \ldots, k$, which implies that $x' \geq x$, contradicting the hypothesis that $x' \in Q(x) - T(x)$. ■

If x^* exists, denote by x^*_{-1} the vector satisfying $p(x^*_{-1}) = p(x^*) - 1$. Note that x^*_{-1} is of the form $(x_1, \ldots, x_{t-1}, 0, 1, \ldots, 1)$ and thus $x^*_{-1} \geq x$. Also let $W(x) = \{y | y \text{ binary}, p(x) \leq p(y) \leq p(x^*_{-1})\}$. A result that allows for skipping of vectors in the enumeration process is

Lemma 1: *If h is a monotone nonincreasing function, then $h(x^*_{-1}) \leq h(y) \leq h(x)$ for all $y \in W(x)$.*

PROOF: By the construction of x^*, $x \leq y \leq x^*_{-1}$, for all $y \in W(x)$. Then, from the monotonicity of h, the lemma follows. ■

Algorithm

STEP 1: If $x = (0, \ldots, 0)$ is feasible to (10), it is also optimal. Otherwise let $x = (0, \ldots, 0, 1)$ and $\underline{z}$ (possibly $-\infty$) be the best known lower bound on $z(x)$. Go to Step 2.

STEP 2: If $z(x) \leq \underline{z}$, go to Step 5. Otherwise go to Step 3.

STEP 3: If x is feasible to (10), let $\underline{z} = z(x)$ and go to Step 5. Otherwise go to Step 4.

STEP 4: If x^* exists and, for some i, $g_{i1}(x^*_{-1}) - g_{i2}(x) > 0$, go to Step 5. Otherwise, if $x = (1, \ldots, 1)$, go to Step 6; if $x \neq (1, \ldots, 1)$, let $\hat{x}$ be the vector satisfying $p(\hat{x}) = p(x) + 1$, put $x = \hat{x}$ and go to Step 2.

STEP 5: If x^* does not exist, go to Step 6. Otherwise let $x = x^*$ and go to Step 2.

STEP 6: Terminate. If $\underline{z} = -\infty$, (10) has no feasible solution. Otherwise, the solution that yielded $\underline{z}$ is optimal.

The skipping steps 2–5 are justified by Lemma 1. Since z is monotone nonincreasing, it follows that $z(y) \leq z(x)$, $y \in W(x)$, which verifies Steps 2 and 3. Step 4 follows from g_{i1}, g_{i2} monotone nonincreasing, and

$$g_{i1}(y) \geq g_{i1}(x^*_{-1}), \qquad g_{i2}(y) \leq g_{i2}(x), \qquad y \in W(x)$$

Thus

$$g_{i1}(y) - g_{i2}(y) \geq g_{i1}(x^*_{-1}) - g_{i2}(x) > 0, \qquad y \in W(x)$$

violating the ith constraint.

Example

$$\begin{aligned} \max z(x) = \; & -2x_1x_3 - 4x_2 - 3x_1x_5 - 2x_4 - 3x_5 \\ & -x_1x_4 - 3x_2x_5 - (-x_1 - 2x_3 - x_4x_5 - 1) \leq 0 \\ & -2x_1 - 5x_3 - (-2x_4 - 3x_5 - 2) \leq 0 \\ & x_1, \cdots, x_5 = 0, 1 \end{aligned}$$

The solution sequence is given in Table 1. Initially, $\underline{z} = -\infty$, and $x = (0, \ldots, 0)$ is infeasible.

Table 1

x_1	x_2	x_3	x_4	x_5	x^*	x^*_{-1}	Comments
0	0	0	0	1	(00010)	(00001)	Step 4, i = 1, 2, skip to x^*.
0	0	0	1	0	(00100)	(00011)	Step 4, i = 1, 2, skip to x^*.
0	0	1	0	0	(01000)	(00111)	Step 4, i = 1, skip to x^*.
0	1	0	0	0	(10000)	(01111)	Step 4, no skipping.
0	1	0	0	1	(01010)	(01001)	Step 4, i = 2, skip to x^*.
0	1	0	1	0	(01100)	(01011)	Step 4, i = 2, skip to x^*.
0	1	1	0	0	(10000)	(01111)	Step 4, no skipping.
0	1	1	0	1	(01110)	(01101)	Step 3, x is feasible; $z(x) = \underline{z} = -7$.
0	1	1	1	0	(10000)	(01111)	Step 4, no skipping.
0	1	1	1	1	(10000)	(01111)	Step 2, $z(x) = -9$, skip to x^*.
1	0	0	0	0	—	—	Step 4, no skipping.
1	0	0	0	1	(10010)	(10001)	Step 4, i = 1, 2, skip to x^*.
1	0	0	1	0	(10100)	(10011)	Step 4, i = 1, 2, skip to x^*.
1	0	1	0	0	(11000)	(10111)	Step 4, i = 1, skip to x^*.
1	1	0	0	0	—	—	Step 4, no skipping.
1	1	0	0	1	(11010)	(11001)	Step 2, $z(x) = -10$, skip to x^*.
1	1	0	1	0	(11100)	(11011)	Step 4, i = 2, skip to x^*.
1	1	1	0	0	—	—	Step 4, no skipping.
1	1	1	0	1	(11110)	(11101)	Step 2, $z(x) = -12$, skip to x^*.
1	1	1	1	0	—	—	Step 2, $z(x) = -8$, Steps 5 and 6, terminate.

The optimal solution (0, 1, 1, 0, 1) was found in 21 steps, compared to 32 for total enumeration. Additional tests, similar to those of Section 4.6, could be devised to accelerate the enumeration and make the algorithm more attractive.

10.3. PSEUDO-BOOLEAN PROGRAMMING

A different class of enumerative algorithms for (7) goes under the name "pseudo-boolean programming." The name comes from logic, where a *boolean function* is a map from the set of binary vectors to {0, 1}.

A *pseudo-boolean function* is simply a real-valued function of a binary n-vector. As shown in Section 10.2, any pseudo-boolean function can be represented by a polynomial. Specifically, from (12) the polynomial can be written

$$f(x) = \sum_{i=1}^{p} a_i \prod_{j=1}^{n} x_j^{k^{ij}} \tag{13}$$

where $k^{ij} = 0, 1$.

Unconstrained Maximization

Here we treat the problem of maximizing unconstrained pseudo-boolean functions. The technique involves elimination of one variable at a time, until finally a trivial problem in one variable is solved.

Consider the problem

$$\max f(x), \qquad x \text{ binary} \tag{14}$$

where $f(x)$ is a polynomial of the form (13). To begin, write $f(x)$ in the form

$$f(x) = x_1 g_1(x_2, \ldots, x_n) + h_1(x_2, \ldots, x_n) \tag{15}$$

It follows from (15) that there is an optimal solution $(x_1^0, \ldots, x_n^0)$ to (14) having

$$x_1^0 = \begin{cases} 1 & \text{if } g_1(x_2^0, \ldots, x_n^0) \geq 0 \\ 0 & \text{otherwise} \end{cases} \tag{16}$$

Thus we can think of x_1 as a function $x_1(x_2, \ldots, x_n)$ that satisfies (16). It is straightforward, although computationally nontrivial, to express $x_1(x_2, \ldots, x_n)$ as a polynomial by enumerating every combination of the variables $x_2, \ldots, x_n$ that satisfies

$$g_1(x_2, \ldots, x_n) \geq 0 \tag{17}$$

If (17) has k solutions, $x^1, \ldots, x^k$, then as in (12)

$$x_1(x_2, \ldots, x_n) = \prod_{j \in J_+^1} x_j \prod_{j \in J_-^1} (1 - x_j) + \cdots + \prod_{j \in J_+^k} x_j \prod_{j \in J_-^k} (1 - x_j) \tag{18}$$

where $j \in J_+^i$ if $x_j^i = 1$ and $j \in J_-^i$ if $x_j^i = 0$. Since the solutions are distinct, not more than one of the terms on the right equals one for any vector $(x_2, \ldots, x_n)$. Thus $x_1(x_2, \ldots, x_n) = 1 \Leftrightarrow g_1(x_2, \ldots, x_n) \geq 0$.

Example

$$f(x) = 2x_1 + x_2 - 7x_3 - 5x_1x_2x_3 + 3x_2x_4 + 9x_4x_5 \tag{19}$$

In the form of (15)

$$f(x) = x_1(2 - 5x_2x_3) + x_2 - 7x_3 + 3x_2x_4 + 9x_4x_5$$

Thus $g_1(x_2, \ldots, x_5) = 2 - 5x_2x_3$ is independent of x_4 and x_5 and the solutions to (17) can be expressed in terms of x_2 and x_3 as $(x_2, x_3) = (0, 0)$, $(0, 1)$, and $(1, 0)$. From (18)

$$\begin{aligned} x_1(x_2, \ldots, x_5) &= (1 - x_2)(1 - x_3) + (1 - x_2)x_3 + x_2(1 - x_3) \\ &= 1 - x_2 - x_3 + x_2x_3 + x_3 - x_2x_3 + x_2 - x_2x_3 \\ &= 1 - x_2x_3 \end{aligned}$$

Since g_1 is not a function of x_4 and x_5, neither is x_1.

Substituting $x_1(x_2, \ldots, x_n)$ into (15) yields

$$f_2(x_2, \ldots, x_n) = x_1(x_2, \ldots, x_n)g_1(x_2, \ldots, x_n) + h_1(x_2, \ldots, x_n) \tag{20}$$

In the example, (20) is

$$\begin{aligned} f_2(x_2, \ldots, x_5) &= (1 - x_2x_3)(2 - 5x_2x_3) + x_2 - 7x_3 + 3x_2x_4 + 9x_4x_5 \\ &= 2 - 7x_2x_3 + 5x_2^2x_3^2 + x_2 - 7x_3 + 3x_2x_4 + 9x_4x_5 \\ &= 2 - 7x_2x_3 + 5x_2x_3 + x_2 - 7x_3 + 3x_2x_4 + 9x_4x_5 \\ &= 2 - 2x_2x_3 + x_2 - 7x_3 + 3x_2x_4 + 9x_4x_5 \end{aligned}$$

General Step

For $i < n$, $f_i(x_i, \ldots, x_n)$ is transformed into

$$f_i(x_i, \ldots, x_n) = x_i g_i(x_{i+1}, \ldots, x_n) + h_i(x_{i+1}, \ldots, x_n)$$

Then the condition

$$x_i^\circ = \begin{cases} 1 & \text{if } g_i(x_{i+1}^\circ, \ldots, x_n^\circ) \geq 0 \\ 0 & \text{otherwise} \end{cases}$$

is used to determine

$$x_i = x_i(x_{i+1}, \ldots, x_n) \tag{21}$$

At the nth step, no transformation is necessary, and the function $f_n(x_n)$ is maximized by inspection. Then (21) is used recursively, beginning with $i = n - 1$, to trace the optimal solution.

The technique given in this section can be generalized in a straightforward way to constrained INLP's.

Example, Continued

$$f_2(x_2, \ldots, x_5) = x_2(1 - 2x_3 + 3x_4) + 2 - 7x_3 + 9x_4x_5$$

The solutions to $1 - 2x_3 + 3x_4 \geq 0$ are $(x_3, x_4) = (0, 0), (0, 1), (1, 1)$. Thus

$$\begin{aligned} x_2(x_3, x_4, x_5) &= (1 - x_3)(1 - x_4) + (1 - x_3)x_4 + x_3x_4 \\ &= 1 - x_3 - x_4 + x_3x_4 + x_4 - x_3x_4 + x_3x_4 \\ &= 1 - x_3 + x_3x_4 \end{aligned}$$

Therefore,

$$\begin{aligned} f_3(x_3, x_4, x_5) &= (1 - x_3 + x_3x_4)(1 - 2x_3 + 3x_4) + 2 - 7x_3 + 9x_4x_5 \\ &= 3 - 8x_3 + 3x_4 - x_3x_4 + 9x_4x_5 \\ &= x_3(-8 - x_4) + 3 + 3x_4 + 9x_4x_5 \end{aligned}$$

Since $-8 - x_4 \geq 0$ has no binary solution, $x_3(x_4, x_5) = 0$. Thus

$$\begin{aligned} f_4(x_4, x_5) &= 3 + 3x_4 + 9x_4x_5 \\ &= x_4(3 + 9x_5) + 3 \end{aligned}$$

Also, $3 + 9x_5 \geq 0$ for any binary x_5, which implies $x_4(x_5) = 1$, so that

$$f_5(x_5) = 6 + 9x_5$$

Thus $x_5^0 = 1$ and, recursively, $x_4^0 = 1$, $x_3^0 = 0$, $x_2^0 = 1 - x_3^0 + x_3^0x_4^0 = 1$, $x_1^0 = 1 - x_2^0x_3^0 = 1$ is an optimal solution to (19).

10.4. BOTTLENECK INTEGER PROGRAMMING

In this section a class of binary INLP's, identified by the form of their objective function, is considered. The bottleneck programming problem is

$$\max z(x),\ x \in S = \{x \mid g_i(x) \leq 0,\ i = 1, \ldots, m,\ x \text{ a binary } n\text{-vector}\} \tag{22}$$

where

$$z(x) = \min f(j), \qquad j \in V(x) = \{j \mid x_j = 1\} \tag{23}$$

and f is a function from $J = \{1, \ldots, n\}$ to the real numbers. It is assumed that if $V(x) = \varnothing$, then $z(x) = -\infty$ in (23).

Let the elements of J be ordered such that

$$j < k \Rightarrow f(j) \leq f(k) \tag{24}$$

From (23), it follows that an optimal solution to (22) is determined by the order (24), independent of the particular values of $f(j)$ [see Exercise 14].

There are many applications in which maximizing the minimum profit (utility) is an appropriate objective. For instance, (23) is a suitable objective function for the political districting problem described in Section 1.5. Another well-known application is the bottleneck assignment problem, which is to assign n men to n jobs with the objective of maximizing the minimum efficiency.

We present two algorithms for bottleneck problems. An important aspect of both is that they require only feasible solutions to subproblems.

A Primal Algorithm

Let

$$J_q = \{j \mid f(j) > q, j \in J\} \qquad \text{and} \qquad S_q = \{x \mid x_j = 0, j \notin J_q\}$$

An algorithm that is primal, in the sense that a sequence of feasible solutions to (23) is generated, proceeds as follows:

STEP 1: Let $\underline{z} = -\infty$, $J_{\underline{z}} = J$. Go to Step 2.

STEP 2: Let $Q_{\underline{z}} = S \cap S_{\underline{z}}$. If $Q_{\underline{z}} = \varnothing$, go to Step 4. Otherwise choose any $x^* \in Q_{\underline{z}}$ and go to Step 3.

STEP 3: Let

$$\underline{z} = \min f(j), \qquad j \in V(x^*)$$

Go to Step 2.

STEP 4: If $\underline{z} = -\infty$, no feasible solution to (22) exists. Otherwise x^* is an optimal solution to (22) with value $\underline{z}$.

Finite termination of the primal algorithm is assured as long as Step 2 is finite, since $\underline{z}$ strictly increases at each occurrence of Step 3. The difficulty occurs in Step 2. For problems in which Step 2 is especially hard, a natural approach would be to try an approximate method first, and resort to an algorithm only if necessary.

Example

$$\begin{aligned} \max z = \min \{-7x_1, -3x_2, 2x_3, 4x_4, 5x_5\} & \\ -2x_1 - x_2 + 3x_3 + 2x_4 - x_5 &\le 2 \\ 3x_1 + x_3 - x_4 + x_5 &\le 2 \\ -4x_1 + x_2 - 4x_3 + 6x_4 + 3x_5 &\le -1 \\ x_1, \ldots, x_5 &= 0, 1 \end{aligned} \tag{25}$$

STEP 1: $\underline{z} = -\infty$, $J_{-\infty} = J = \{1, \ldots, 5\}$.

STEP 2: An element of $Q_{-\infty}$ is $x^* = (0, 1, 1, 0, 0)$.

STEP 3: $\underline{z} = -3$, $J_{-3} = \{3, 4, 5\}$.

STEP 2: An element of Q_{-3} is $x^* = (0, 0, 1, 0, 1)$.

STEP 3: $\underline{z} = 2$, $J_2 = \{4, 5\}$.

STEP 2: Clearly, the third constraint cannot be satisfied and $Q_2 = \varnothing$.

STEP 4: An optimal solution to (25) is $x^* = (0, 0, 1, 0, 1)$ with value 2.

A Threshold Algorithm

Another approach, in a sense the converse of the primal algorithm, is given here. Let

$$J_q^+ = \{j \mid f(j) \geq q, j \in J\} \qquad \text{and} \qquad S_q^+ = \{x \mid x_j = 0, j \notin J_q^+\}$$

STEP 1: Let $\bar{z} = f(n)$. Go to Step 2.

STEP 2: Let $T_{\bar{z}} = S \cap S_{\bar{z}}^+$. If $T_{\bar{z}} = \varnothing$, go to Step 3. Otherwise choose any $x^* \in T_{\bar{z}}$ and go to Step 4.

STEP 3: If $J_{\bar{z}}^+ = J$, go to Step 5. Otherwise let

$$\bar{z} = \max f(j), \qquad j \in J - J_{\bar{z}}^+$$

Go to Step 2.

STEP 4: Terminate; x^* is an optimal solution to (22) with value $\bar{z}$.

STEP 5: Terminate; there is no feasible solution to (22).

Example

The example (25) is solved by the threshold algorithm.

STEP 1: $\bar{z} = 5, J_5^+ = \{5\}$.
STEP 2: $T_5 = \varnothing$.
STEP 3: $\bar{z} = 4, J_4^+ = \{4, 5\}$.
STEP 2: $T_4 = \varnothing$.
STEP 3: $\bar{z} = 2, J_+^2 = \{3, 4, 5\}$.
STEP 2: $x^* = (0, 0, 1, 0, 1)$ is an element of T_2.
STEP 4: $(0, 0, 1, 0, 1)$ is an optimal solution to (25) with value 2.

Modifications

A better initial estimate of $\bar{z}$ can be determined by relaxing (22). Let

$$\bar{z}_i = \max z(x), x \in \{x \mid g_i(x) \leq 0, x \text{ binary}\} \tag{26}$$

Since (26) is a relation of (22),

$$\bar{z}_i \geq z(x), \qquad x \in S, \qquad i = 1, \ldots, m \tag{27}$$

It follows from (27) that if (26) is infeasible for any i, (22) is infeasible. If (26) is feasible for every i, then from (27) the threshold algorithm can begin with $\bar{z} = \min_{i=1}^{m} \bar{z}_i$. In (25), $\bar{z}_1 = 5$, $\bar{z}_2 = 5$, $\bar{z}_3 = 2$, $\bar{z} = 2$, and the first two iterations are not needed.

Finally, one should observe that if Step 2 is difficult, the algorithm may be very slow unless a sizable improvement in $\bar{z}$ can be achieved in Step 3. In some applications (see Section 10.7 for reference to the bottleneck assignment problem), special structure leads to better estimates of $\bar{z}$ and renders the algorithm more efficient.

10.5. THE TRAVELING SALESMAN PROBLEM

Consider the problem of the traveling salesman who must make a tour of n cities $(1, \ldots, n)$ beginning and ending at city 1. His objective is to minimize the total cost or distance of his tour.

The problem can be represented in terms of a directed graph $G = (V, E)$. Let V correspond to the set of cities, and let there be an edge (i, j) if it is possible to go directly from city i to city j. Formally, a *tour* is a cycle in G that contains every vertex exactly once (hamiltonian cycle). A vertex representation of a tour is

$$t = (1, i_1, \ldots, i_{n-1}, 1) \tag{28}$$

where $(i_1, \ldots, i_{n-1})$ is a permutation of the integers $(2, \ldots, n)$. Thus there are at most $(n - 1)!$ tours.

In terms of edges, the tour (28) is

$$t = ((1, i_1), (i_1, i_2), \ldots, (i_{n-2}, i_{n-1}), (i_{n-1}, 1))$$

Letting the cost (length) of an edge (i, j) be c_{ij}, and the edge set of t be $E(t)$, the cost of tour t is defined to be

$$z(t) = \sum_{(i,j)\in E(t)} c_{ij}$$

If no edge joins vertices i and j, $c_{ij} = \infty$. In particular, it is assumed that $c_{ii} = \infty$, $i = 1, \ldots, n$, and that the graph G contains at least one tour. Thus the traveling salesman problem can be written as

$$\min z(t) = \sum_{(i,j)\in E(t)} c_{ij}, \qquad t \text{ is a tour} \tag{29}$$

It is not necessarily the case that $c_{ij} = c_{ji}$, since c_{ij} can have other interpretations besides distance.

Relationship to the Assignment Problem

Define the variable

$$x_{ij} = \begin{cases} 1 & \text{if edge } (i, j) \text{ is in the tour} \\ 0 & \text{otherwise} \end{cases} \tag{30}$$

For every vertex i, exactly one edge (i, j) must be in every tour. Thus

$$\sum_{j=1}^{n} x_{ij} = 1, \qquad i = 1, \ldots, n \tag{31}$$

Also, for every vertex j, exactly one edge (i, j) must be in every tour. This implies that

$$\sum_{i=1}^{n} x_{ij} = 1, \qquad j = 1, \ldots, n \tag{32}$$

The cost of a tour is

$$\sum_{i=1}^{n} \sum_{j=1}^{n} c_{ij} x_{ij} \tag{33}$$

The problem of minimizing (33) subject to (30)–(32) is the assignment problem introduced in Section 3.1. Unfortunately, as illustrated in Figure 4, the assignment and traveling salesman problems are not equivalent.

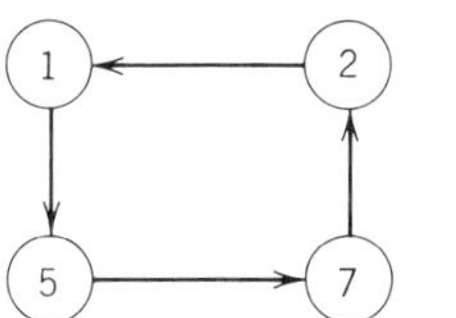

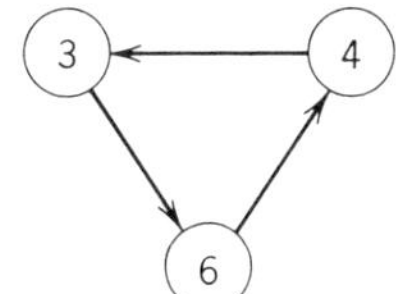

Figure 4

Figure 4 represents a solution to an assignment problem, where $x_{15} = x_{57} = x_{72} = x_{21} = x_{36} = x_{64} = x_{43} = 1$, and $x_{ij} = 0$ otherwise. However, it does not represent a solution to the traveling salesman problem. The difficulty is that this solution represents the two cycles (1, 5, 7, 2, 1) and (3, 6, 4, 3), neither of which is hamiltonian. Such cycles are called *subtours*. It follows that a valid, although not very satisfactory, statement of the traveling salesman problem is to minimize (33), subject to (30)–(32), with no subtours allowed.

To express the subtour constraint mathematically, observe that x^* yields a subtour if and only if for some nonempty subset $Q \subset V$,

$$\sum_{i \in Q} \sum_{j \in \bar{Q}} x_{ij}^* = 0 \tag{34}$$

where $\bar{Q} = V - Q$. Thus tours are the only solutions to (30), (31) and (32) that satisfy, for all $Q \subset V$, $Q \neq \varnothing$

$$\sum_{i \in Q} \sum_{j \in \bar{Q}} x_{ij} \geq 1 \tag{35}$$

and the traveling salesman problem can be written as the ILP

$$\min z = \sum_{i=1}^{n} \sum_{j=1}^{n} c_{ij}x_{ij}$$

$$\begin{aligned} \sum_{j=1}^{n} x_{ij} &= 1, \qquad i = 1, \ldots, n \\ \sum_{i=1}^{n} x_{ij} &= 1, \qquad j = 1, \ldots, n \\ \sum_{i \in Q} \sum_{j \in \bar{Q}} x_{ij} &\geq 1, \qquad \text{every } Q \subset V, Q \neq \varnothing \\ x_{ij} &= 0, 1, \qquad i, j = 1, \ldots, n \end{aligned} \tag{36}$$

However, it is not practical to solve (36) by a general algorithm because of the large number of subtour constraints (35). In fact, for an n-city problem, there are $2^n - 2$ subtour constraints.

The assignment problem is a relaxation of (36) and its solution yields a lower bound on z. Thus it is natural to consider solving (36) by a branch and bound algorithm.

A Branch and Bound Algorithm

At vertex k of the enumeration tree, the problem to be solved is the traveling salesman problem over the graph $G^k = (V, E^k)$, where $E^k \subseteq E^0 = E$. The bounding problem is the related assignment problem. Using the terminology of Section 4.2, the subset of tours at v_k of the enumeration tree is denoted by S_k. If the solution to the corresponding assignment problem does not represent a tour, and v_k is not fathomed by bounds, then a separating set S_k^* must be determined.

To separate S_k, choose any vertex set $Q \subset V$ (there exist at least two), which corresponds to a subtour. If $Q = \{i_1, \ldots, i_p\}$, we impose the constraint (35) by separating S_k into p subsets. Each subset corresponds to a graph having some of the edges of E^k deleted.

For an arbitrary vertex $i \in Q$, define

$$Q_i = \{(i, j) | j \in Q\} \qquad \text{and} \qquad Q_i' = \{(i, j) | j \in \bar{Q}\}$$

Then the edge sets corresponding to the separation S_k^* are

$$\begin{aligned} E_{i_1}^k &= E^k - Q_{i_1} \\ E_{i_2}^k &= E^k - Q_{i_1}' - Q_{i_2} \\ &\vdots \\ E_{i_p}^k &= E^k - Q_{i_1}' - \cdots - Q_{i_{p-1}}' - Q_{i_p} \end{aligned} \tag{37}$$

For example, in a seven-city problem, consider the two subtours of Figure 5 that are assumed to be part of the edge set E^k. If $i_1 = 1, i_2 = 2, i_3 = 3$, the set S_k^* is determined from

$$E_1^k = E^k - \{(1, 2), (1, 3)\}$$
$$E_2^k = E^k - \{(1, 4), (1, 5), (1, 6), (1, 7), (2, 1), (2, 3)\}$$
$$E_3^k = E^k - \{(1, 4), (1, 5), (1, 6), (1, 7), (2, 4), (2, 5), (2, 6), (2, 7), (3, 1), (3, 2)\}$$

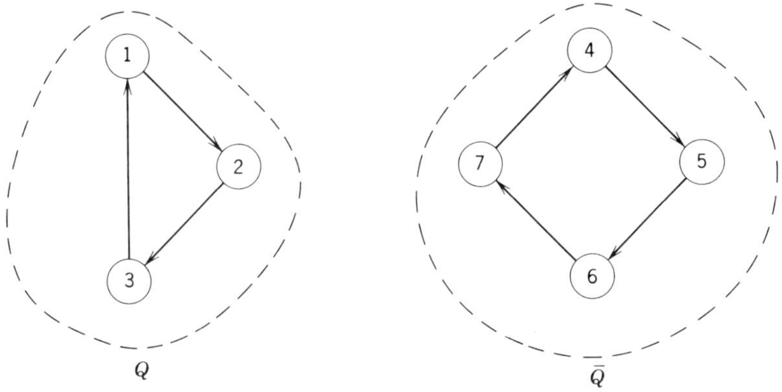

Figure 5

Lemma 2: *Every tour t that is contained in $G^k = (V, E^k)$ is contained in exactly one of the graphs $G_{i_j}^k = (V, E_{i_j}^k), j = 1, \ldots, p$.*

PROOF: We have already shown that t must contain at least one edge of the set $Q'_{i_1} \cup Q'_{i_2} \cup \cdots \cup Q'_{i_p}$. Suppose that $E(t) \cap Q'_{i_1} \neq \varnothing$, then from (37) $E(t) \subseteq E_{i_1}^k$ and $E(t) \nsubseteq E_{i_j}^k, j = 2, \ldots, p$. If $E(t) \cap Q'_{i_1} = \varnothing$ and $E(t) \cap Q'_{i_2} \neq \varnothing$, then $E(t) \nsubseteq E_{i_1}^k$, since $E(t) \cap Q_{i_1} \neq \varnothing$. It follows that $E(t) \subseteq E_{i_2}^k$ and $E(t) \nsubseteq E_{i_j}^k, j \neq 2$. By the same reasoning, for $2 \leq r \leq p$, if $E(t) \cap Q'_{i_j} = \varnothing, j = 1, \ldots, r - 1$, and $E(t) \cap Q'_{i_r} \neq \varnothing$, then $E(t) \subseteq E_{i_r}^k$ and $E(t) \nsubseteq E_{i_j}^k, j \neq r$. ■

To summarize the branch and bound procedure, at vertex k of the enumeration tree the bounding problem

$$\min z_k = \sum_{i=1}^{n} \sum_{j=1}^{n} c_{ij} x_{ij}$$

$$\sum_{j=1}^{n} x_{ij} = 1, \qquad i = 1, \ldots, n \tag{38}$$

$$\sum_{i=1}^{n} x_{ij} = 1, \qquad j = 1, \ldots, n$$

$$x_{ij} = 0, 1, \qquad i, j = 1, \ldots, n$$

is solved over $G^k = (V, E^k)$. In other words, edges in $E - E^k$ have infinite cost.

Let the optimal solution to (38) be $(z_k^o, x^o(k))$, then z_k^o is a lower bound on the value of an optimal tour at v_k. If $x^o(k)$ represents a tour, then v_k is fathomed and $\bar{z}_0$ (the value of the best known solution) is replaced by min $\{\bar{z}_0, z_k^o\}$. If $z_k^o \geq \bar{z}_0$, then v_k is fathomed by bounds. If $x^o(k)$ represents a solution containing subtours, and v_k is not fathomed, then S_k is partitioned as in (37).

Example

$$C = \begin{bmatrix} \infty & 7 & 3 & 12 & 5 & 8 \\ 4 & \infty & 2 & 10 & 9 & 3 \\ 6 & 7 & \infty & 11 & 1 & 7 \\ 7 & 3 & 1 & \infty & 8 & 8 \\ 2 & 10 & 2 & 7 & \infty & 3 \\ 4 & 11 & 7 & 6 & 3 & \infty \end{bmatrix}$$

Solving the associated assignment problem yields an optimal solution of $x_{13} = x_{35} = x_{51} = x_{26} = x_{64} = x_{42} = 1$, and $z_0^o = 18$. The two subtours are shown in Figure 6, and the enumeration tree in Figure 7, where $Q = \{1, 3, 5\}$ has been arbitrarily chosen for partitioning, and $i_1 = 1$, $i_2 = 3$, $i_3 = 5$.

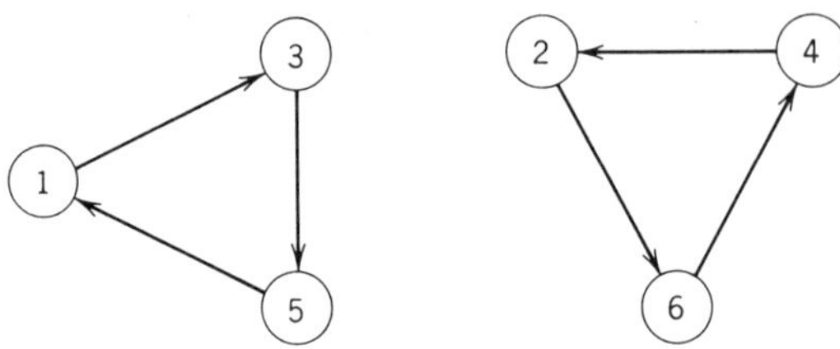

Figure 6

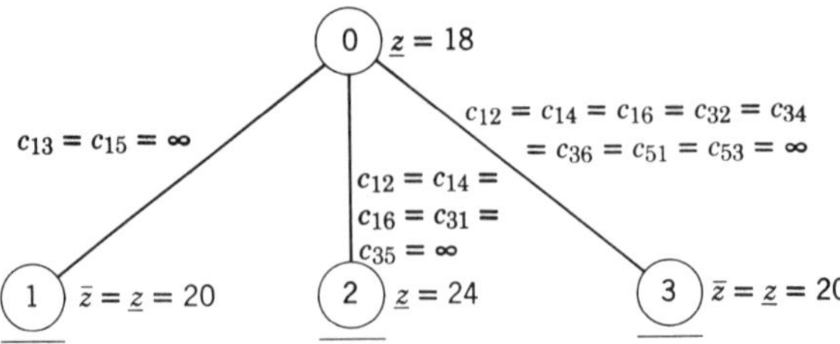

Figure 7

The assignment problem at v_1 has optimal solution $x_{12} = x_{26} = x_{64} = x_{43} = x_{35} = x_{51} = 1$, yielding a tour of length 20. Thus v_1 is fathomed and $\bar{z}_0 = 20$.

At v_2 an optimal solution to the assignment problem is $x_{15} = x_{51} = x_{26} = x_{64} = x_{43} = x_{32} = 1$, $z_2 = 24$, and v_2 is fathomed, since $z_2 > \bar{z}_0$.

At v_3 an optimal solution is $x_{13} = x_{35} = x_{56} = x_{64} = x_{42} = x_{21} = 1$, yielding another tour of length 20. Thus v_3 is fathomed, and the tours of v_1 and v_3 are alternative optimal solutions.

Additional Considerations

Finiteness of the algorithm is easily shown by the arguments of Section 4.3. The formal proof is left to the reader in Exercise 22. As usual, an efficient branching rule is to branch to one of the successor vertices of a vertex just partitioned. Finally, the modified assignment problems are very easily solved, since $n - 1$ assignments are available and a dual feasible solution is at hand. However, the choice of Q and which vertices of Q to designate as $i_1, i_2, \ldots, i_p$, is an open question.

An Approximate Method

For very large traveling salesmen problems, a simple approximate method has worked quite well. It is an adaptation of AM2 given in Section 9.2. An iteration begins with an arbitrary permutation $t^* = (i_1, \ldots, i_n)$ of $(1, \ldots, n)$. If edges $(i_1, i_2), \ldots, (i_n, i_1)$ exist, then t^* is a tour. In our development, it will be convenient to speak of t^* as a tour in either case, although its cost may be infinite. Thus phase 1 calculations will be unnecessary and neighborhoods may be given as functions of tours. The neighborhoods to be used are

$$Q_i(t) = \{t' | t' \text{ differs from } t \text{ in } i \text{ edges or fewer}\}$$

Note that $Q_0(t) = Q_1(t) = \{t\}$.

The Nature of $Q_i(t)$

In general $Q_2(t) = \{t\}$ since, if $t = (1, \ldots, i, j, \ldots, k, l, \ldots, 1)$ as shown in Figure 8, then the only possible substitutions for (i, j) and (k, l) would be (i, l) and (k, j), which would yield the subtour $(1, \ldots, i, l, \ldots, 1)$.

However, if $c_{ij} = c_{ji}$ for all i and j, the problem can be represented on an undirected graph and $Q_2(t) \neq \{t\}$. If $t = (1, \ldots, i, j, \ldots, k, l, \ldots, 1)$, there is a tour $t' = (1, \ldots, i, k, \ldots, j, l, \ldots, 1)$ such that

$$E(t') = E(t) \cup \{(i, k), (j, l)\} - \{(i, j), (k, l)\}$$

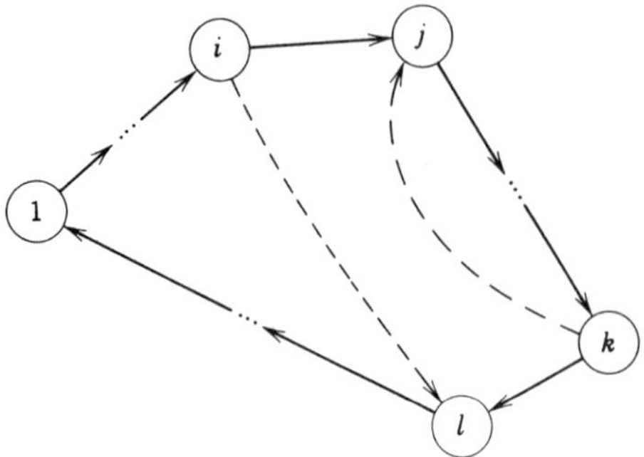

Figure 8

as shown in Figure 9. Thus $|Q_2(t)| = \binom{n}{2} + 1$.

For a directed graph, $|Q_3(t)| = \binom{n}{3} + 1$. If

$$t = (1, \ldots, i, j, \ldots, k, l, \ldots, r, s, \ldots, 1),$$

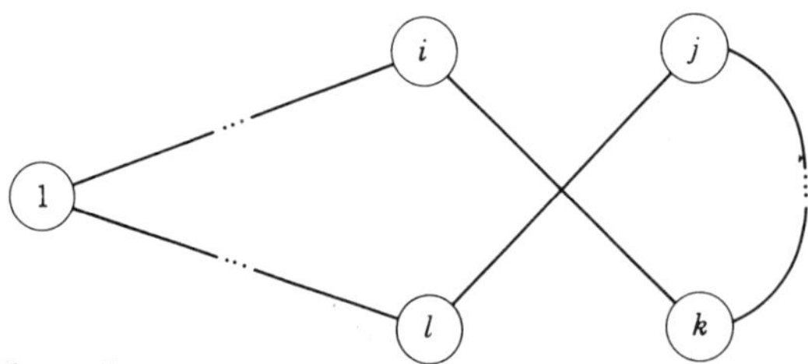

Figure 9

where (i, j), (k, l), and (r, s) are to be replaced, then the unique replacement arcs are (i, l), (k, s), and (r, j), as shown in Figure 10.

The cases given are the only ones in which the set of replacement arcs are uniquely determined.

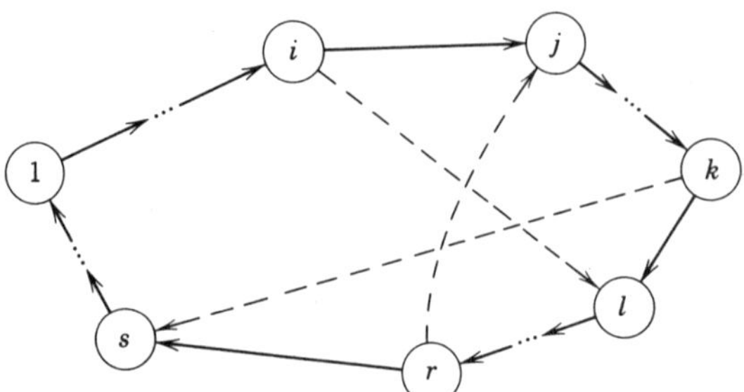

Figure 10

Algorithm

STEP 1: Select p and q such that $2 \le p \le q \le n$. Choose any tour t^*. Let $\bar{z} = z(t^*)$. Go to Step 2.

STEP 2: Attempt to find a solution to

$$z(t) < z(t^*), \qquad t \in Q_p(t^*) \tag{39}$$

Go to Step 3.

STEP 3: If a solution t' to (39) is found in Step 2, let $t^* = t'$ and go to Step 2. Otherwise go to Step 4.

STEP 4: Let $\bar{z} = \min \{\bar{z}, z(t^*)\}$. For each neighborhood $Q_i(t)$, $i = p + 1, \ldots, q$, attempt in turn to find a solution to

$$z(t) < z(t^*), \qquad t \in Q_i(t^*) \tag{40}$$

If t' satisfying (40) is found for some i, let $t^* = t'$ and go to Step 2. If no solution to (40) is found, go to Step 5.

STEP 5: If $\bar{z} = \infty$, no tour has been found. If $\bar{z} < \infty$, the tour that yielded $\bar{z}$ is the best known. Either choose a new t^* and go to Step 2, or terminate.

Example

We solve the example of this section letting $p = q = 3$.

STEP 1: Set $p = q = 3$. Let $t^* = (1, 2, 3, 4, 5, 6, 1)$ and $\bar{z} = z(t^*) = 35$.

STEP 2: Replace (1, 2), (2, 3), and (4, 5) by (1, 3), (2, 5), and (4, 2).

STEP 3: $t^* = (1, 3, 4, 2, 5, 6, 1)$, $z(t^*) = 33$.

STEP 2: Replace (1, 3), (3, 4), and (2, 5) by (1, 4), (3, 5), and (2, 3).

STEP 3: $t^* = (1, 4, 2, 3, 5, 6, 1)$, $z(t^*) = 25$.

STEP 2: Replace (1, 4), (2, 3), and (6, 1) by (1, 3), (2, 1), and (6, 4).

STEP 3: $t^* = (1, 3, 5, 6, 4, 2, 1)$, $z(t^*) = 20$.

STEP 2: No solution is found.

STEP 4: $\bar{z} = 20$

STEP 5: Terminate. For this small example, an optimal solution has been found.

10.6. EXERCISES

1. Solve the example of Section 10.1 by branch and bound, using bounding problems defined by (2b). Solve the same problem using (2c), where an appropriate set of linear constraints must be constructed at every vertex.

2. Consider the problem

$$\max z(x) = x^T Dx$$
$$Ax \leq b \tag{41}$$
$$x \text{ binary}$$

where $D = (d_{ij})$ is a symmetric negative semidefinite matrix. Define S_k^+, S_k^-, and F_k as in the implicit enumeration scheme of Section 4.5. Show that an upper bound on the increment to $z(x)$ caused by transferring j from F_k to S_k^+ is

$$d_{jj} + 2 \sum_{i \in S_k^+} d_{ij} + 2 \sum_{\substack{i \in F_k \\ i \neq j}} \max \{d_{ij}, 0\} \tag{42}$$

3. Extend the implicit enumeration algorithm of Section 4.5 to solve (41) based on (42).

4. Using the algorithm of Exercise 3, solve the problem (41), where

$$D = \begin{bmatrix} -2 & 3 & -1 & -2 & 0 \\ 3 & -5 & 2 & 0 & 0 \\ -1 & 2 & -15 & 1 & 0 \\ -2 & 0 & -1 & -30 & 0 \\ 0 & 0 & 0 & 0 & -2 \end{bmatrix}, \quad A = \begin{bmatrix} 2 & -4 & -3 & 1 & -1 \\ 0 & 2 & 1 & -3 & 1 \\ -1 & -2 & 0 & 3 & -1 \end{bmatrix}$$

$$b = \begin{bmatrix} 2 \\ 0 \\ -1 \end{bmatrix}$$

5. Determine which tests of Section 4.6 apply directly to the problem (41). Attempt to modify those tests which do not apply directly.

6. For an ILP, how does the lexicographic algorithm of Section 10.2 differ from the implicit enumeration algorithm of Section 4.5?

7. Show that the ordering of binary vectors by the function p in Section 10.2 corresponds to a lexicographic ordering.

8. Show that an alternative construction of x^* in Section 10.2 is

(a) Let x_{-1} satisfy $p(x_{-1}) = p(x) - 1$.

(b) Let $\hat{x} = x \diamond x_{-1}$, where $\diamond$ denotes the OR operation described in Section 8.4.

(c) x^* satisfies $p(x^*) = p(\hat{x}) + 1$.

9. Put the following example in the form of (10), and solve by the lexicographic enumeration algorithm:

$$\max z(x) = -3x_1x_3x_5 - 2x_2 - 4x_2x_4 - 3x_5$$
$$2x_1 - 3x_1x_3 + x_4 - 2x_4x_5 \leq -2$$
$$-x_1x_4 - x_2 + 2x_3 - x_4x_5 + 3x_2x_5 \leq 0$$
$$x_1, \ldots, x_5 = 0, 1$$

10. Consider the problem of finding all optimal solutions to a pseudo-boolean program. Then optimality conditions would generalize to

$$x_i^o = \begin{cases} 1 & \text{if } g_i(x_{i+1}^o, \ldots, x_n^o) > 0 \\ 0 & \text{if } g_i(x_{i+1}^o, \ldots, x_n^o) < 0 \\ p_i & \text{if } g_i(x_{i+1}^o, \ldots, x_n^o) = 0 \end{cases}$$

where p_i can be either zero or one. Show how to incorporate the parameter p_i into the algorithm of Section 10.3.

11. Consider the problem

$$\max f(x) = 3x_1 - x_2 - 2x_1x_3x_5 + 2x_2x_6 - x_1x_4x_6 + 2x_4$$
$$x_1, \ldots, x_6 = 0, 1$$

(a) Solve by the algorithm of Section 10.3.

(b) Find all optimal solutions, using the algorithm of Exercise 10.

12. Extend the algorithm of Section 10.3 to solve the problem

$$\begin{aligned} &\max f_0(x) \\ &f_i(x) \leq 0, \qquad i = 1, \ldots, m \\ &x \text{ binary} \end{aligned}$$

where $f_0, \ldots, f_m$ are polynomials.

13. Show that if $x^1, x^2 \neq 0$, are in S of (22), and if $x^1 > x^2$, then an optimal solution to (22) is in $S - \{x^1\}$.

14. Show that the bottleneck problems

$$\max_{x \in S} \min_{j \in V(x)} f(j)$$

and

$$\max_{x \in S} \min_{j \in V(x)} g(j)$$

have the same set of optimal solutions if

$$f(j) < f(k) \Leftrightarrow g(j) < g(k)$$

15. Put the bottleneck assignment problem in the form (22), and devise primal and threshold algorithms for it.

16. Solve the problem

$$\begin{aligned} \max z = \min\{f(1), \ldots, f(6)\} \\ -3x_1 + x_2 + x_3 - 2x_4 + x_6 &\leq -1 \\ 2x_1 - 3x_2x_4 - x_3 + x_4 + x_5x_6 &\leq 0 \\ x_1, \ldots, x_6 &= 0, 1 \end{aligned}$$

where $f(1) < \cdots < f(6)$.

(a) Use the primal algorithm.
(b) Use the threshold algorithm.

17. Consider a bottleneck ILP in the form (22), where $S = \{x | Ax \leq b\}$, and $f(j) = c_j$. Show how this problem can be converted into an ILP. Would it be more effective computationally to make this conversion, or to treat the bottleneck ILP directly?

18. Show that if the "triangle inequality"

$$c_{ij} \leq c_{ik} + c_{kj}, \qquad \text{all } i, j, k$$

is not satisfied, the traveling salesman may find it optimal to visit a city more than once.

19. Extend the branch and bound algorithm of Section 10.5 to the case in which cities may be visited more than once.

20. Show that the number of subtour constraints in (36) is $2^n - 2$.

21. Suppose that an optimal solution to the assignment problem at vertex k of the enumeration tree yields a subtour with edge set Q. The subtour could be eliminated by adding the constraint

$$\sum_{(i,j) \in Q} x_{ij} \leq |Q| - 1 \tag{43}$$

Constraint (43) could be implemented by partitioning S_k into $|Q|$ subsets, each of which has $c_{ij} = \infty$ for some $(i, j) \in Q$.

(a) Show that the constraint (43) is dominated by (35), in the sense that every solution to the assignment problem eliminated by (43) is also eliminated by (35), but not conversely.
(b) What would be the consequence of using (43) instead of (35) for a problem having 10 cities, 5 in New York and 5 in California?

22. Prove that the branch and bound algorithm of Section 10.5 is finite.

23. Solve the traveling salesman problem with the following cost matrix

$$\begin{bmatrix} \infty & 11 & 11 & 9 & 6 & 10 \\ 7 & \infty & 8 & 2 & 3 & 5 \\ 7 & 3 & \infty & 3 & 7 & 6 \\ 10 & 9 & 4 & \infty & 4 & 5 \\ 16 & 13 & 16 & 13 & \infty & 9 \\ 5 & 10 & 6 & 6 & 8 & \infty \end{bmatrix}$$

(a) By the branch and bound algorithm of Section 10.5.
(b) By the branch and bound algorithm suggested in Exercise 21.
(c) By the approximate method of Section 10.5 using $p = q = 3$.

24. Suggest some other neighborhood definitions beside those of Section 10.5 that might prove useful for the traveling salesman problem.

10.7. NOTES

10.1. Laughhunn (1970) extends the basic implicit enumeration algorithm of Section 4.5 to the case in which the constraints are linear and the objective function is a negative semidefinite quadratic form. Also, see Rubin and Hammer (1969) and Hansen (1971). Gomory's method of integer forms has been extended by Witzgall (1963) to handle quadratic constraints of a specific form. Witzgall's procedure is finite; however, Jeroslow (1972) has shown that a finite algorithm of this type cannot be obtained for arbitrary quadratic constraints. Kunzi and Oettli (1963) describe a different cutting plane algorithm for problems with a quadratic objective function and linear constraints. Huard (1970) extends the method of centers given in Huard (1967) to develop a cutting plane algorithm for binary INLP's. Miller (1971) gives an approach to solving unconstrained INLP's using the notion of "discrete convexity."

10.2. The basic reference for this section is Lawler and Bell (1966). Dragan (1968) gives a variation. Mao and Wallingford (1968) extend the basic algorithm to the case in which the objective function is quadratic but not necessarily monotone. Korte, Krelle, and Oberhoffer (1969) develop an algorithm that combines lexicographic search with branch and bound.

10.3. The fundamental reference in pseudo-boolean programming is Hammer and Rudeanu (1968). A survey of their methods is given in Hammer and Rudeanu (1969). Some extensions are given by Rudeanu (1969). The special case of hyperbolic programming, in which the objective function is a ratio of linear functionals, is treated by Robillard (1971). Hansen (1970), Granot and Hammer (1970) and Hammer (1971a) and (1971b) develop algorithms that combine pseudo-boolean programming with branch and bound.

10.4. A general reference for the bottleneck programming problem is Edmonds and Fulkerson (1970). Garfinkel and Nemhauser (1970) give a bottleneck model for political districting and a primal algorithm for solving it. Gross (1958) gives a primal algorithm for the bottleneck assignment problem; a threshold algorithm is given by Garfinkel (1971). Hammer (1969b) gives a primal algorithm for a bottleneck variation of the transportation problem; also see Swarc (1971). Garfinkel and Rao (1971) give a threshold algorithm for the same problem.

10.5. A comprehensive survey of methods for solving the traveling salesman problem is given in Bellmore and Nemhauser (1968); most of their bibliography is not repeated here.

The constraint (35) for blocking subtours is from Dantzig, Fulkerson, and Johnson (1954) and (1959). Different linear constraints for blocking subtours are given by Miller, Tucker, and Zemlin (1960).

The basic branch and bound algorithm of this section is suggested for problems with asymmetric costs in Bellmore and Malone (1971). The partitioning procedure is from Garfinkel (1972). Branch and bound techniques that differ from the algorithm of this section only in the method of breaking subtours (see Exercise 21) and partitioning are given by Eastman (1958) and Shapiro (1966). For problems with symmetric costs, Bellmore and Malone establish by statistical arguments that Shapiro's algorithm is likely to encounter many subtours containing exactly two arcs. For this case, an alternative algorithm, based on weighted b-matching on graphs (see Section 3.7) is developed.

Many procedures are available for solving assignment problems; for instance, see Ford and Fulkerson (1962). Little et al. (1963), give a branch and bound technique in which the separation is not based on breaking subtours, and which uses lower bounds on the assignment problem optimal solution for fathoming.

Held and Karp (1970) and (1971) describe a branch and bound algorithm for the traveling salesman problem that uses a relaxation based on finding minimal weight spanning trees.

The approximate method of this section is from Lin (1965). Lin discusses determining the probability of finding an optimal tour as a function of the number of starting points. Other approximate methods are given by Reiter and Sherman (1965); Webb (1971); and Krolak, Felts, and Marble (1971).

11 Computational Experience

11.1. INTRODUCTION

In this chapter we give an indication of the computational efficiency of various general and special purpose integer programming algorithms. The algorithms cited include those described in the rest of the book and others. No attempt is made to describe in detail algorithms not previously discussed.

Almost every author who reports computational results also invents his own parameters, which he feels are important in generating meaningful problems. Since there is no obvious consistency in these parameters, they have generally been omitted from this chapter. In general, from the many data points listed in an article, a few are chosen in order to convey some typical computer runs and the maximum-size problem attempted.

The exception, where comparisons are possible, is a set of problems developed by Haldi (1964). A number of authors have attempted these problems, which are described in detail in the next section. Wherever possible, the results of these problems are reported. Unfortunately, the Haldi problems are rather small for making meaningful comparisons concerning many current algorithms. A new set of larger test problems is needed.

Another difficulty in reporting computational results is the number of different computers used. Every computer has its own characteristics, and it is not easy to make meaningful statements like "computer A is k times faster than computer B." We will simply list the computer used and let the reader draw his own conclusions. Also, the efficiency of the program is relevant but impossible to evaluate.

Finally, there is a problem in reporting results on approximate methods. There is virtually no point in reporting times to find a suboptimal solution without also giving its deviation from optimality. Unfortunately, for the large problems for which approximate methods are designed, the optimal solutions are generally unknown.

One source of our information has been the survey of Geoffrion and Marsten (1972), which contains a substantial amount of computational experience. In a few cases we cite results from Geoffrion and Marsten that are from unpublished papers or private communications to which we have not had direct access.

11.2. TEST PROBLEMS

These problems were reported by Haldi (1964). Solutions are given in Trauth and Woolsey (1969) and Wahi and Bradley (1969). In addition, Wahi and Bradley give other test problems and relevant information about the solutions.

The first 10 problems, which will be referred to as "Haldi 1–10" are fixed charge problems. Although fairly small, they are often difficult to solve.

Problems 1–4 are of the form

$$\begin{aligned}
\max x_0 = {} & x_3 + x_4 + x_5 \\
& 2x_1 + 3x_2 + x_3 + 2x_4 + 2x_5 \leq b_1 \\
& 3x_1 + 2x_2 + 2x_3 + x_4 + 2x_5 \leq b_2 \\
& -a_{31}x_1 + x_3 \leq 0 \\
& -a_{42}x_2 + x_4 \leq 0 \\
& x_1, x_2 = 0, 1 \\
& x_3, x_4, x_5 \geq 0 \text{ integer}
\end{aligned}$$

The values of the parameters for the four problems are given in Table 1.

Table 1

Problem	a_{31}	a_{42}	b_1	b_2
1	6	7	18	15
2	9	7	18	17
3	9	9	21	21
4	6	8	19	15

Problems 5 and 6 are of the form

$$\begin{aligned}
\max x_0 = {} & x_3 + x_4 + x_5 \\
& 20x_1 + 30x_2 + x_3 + 2x_4 + 2x_5 \leq b_1 \\
& 30x_1 + 20x_2 + 2x_3 + x_4 + 2x_5 \leq b_2 \\
& -a_{31}x_1 + x_3 \leq 0 \\
& -a_{42}x_2 + x_4 \leq 0 \\
& x_1, x_2 = 0, 1 \\
& x_3, x_4, x_5 \geq 0 \text{ integer}
\end{aligned}$$

The parameters are given in Table 2.

Table 2

Problem	a_{31}	a_{42}	b_1	b_2
5	60	75	180	150
6	90	90	210	210

Problems 7 and 8 are obtained from problems 5 and 6, respectively, by removing the third and fourth constraints.

Problem 9 is

$$
\begin{array}{rl}
\max x_0 = & x_4 + x_5 + x_6 \\
& 2x_1 + 2x_2 + x_4 + x_5 \leq 10 \\
& 2x_1 + 2x_3 + x_4 + x_6 \leq 10 \\
& 2x_2 + 2x_3 + x_5 + x_6 \leq 10 \\
& -8x_1 + x_4 \leq 0 \\
& -8x_2 + x_5 \leq 0 \\
& -8x_3 + x_6 \leq 0 \\
& x_1, x_2, x_3 = 0, 1 \\
& x_4, x_5, x_6 \geq 0 \text{ integer}
\end{array}
$$

Problem 10 is of the form max cx, $Ax \leq b$, $x_1, \ldots, x_6$ binary, $x_7, \ldots, x_{12} \geq 0$ integer. The data are given in Table 3 with all the missing entries zero.

Table 3

	x_1	x_2	x_3	x_4	x_5	x_6	x_7	x_8	x_9	x_{10}	x_{11}	x_{12}	
c							1	1	1	1	1	1	b
	9	7	16	8	24	5	3	7	8	4	6	5	110
	12	6	6	2	20	8	4	6	3	1	5	8	95
	15	5	12	4	4	5	5	5	6	2	1	5	80
A	18	4	4	18	28	1	6	4	2	9	7	1	100
	−12						1						0
		−15						1					0
			−12						1				0
				−10						1			0
					−11						1		0
						−11						1	0

The next set of nine problems is referred to as "IBM 1–9." Problems 1–5 are of the form min cx, $Ax \geq b$, $x \geq 0$ integer.

Problems 1 and 2 have $c = (1, \ldots, 1)$, and

$$A = \begin{bmatrix} 1 & 0 & 0 & 1 & 0 & 0 & 1 \\ 0 & 1 & 0 & 1 & 0 & 1 & 1 \\ 0 & 0 & 1 & 0 & 1 & 1 & 1 \\ 1 & 1 & 0 & 0 & 1 & 1 & 0 \\ 1 & 0 & 1 & 1 & 0 & 1 & 0 \\ 0 & 1 & 1 & 1 & 1 & 0 & 0 \\ 1 & 1 & 1 & 0 & 0 & 0 & 1 \end{bmatrix}$$

Problem 1 has $b = (5, 5, 5, 4, 4, 4, 3)$ and problem 2 has $b = (4, 4, 4, 3, 3, 3, 2)$.

Problem 3 has $c = (13, 15, 14, 11, 0, 0, 0)$, $b = (96, 200, 101)$, and

$$A = \begin{bmatrix} 4 & 5 & 3 & 6 & -1 & 0 & 0 \\ 20 & 21 & 17 & 12 & 0 & -1 & 0 \\ 11 & 12 & 12 & 7 & 0 & 0 & -1 \end{bmatrix}$$

Problems 4 and 5 both have $c = (1, \ldots, 1)$. They also have the same binary A matrix. Each column has exactly eight 1's, whose locations are listed in Table 4. The b vector for problem 4 is

$$b^4 = (6, 6, 6, 6, 5, 5, 5, 5, 5, 5, 4, 4, 4, 4, 3).$$

Table 4

Column	1	2	3	4	5	6	7	8	9	10	11	12	13	14	15
Row index of location of 1's	1	2	3	4	1	1	1	2	2	3	1	1	1	2	1
	5	5	6	7	2	3	4	3	4	4	2	2	3	3	2
	6	8	8	9	6	5	5	5	5	6	3	4	4	4	3
	7	9	10	10	7	7	6	6	7	7	7	6	5	5	4
	11	11	11	12	8	8	9	9	8	8	9	8	8	6	11
	12	12	13	13	9	10	10	10	10	9	10	10	9	7	12
	13	14	14	14	13	12	11	12	11	11	11	12	13	14	13
	15	15	15	15	14	14	14	13	13	12	15	15	15	15	14

For problem 5, $b_i^5 = b_i^4 + 2$, $i = 1, \ldots, 15$.

Problem 6 also has a binary A matrix and $c = (1, \ldots, 1)$. The A matrix and b are given in Table 5. For Problems 7 and 8, c, b, and the transpose of A are given in Tables 6 and 7 respectively.

Problem 9 is a set covering problem having $c = (1, \ldots, 1)$ and exactly seven 1's per column. The (35×15) A matrix is described in Table 8.

Table 5

b	A																														
10	1	0	1	0	1	0	1	0	1	0	1	0	1	0	1	0	1	0	1	0	1	0	1	0	1	0	1	0	1	0	1
10	0	1	1	0	0	1	1	0	0	1	1	0	0	1	1	0	0	1	1	0	0	1	1	0	0	1	1	0	0	1	1
9	1	1	0	0	1	1	0	0	1	1	0	0	1	1	0	0	1	1	0	0	1	1	0	0	1	1	0	0	1	1	0
10	0	0	0	1	1	1	1	0	0	0	0	1	1	1	1	0	0	0	0	1	1	1	1	0	0	0	0	1	1	1	1
9	1	0	1	1	0	1	0	0	1	0	1	1	0	1	0	0	1	0	1	1	0	1	0	0	1	0	1	1	0	1	0
9	0	1	1	1	1	0	0	0	0	1	1	1	1	0	0	0	0	1	1	1	1	0	0	0	0	1	1	1	1	0	0
8	1	1	0	1	0	0	1	0	1	1	0	1	0	0	1	0	1	1	0	1	0	0	1	0	1	1	0	1	0	0	1
10	0	0	0	0	0	0	0	1	1	1	1	1	1	1	1	0	0	0	0	0	0	0	0	1	1	1	1	1	1	1	1
9	1	0	1	0	1	0	1	1	0	1	0	1	0	1	0	0	1	0	1	0	1	0	1	1	0	1	0	1	0	1	0
9	0	1	1	0	0	1	1	1	1	0	0	1	1	0	0	0	0	1	1	0	0	1	1	1	1	0	0	1	1	0	0
8	1	1	0	0	1	1	0	1	0	0	1	1	0	0	1	0	1	1	0	0	1	1	0	1	0	0	1	1	0	0	1
9	0	0	0	1	1	1	1	1	1	1	1	0	0	0	0	0	0	0	0	1	1	1	1	1	1	1	1	0	0	0	0
8	1	0	1	1	0	1	0	1	0	1	0	0	1	0	1	0	1	0	1	1	0	1	0	1	0	1	0	0	1	0	1
8	0	1	1	1	1	0	0	1	1	0	0	0	0	1	1	0	0	1	1	1	1	0	0	1	1	0	0	0	0	1	1
7	1	1	0	1	0	0	1	1	0	0	1	0	1	1	0	0	1	1	0	1	0	0	1	1	0	0	1	0	1	1	0
10	0	0	0	0	0	0	0	0	0	0	0	0	0	0	0	1	1	1	1	1	1	1	1	1	1	1	1	1	1	1	1
9	1	0	1	0	1	0	1	0	1	0	1	0	1	0	1	1	0	1	0	1	0	1	0	1	0	1	0	1	0	1	0
9	0	1	1	0	0	1	1	0	0	1	1	0	0	1	1	1	1	0	0	1	1	0	0	1	1	0	0	1	1	0	0
8	1	1	0	0	1	1	0	0	1	1	0	0	1	1	0	1	0	0	1	1	0	0	1	1	0	0	1	1	0	0	1
9	0	0	0	1	1	1	1	0	0	0	0	1	1	1	1	1	1	1	1	0	0	0	0	1	1	1	1	0	0	0	0
8	1	0	1	1	0	1	0	0	1	0	1	1	0	1	0	1	0	1	0	0	1	0	1	1	0	1	0	0	1	0	1
8	0	1	1	1	1	0	0	0	0	1	1	1	1	0	0	1	1	0	0	0	0	1	1	1	1	0	0	0	0	1	1
7	1	1	0	1	0	0	1	0	1	1	0	1	0	0	1	1	0	0	1	0	1	1	0	1	0	0	1	0	1	1	0
9	0	0	0	0	0	0	0	1	1	1	1	1	1	1	1	1	1	1	1	1	1	1	1	0	0	0	0	0	0	0	0
8	1	0	1	0	1	0	1	1	0	1	0	1	0	1	0	1	0	1	0	1	0	1	0	0	1	0	1	0	1	0	1
8	0	1	1	0	0	1	1	1	1	0	0	1	1	0	0	1	1	0	0	1	1	0	0	0	0	1	1	0	0	1	1
7	1	1	0	0	1	1	0	1	0	0	1	1	0	0	1	1	0	0	1	1	0	0	1	0	1	1	0	0	1	1	0
8	0	0	0	1	1	1	1	1	1	1	1	0	0	0	0	1	1	1	1	0	0	0	0	0	0	0	0	1	1	1	1
7	1	0	1	1	0	1	0	1	0	1	0	0	1	0	1	1	0	1	0	0	1	0	1	0	1	0	1	1	0	1	0
7	0	1	1	1	1	0	0	1	1	0	0	0	0	1	1	1	1	0	0	0	0	1	1	0	0	1	1	1	1	0	0
6	1	1	0	1	0	0	1	1	0	0	1	0	1	1	0	1	0	0	1	0	1	1	0	0	1	1	0	1	0	0	1

Table 6

A Transpose

c												
20	46	0	0	0	0	0	0	0	0	0	0	0
25	68	0	0	0	0	0	0	0	0	0	0	0
30	91	0	0	0	0	0	0	0	0	0	0	0
38	137	0	0	0	0	0	0	0	0	0	0	0
46	182	0	0	0	0	0	0	0	0	0	0	0
61	273	0	0	0	0	0	0	0	0	0	0	0
74	364	0	0	0	0	0	0	0	0	0	0	0
101	545	0	0	0	0	0	0	0	0	0	0	0
137	818	0	0	0	0	0	0	0	0	0	0	0
15	0	46	0	0	0	0	0	0	0	0	0	0
17	0	68	0	0	0	0	0	0	0	0	0	0
21	0	91	0	0	0	0	0	0	0	0	0	0
29	0	137	0	0	0	0	0	0	0	0	0	0
31	0	182	0	0	0	0	0	0	0	0	0	0
41	0	273	0	0	0	0	0	0	0	0	0	0
50	0	364	0	0	0	0	0	0	0	0	0	0
67	0	545	0	0	0	0	0	0	0	0	0	0
10	0	0	46	0	0	0	0	0	0	0	0	0
11	0	0	68	0	0	0	0	0	0	0	0	0
14	0	0	91	0	0	0	0	0	0	0	0	0
19	0	0	137	0	0	0	0	0	0	0	0	0
20	0	0	182	0	0	0	0	0	0	0	0	0
27	0	0	273	0	0	0	0	0	0	0	0	0
15	0	0	0	23	0	0	0	0	0	0	0	0
18	0	0	0	46	0	0	0	0	0	0	0	0
22	0	0	0	68	0	0	0	0	0	0	0	0
26	0	0	0	91	0	0	0	0	0	0	0	0
33	0	0	0	137	0	0	0	0	0	0	0	0
41	0	0	0	182	0	0	0	0	0	0	0	0
9	0	0	0	0	23	0	0	0	0	0	0	0
11	0	0	0	0	46	0	0	0	0	0	0	0
14	0	0	0	0	68	0	0	0	0	0	0	0
16	0	0	0	0	91	0	0	0	0	0	0	0
20	0	0	0	0	137	0	0	0	0	0	0	0
25	0	0	0	0	182	0	0	0	0	0	0	0
8	0	0	0	0	0	23	0	0	0	0	0	0
10	0	0	0	0	0	46	0	0	0	0	0	0
12	0	0	0	0	0	68	0	0	0	0	0	0
5	20	0	0	0	0	0	−21	0	0	0	0	0
14	70	0	0	0	0	0	−85	0	0	0	0	0
5	20	20	0	0	0	0	0	−21	0	0	0	0
14	70	70	0	0	0	0	0	−85	0	0	0	0
5	20	20	20	0	0	0	0	0	−21	0	0	0
14	70	70	70	0	0	0	0	0	−85	0	0	0
5	20	20	20	20	0	0	0	0	0	−21	0	0
14	70	70	70	70	0	0	0	0	0	−85	0	0
5	20	20	20	20	20	0	0	0	0	0	−21	0
14	70	70	70	70	70	0	0	0	0	0	−85	0
5	20	20	20	20	20	20	0	0	0	0	0	−21
14	70	70	70	70	70	70	0	0	0	0	0	−85
b	717	498	266	150	165	56	−630	−211	−51	−158	−137	−88

Table 7

A Transpose

c												
1	1	1	1	1	1	1	0	0	0	0	0	0
0	−12	0	0	0	0	0	1	0	0	0	0	0
0	0	−12	0	0	0	0	1	0	0	0	0	0
0	0	0	−12	0	0	0	1	0	0	0	0	0
0	0	0	0	−12	0	0	1	0	0	0	0	0
0	0	0	0	0	−12	0	1	0	0	0	0	0
0	0	0	0	0	0	−12	1	0	0	0	0	0
0	−14	0	0	0	0	0	0	1	0	0	0	0
0	0	−14	0	0	0	0	0	1	0	0	0	0
0	0	0	−14	0	0	0	0	1	0	0	0	0
0	0	0	0	−14	0	0	0	1	0	0	0	0
0	0	0	0	0	−14	0	0	1	0	0	0	0
0	0	0	0	0	0	−14	0	1	0	0	0	0
0	−17	0	0	0	0	0	0	0	1	0	0	0
0	0	−17	0	0	0	0	0	0	1	0	0	0
0	0	0	−17	0	0	0	0	0	1	0	0	0
0	0	0	0	−17	0	0	0	0	1	0	0	0
0	0	0	0	0	−17	0	0	0	1	0	0	0
0	0	0	0	0	0	−17	0	0	1	0	0	0
0	−6	0	0	0	0	0	0	0	0	1	0	0
0	0	−6	0	0	0	0	0	0	0	1	0	0
0	0	0	−6	0	0	0	0	0	0	1	0	0
0	0	0	0	−6	0	0	0	0	0	1	0	0
0	0	0	0	0	−6	0	0	0	0	1	0	0
0	0	0	0	0	0	−6	0	0	0	1	0	0
0	−20	0	0	0	0	0	0	0	0	0	1	0
0	0	−20	0	0	0	0	0	0	0	0	1	0
0	0	0	−20	0	0	0	0	0	0	0	1	0
0	0	0	0	−20	0	0	0	0	0	0	1	0
0	0	0	0	0	−20	0	0	0	0	0	1	0
0	0	0	0	0	0	−20	0	0	0	0	1	0
0	−9	0	0	0	0	0	0	0	0	0	0	1
0	0	−9	0	0	0	0	0	0	0	0	0	1
0	0	0	−9	0	0	0	0	0	0	0	0	1
0	0	0	0	−9	0	0	0	0	0	0	0	1
0	0	0	0	0	−9	0	0	0	0	0	0	1
0	0	0	0	0	0	−9	0	0	0	0	0	1
b	−105	−105	−105	−105	−105	−105	8	10	5	4	6	18

Table 8

Column	1	2	3	4	5	6	7	8	9	10	11	12	13	14	15
Row index of location of 1's	3	4	1	1	2	1	2	3	4	5	11	12	13	14	15
	4	5	5	2	3	6	7	8	9	10	16	17	18	19	20
	7	6	7	8	6	13	14	11	11	12	23	24	21	21	22
	10	8	9	10	9	14	15	15	12	13	24	25	25	22	23
	21	22	23	24	25	17	16	17	18	16	27	26	27	28	26
	26	27	28	29	30	20	18	19	20	19	30	28	29	30	29
	31	32	33	34	35	31	32	33	34	35	31	32	33	34	35

11.3. COMPUTATIONAL RESULTS. GENERAL METHODS

Enumeration Algorithms

Throughout this section and Section 11.4, we use the notation (m, n) to describe an ILP with m constraints and n variables, and (m, n, p) for an MILP with m constraints, n integer variables and p continuous variables.

Roy, Benayoun, and Tergny (1970) have devised a commercial code using a branch and bound algorithm based on some of the penalties of Section 4.4. Some results for runs on the CDC 6600 are given in Table 9.

Table 9

m Constraints	p Continuous Variables	n Integer Variables	Time (sec)
288	244	59	27.9
68	0	536	6.6
604	1955	25	1146.0
1244	3884	24	361.6

Another commercial algorithm along similar lines is described in Benichou et al. (1970). In terms of integer variables, the largest problem solved had $(m, n) = (69, 590)$, and was solved in 25.7 minutes on an unspecified model of the IBM 360. An MILP with $(m, n, p) = (721, 39, 1117)$ was solved in about 18 minutes.

Tomlin (1971) reports that use of all of the penalties of Section 4.4 can reduce running times in predominately integer problems by 50 percent or more over a similar algorithm not using the penalties based on Gomory cuts. For instance, for an ILP with $(m, n) = (5, 50)$ the running time on the UNIVAC 1108 was reduced from 93 to 42 seconds.

Computational experience with some very large problems is reported in Tomlin (1970). These include a binary MILP with $(m, n, p) = (200, 40, 700)$ and two binary ILP's of size about $(m, n) = (800, 200)$ and $(1000, 200)$ respectively. Integer solutions within 3 percent of the continuous optima were obtained, although they could not be proved optimal. Computations were done on a UNIVAC 1108. The first problem took about 10 minutes, the latter two more than 45 minutes.

Davis, Kendrick, and Weitzman (1971) have also developed an algorithm of the same general type for binary MILP's. Running on the IBM 7094, the largest problem attempted had $(m, n, p) = (197, 100, 217)$ and was solved in 5 minutes.

Davis (1969) has also developed an algorithm for the binary MILP. It is of a branch and bound variety but also uses the constraints derived in Section 4.10. Running on the UNIVAC 1108, the code solved all of the Haldi fixed charge problems in less than 1 second. Two scheduling problems with $(m, n) = (40, 131)$ were solved within 20 seconds.

Aldrich (1969) has coded a version of the algorithm of Section 4.9 for the IBM 360/65. The ILP's are solved by a variation of the algorithm of Section 4.5. A problem with $(m, n, p) = (39, 14, 53)$ was solved in 4.5 minutes. The largest problem solved had $(m, n, p) = (378, 24, 136)$, but computation time was not reported.

Childress (1969) has used a similar algorithm to obtain good solutions to very large MILP's. A feasible solution with 1.3 percent of optimum was obtained on a problem with $(m, n, p) = (515, 48, 851)$ in about 20 minutes on a UNIVAC 1108. This type of method has also been used by Manne (1971), who reports solving a binary MILP with $(m, n, p) = (531, 22, 622)$ in about 30 minutes on an IBM 360/67.

Geoffrion (1969) reports contrasting experience with the basic algorithm of Section 4.5 and one incorporating surrogate constraints via linear programming at every partial solution, as described in Section 4.7. The computer used was the IBM 7044. Table 10 contains some of the data points for general test problems, where upper bounds of 3 and 1 have been imposed on the variables for IBM 5 and 6, respectively. Table 11 gives results for set covering problems. Times in Table 11 are averages for five problems using surrogate constraints at every vertex. Seven percent of the entries in the A matrices are 1's (density = 0.07). Two 30-variable problems were solved without the surrogate constraints in 1.2 and 1.6 minutes; however, all attempted 40-variable problems exceeded their 16-minute time limit.

Table 10

Problem	m	n	Time (sec) No LP	Time (sec) LP
	5	39	>600	25.8
	5	50	>600	27.6
Haldi 8	4	20	6.0	3.0
10	10	30	24.6	3.6
IBM 1	7	21	8.4	0.6
5	15	30	>600	114.0
6	31	31	>600	120.0

Table 11

m	n	Time (sec)
30	30	1.8
30	40	4.2
30	50	4.8
30	60	8.4
30	70	9.0
30	80	12.6
30	90	10.2

Trotter and Shetty (1971) have extended the surrogate constraint algorithm of Section 4.7 to handle general integer variables directly. Times on the CDC 6600 for the Haldi and IBM problems, treated directly and by binary expansion, are given in Table 12. These results indicate that direct treatment is more efficient.

Table 12

	Time (sec)	
Problem	Direct	Binary Expansion
Haldi 1	0.03	0.12
2	0.03	0.13
3	0.03	0.20
4	0.02	0.08
7	0.11	2.03
8	0.14	1.57
9	0.03	0.23
10	0.29	4.22
IBM 1	0.06	0.27
2	0.14	0.89
3	0.02	0.24
4	1.50	9.23
5	53.37	519.66
6	62.92	747.73
7	3.33	19.10
8	79.33	22.69
9	4.75	4.75

Thangavelu and Shetty (1971) have used the algorithm of Section 4.7 to solve assembly line balancing problems. The largest problem considered had $(m, n) = (73, 450)$ and was solved in less than 5 seconds on a UNIVAC 1108. Nine other problems were solved in less than 1 second. Additional computational experience with implicit enumeration algorithms on combinatorial problems is given by Ashour and Char (1971).

In Salkin (1970) the importance of a good starting point (see Section 4.5) for implicit enumeration algorithms is illustrated. In fact, it is seen to be often advantageous to restart the enumeration process when a good feasible solution is discovered. The time sacrificed in possibly repeating some of the previous enumeration seems to be more than offset by the flexibility obtained by allowing those variables previously fixed at zero to be free. A good deal of computational experience is reported but in a form difficult to summarize meaningfully.

Hillier (1969b) has derived another enumerative algorithm for ILP's. He uses an efficient heuristic procedure [see Hillier (1969a) and Section 9.5] to generate good starting points. Computations are on the IBM 360/67, and the problems are randomly generated using various parameters. Two problems with $(m, n) = (30, 15)$ each took less than 4 seconds. Two others with $(m, n) = (15, 30)$ took 5 and 160 seconds, respectively. The two largest problems had $(m, n) = (30, 30)$, but neither terminated in 5 minutes.

Korte, Krelle, and Oberhoffer (1970) report results with an enumeration algorithm based on lexicographic search. They contrast their results with those of many other authors. All of the Haldi and IBM problems, except IBM 6 and 7, were attempted with times ranging from 0.5 seconds (Haldi 1–4, 9, IBM 5) to 484 seconds (IBM 8) on an IBM 7090.

Finally, it should be noted that many authors have reported that enumerative algorithms have the desirable property that an optimal solution is often found quite early in the search so that the majority of the time is spent verifying optimality. Thus premature termination often results in a very good feasible, if not optimal, solution.

Cutting Plane Algorithms

Trauth and Woolsey (1969) report computational results that contrast Gomory's method of integer forms (Sections 5.3–5.5) with his dual all-integer algorithm (Sections 5.8, 5.9). A total of five codes based on the two algorithms are compared. Here we give results for two of them. The first, known as LIP1, is a version of the method of integer forms developed by Haldi and Isaacson (1965). It is written in Fortran and FAP (IBM 7090 assembly language) and has a maximum problem size of $(m, n) = (60, 240)$. The second code, called ILP2, is based on the dual all-integer algorithm. Runs of LIP1 were on the IBM 7090, for ILP2 on the CDC 3600. An upper bound of 7000 was used on the number of iterations for both codes where iterations denote constraints added. Note that there will, in general, be many dual simplex iterations for each new constraint in LIP1, but only one in ILP2.

The first set of problems attempted were zero-one knapsack problems of the form

$$\max 20x_1 + 18x_2 + 17x_3 + 15x_4 + 15x_5 + 10x_6 + 5x_7 + 3x_8 + x_9 + x_{10}$$
$$30x_1 + 25x_2 + 20x_3 + 18x_4 + 17x_5 + 11x_6 + 5x_7 + 2x_8 + x_9 + x_{10} \le b$$
$$x_1, \ldots, x_{10} = 0, 1$$

Some Haldi and IBM problems were also run. Results are in Table 13.

Table 13

Problem	LIP 1		ILP 2	
	Time (sec)	Iterations	Time (sec)	Iterations
Knapsack $b = 35$	2.4	19	1.9	51
$b = 70$	2.6	19	1.9	48
$b = 90$	4.5	51	2.0	57
$b = 100$	1.9	12	1.6	34
Haldi 6	7.6	123	3.3	311
7	7.8	159		>7000
8	6.4	126	3.0	306
9	3.2	42	3.6	298
10	9.2	102		>7000
IBM 1	1.9	11	1.1	11
2	3.0	32	1.1	15
3	2.9	53	0.6	14
4	11.7	73	3.1	18
5	66.4	351	26.2	842
9	473.1	953	75.1	1105

Srinivasan (1965) investigates various source row selection rules in the method of integer forms (Section 5.4). He considers 8 ILP's with $11 \le m \le 48$ and $14 \le n \le 98$ and various combinations of selection rules too complex to describe here. Computations were done on the IBM 7094. The best rule varied from problem to problem. One of the more difficult of the eight problems with $(m, n) = (37, 44)$, was solved in about 32 seconds and required 289 pivots using the best rule for it. To contrast the different rules, a problem with $(m, n) = (10, 20)$ was solved in about 4 seconds and 71 pivots, using its best rule, but could not be solved in 600 pivots using other reasonable rules. Finally, another group of 6 ILP's that could not be solved with any of the rules in 600–900 pivots is given. One of these problems is as small as $(m, n) = (10, 7)$.

Geoffrion and Marsten (1972) state that Martin reports success in solving a number of large set covering problems on the CDC 3600 using the algorithm of Section 5.14. Crew scheduling problems with $(m, n) = (100, 4000)$ typically take about 10 minutes; larger ones with $(m, n) = (150, 7000)$ are often solved using only a few cuts in 40 minutes.

Other authors also report that set covering and partitioning problems often require very few cuts after an optimal solution to the corresponding LP has been obtained. Of the 10 partitioning problems, ranging in size from $(m, n) = (5, 30)$ to $(m, n) = (15, 388)$, considered by Balinski and Quandt (1964), only 3 required more than two cuts and only 1 required more than seven cuts.

Toregas, et al. (1971) have solved more than 150 structured set covering problems arising from location applications. These problems have square A matrices, unit costs, and as many as 50 constraints. In most cases no cuts were required and more than one cut was never needed.

Baugh, Ibaraki, and Muroga (1971) report solving problems as large as $(m, n) = (91, 240)$, using the dual all-integer algorithm of Section 5.9. Computation times are not given.

Almost no computational experience has been obtained for the primal all-integer algorithm of Sections 5.11 and 5.12. However, Arnold and Bellmore (1971c) have attempted 20 randomly generated problems having $(m, n) = (10, 10)$. Fourteen of those problems required 58 or fewer iterations. However, none of the others was solved in 500 iterations.

ILPC-Based Algorithms

Shapiro (1968b), and Gorry and Shapiro (1971) report computational results for the algorithm of Section 7.6. The algorithm of the 1971 paper uses a different separation in the branch and bound phase than the one given in Section 7.6 and also finds all alternative optima to the ILPC. Nine of the 17 problems reported in the 1968 paper were solved without enumeration. Three others required enumeration only because all alternative optima to the ILPC were not generated before the branch and bound phase. Of 25 problems considered in the 1971 paper, 19 did not require enumeration. On the negative side, it is indicated that some ILPC's could not be solved because of very large determinants. Bradley and Wahi (1969) solve ILPC's for 28 test problems using the transformation to Hermite Normal Form discussed in Chapter 7, Exercises 10–14. In 23 of the 28 problems some optimal solution to the ILPC was feasible to the ILP. Although there was some overlap in the problem sets considered by Shapiro, Gorry and Shapiro, and Bradley and Wahi, in each case about 75 percent of the ILP's could be solved as ILPC's.

Some results in Shapiro (1968), on the Haldi and IBM problems are shown in Table 14. The computer is the IBM 360/65 and D is the absolute value of the determinant of the optimal basis matrix. For some of these problems, slightly better times are reported in Gorry and Shapiro (1971).

Table 14

Problem		D	Time (sec)	ILPC Solves ILP
Haldi	1	183	1.3	Yes
	2	258	1.0	Yes
	3	320	1.3	Yes
	4	205	0.9	Yes
	9	2000	6.6	Yes
IBM	1	32	2.2	Yes
	2	32	2.0	Yes
	3	72	0.8	No
	8	2856	9.5	Yes
	9	96	20.2	No

Some of the larger problems considered in Gorry and Shapiro (1971) were solved on a UNIVAC 1108. Included in these were a group of media selection problems having $(m, n) = (75, 150)$, which were solved in under a minute. The largest problem considered was a crew scheduling problem with $(m, n) = (104, 236)$; it was solved in about 7 minutes. Gorry and Shapiro also indicate that cutting planes and other procedures for reducing the size of determinants have been incorporated into their algorithm so that it is now possible to solve ILPC's when the optimal basis matrix has a very large determinant [see also, Gorry, Shapiro, and Wolsey (1972)]. For problems in which the branch and bound phase of their algorithm has been very time-consuming, an improved fathoming rule has yielded substantial savings [see Gorry et al. (1970)].

Approximate Methods

Roth (1970) gives computational experience for a variation of AM2 of Section 9.2. His technique, specific to binary problems, essentially uses the neighborhoods $N_k(x)$ and tries to find any improving point at Step 2. Some results for standard problems are given in Table 15. Times specify the average for 25 iterations and f indicates the proportion of iterations in which the optimal solution was found. Runs are on the GE 635.

Echols and Cooper (1968) give a direct search approximate method for the general ILP, which is also similar to AM2. Computations were done on the IBM 7072, with problems ranging up to $(m, n) = (27, 21)$. For these

Table 15

Problem	$k = 5$		$k = 3$	
	f	time (sec)	f	time (sec)
IBM 4	0.92	8.2	0.88	0.9
6	0.8	360.0	0.12	8.0
9	0.4	5.0	0.32	0.5

small problems the optimal solution was found in almost every case. Times ranged from 0.006 to 9.3 minutes.

Hillier (1969a) reports results for a variation of a direct search approximate method using phase 1-method 2. Solving problems of size $(m, n) = (15, 15)$ and $(30, 15)$, using various points on the generated line segment as starting points, he is generally able to find an optimal solution. Times are in the order of a few seconds on the IBM 360/67.

11.4. COMPUTATIONAL RESULTS. SPECIAL PURPOSE METHODS

Knapsack Problem

Cabot (1970) devised an enumerative algorithm for the knapsack problem which is described in Exercises 14–17, Chapter 6. The a_j's and c_j's were generated from the uniform distribution between 10 and 110, and $b_j = 2\sum_{j=1}^{n} a_j$. The computer was the CDC 3600. Results are given in Table 16. For each value of n, 50 problems were run.

Table 16

n	Average time (sec)	Maximum time (sec)
60	0.41	3.5
70	0.66	3.5
80	0.86	7.2

Greenberg and Hegerich (1970) use a variation of the branch and bound algorithm of Section 4.3 to solve the binary knapsack problem. The a_j's and c_j's are randomly generated (the distribution used is not reported). The computer is the IBM 360/67. Problems with $b = 100$ and $n = 50$ took on the average 0.163 seconds. It is also reported that several problems with 5000 variables were solved in approximately 4 minutes.

Nemhauser and Ullmann (1969) give results for an algorithm based on (4) in Chapter 6. The variables are bounded. The largest run had a_j's and c_j's uniformly random between 0 and 99, $b = 2500$, $n = 50$, and $x_j \leq 2$ for all j. Computation time was 91.2 seconds on the IBM 7094.

Unfortunately, we cannot summarize any computational experience with the Gilmore and Gomory [(1963), (1965), and (1966)] algorithms. The 1963 paper contains substantial experience, but the data are based on cutting stock problems so that it is difficult to extract the data and times for the knapsack problems. The 1966 paper contains results for the recursion of Section 6.4, modified as described in Exercise 21, Chapter 6. These results demonstrate that the periodic property generally makes it possible to terminate calculations after finding an optimal solution for a much smaller knapsack than the one being considered.

Edge Matching

Gordon (1971) has coded versions of the Balinski (1969a) and Edmonds (1965a) algorithms for finding maximum cardinality edge matchings on graphs. His results indicate slightly faster running times for Balinski's algorithm. The largest problem considered has 60 vertices and 263 edges. It was solved in less than 2 seconds on an IBM 360/65 with both methods.

Edmonds and Johnson (1970) report results for their algorithm for b-weighted edge matching problems. A typical run for a problem with 300 vertices, 1500 edges, $b_i = 1, 2$, and $1 \leq c_j \leq 10$ takes about 30 seconds on an IBM 360/91.

Of all the problems discussed in this chapter, edge matching is the only one for which there is an algorithm with an upper bound on the number of calculations that can be expressed as a polynomial in the parameters (m, n). The general question of whether such "polynomial algorithms" exist for integer programming problems is discussed in Karp (1972).

Set Covering and Partitioning

Bellmore and Ratliff (1971a) coded the algorithm of Section 8.6 for the IBM 7094. For the problems of Table 17, costs are random between 1 and 99. Failures to solve problems were caused by the core limitations of the computer.

A variation of the algorithm of Section 8.7 was run on the IBM 360/50 by Lemke, Salkin, and Spielberg (1971). Their computational results are hard to summarize, since a large number of parameters are introduced. One run with $(m, n) = (30, 90)$ and density of 0.07 took 36 seconds, and another with $(m, n) = (50, 450)$ and density of 0.046 took 2.15 minutes.

Table 17

m	n	Density	Problems Attempted	Problems Solved	Average Time (sec)
50	80	0.07	10	9	11.5
50	90	0.07	10	8	19.7
30	80	0.21	10	9	8.5
30	90	0.21	10	10	7.6

The set partitioning algorithm of Section 8.8 has been coded in FORTRAN and MAP (IBM 7094 assembly language) by Garfinkel and Nemhauser (1969). Some results are given in Table 18. The first 15 problems were randomly generated, while the sixteenth was adapted from a redistricting problem. Problems 10 and 11 illustrate the effect of density on the algorithm. Problems with higher densities are much easier, since more list skipping occurs.

Table 18

Problem	m	n	Density	Time (sec)
1–9	20	100	0.1–0.5	2–5
10	37	200	0.3	5
11	37	200	0.1	257
12	26	385	0.16	25
13	26	777	0.16	39
14	100	1400	0.15	895
15	37	1790	0.25	>900
16	37	1790	0.25	49

A similiar algorithm has also been run by Pierce (1968) with comparable results. Pierce and Lasky (1970) have incorporated the surrogate constraints of Section 8.8, linear programming (see Exercise 17 of Chapter 8) and other modifications into the algorithm and have obtained some better results. A different enumeration scheme, Marsten (1971), enumerates on the rows rather than the columns. Marsten's algorithm is considerably more efficient for low density problems, but does not perform as well on high density ones.

Fixed Charge Problem

Steinberg (1970) has developed several approximate methods, as well as a branch and bound algorithm for the general fixed charge problem. The algorithm, run on the IBM 7072, solved binary MILP's with $(m, n, p) = (15, 30, 30)$ in from 4.8 to 47.3 minutes. For the same size problems, however, one of the heuristics was generally able to get within 3 percent of the optimal solution.

Spielberg (1969a) has developed a specialization of the algorithm of Section 4.10 to a version of the plant location model described in Section 1.5. His model is a binary MILP with m being the number of plants plus the number of customers, n the number of plants, and $p = mn$. He gives a multitude of results that are difficult to summarize. Two of the larger problems had $(m, n, p) = (250, 100, 15000)$ and $(m, n, p) = (184, 92, 8464)$. They were solved in 30 and 46 minutes respectively on the IBM 360/50.

Traveling Salesman Problem

The branch and bound algorithm of Section 10.5 (with a somewhat different separation procedure) has been coded by Bellmore and Malone (1971). Asymmetrical problems with $n \leq 70$ were run on the IBM 7094 and with $n \geq 80$ on the IBM 360/65. Results are in Table 19.

Table 19

n	Number of Problems	Average Time (sec)
10	20	0.1
20	20	1.7
30	20	12.7
40	10	18.7
50	5	27.4
60	2	79.5
70	2	167.7
80	2	165.4
100	1	> 300.0
120	1	> 300.0

Held and Karp (1971) report computational results with an improvement of the branch and bound algorithm of Held and Karp (1970). The bounds are obtained from a relationship between the symmetric traveling salesman problem and the minimum spanning tree problem. It is indicated that the bounds are generally sharp so that very little enumeration is required. Nineteen problems having symmetric costs and ranging from $n = 20$ to $n = 64$ were solved. The longest time was 15 minutes for a problem with $n = 46$ on the IBM 360/91. Most problems were solved in less than 5 minutes and one 64-city problem was solved in 3 minutes.

Martin (1966) has applied the algorithm of Section 5.14 to a 42-city problem. The model has 861 binary variables and 84 assignment constraints. Nine subtour constraints and 10 cuts were added; the problem was solved in 5 minutes on the IBM 7094.

Lin (1965) gives results for the approximate method of Section 10.5 using $p = q = 3$. Table 20 summarizes results of runs on the IBM 7094.

Table 20

n	Runs	Number of Times Optimal Tours Were Produced	Total Time (sec)
33	50	19	10.7
42	40	11	36.3
48	100	21	63.0
57	200	6	256.4
105	20	0	476.0

Webb (1971) gives a very simple, "one-pass" heuristic for large traveling salesman problems. The method is very fast, requiring only seconds of CDC 6600 time for 500-city problems. Tours produced are generally not more than 50 percent longer than a lower bound.

Krolak, Felts, and Marble (1971) describe a man–machine iterative system for obtaining approximate solutions to large traveling salesman problems. Tours claimed to be good are found for 100-, 150-, and 200-city problems in 3–9 minutes of Sigma 7 time and $1\text{–}1\frac{1}{2}$ hours of human time.

11.5. SUMMARY AND SYNTHESIS

Developments in the use, computational aspects, and theory of integer programming have been substantial. There is a rich variety of integer programming models of real problems, and some measure of success has been achieved in solving them. However, problem size is still a major limitation.

For general MILP's, branch and bound programs, based on the developments in Sections 4.3 and 4.4, are most widely used and have been most successful. Less substantial, but comparable, experience has been obtained using algorithms based on the partitioning concept of Section 4.9. The number of integer variables appears to be the most significant measure of difficulty using these approaches. Experience with commercial codes indicates that MILP's having fewer than 50 integer variables are extremely likely to be solvable, problems with 50–100 integer variables have a good chance of being solved, and good feasible solutions can be obtained for substantially larger problems. All of these commercial codes have highly sophisticated linear programming subroutines. The limitation in the number of constraints and continuous variables depends heavily on the linear programming code. Successful runs have been obtained for problems having more than 500 continuous variables and constraints. Nevertheless it is still possible to encounter quite small problems that defy solution, since no reasonable upper bound on the number of calculations is known.

The use of both linear programming and some form of enumeration appears to be necessary to achieve good computational results for general ILP's and MILP's. Algorithms based exclusively on cutting planes or implicit enumeration without surrogate constraints (derived from optimal linear programming solutions) have not, in general, been effective for medium and large problems.

Although pure cutting plane algorithms appear to be of limited practical value, cuts can be very important in branch and bound algorithms. It has already been shown that improved penalties can be obtained from cuts. Moreover, cuts can be incorporated into a branch and bound algorithm in a more direct way. In the process of solving an ILP or MILP by branch and bound, an alternative to making the separation $x_{B_r} \leq [y_{r0}]$, $x_{B_r} \geq \langle y_{r0} \rangle$ is to use row r as a source row for a cut of the type developed in Section 5.17. If there is no dual degeneracy, progress in reducing the objective function can be achieved without resorting to separation, thereby eliminating the resulting growth in tree size. Presumably, empirical rules could be established for choosing between these two options.

Another option is available for ILP's. If an optimal solution to an LP is not all-integer and the determinant of the basis matrix is not too large, a better bound and possibly an optimal ILP solution can be obtained by solving the corresponding ILPC. The point is that the general methods described in Chapters 4, 5, and 7 are compatible and could all be subroutines in one master algorithm. It is not necessary to use only one of these methods to solve a problem. At each iteration a supervisor routine or a human, if an interactive system is being used, could intervene and choose a tactic based upon the current status of the problem. [See Weingartner and Ness (1967), Gorry and Shapiro (1971), and Geoffrion and Marsten (1972).] Although an intelligent choice might be difficult and might require considerable adaptive learning, a plausible start along these lines could be as follows:

1. If the number of variables required to be integer but which are not integer in the current solution is less than K_1, where K_1 is a suitably small specified constant, separation is a good strategy. In this case the enumeration from the current solution should not be excessive.

2. If the problem is an ILP and the magnitude of the determinant of the current basis matrix is no greater than K_2, where K_2 is a suitably small specified constant, solving the corresponding ILPC is a good strategy. In

this case the ILPC will be solved efficiently. If no optimal solution to the ILPC solves the ILP, we could enumerate as in Section 7.6 or just use the ILPC solution as an upper bound. The decision regarding enumeration could be based on the "amount" of infeasibility in the ILPC solution. Note that even if the current problem is an MILP, an ILPC can be derived by setting all of the nonbasic continuous variables to zero. If the number of continuous nonbasic variables is small, this approach might be used to find a good feasible solution to the MILP.

3. The option of adding a cutting plane and reoptimizing the LP is always available. This option is the simplest of the three in the sense that it requires much less work than (2) and does not cause an increase in the number of problems to be solved, as does (1). If a cut produces a satisfactory decrease in the dual objective function, adding it should be a good strategy.

Of course, it is to be expected that situations will be encountered in which none of the three strategies will appear to be attractive. However, there are other alternatives:

4. Generate a surrogate constraint (see Section 4.7) in the hope that new information will become available; for example, a good source row for a cut.

5. Use an approximate method to obtain a good feasible solution.

The synthesis that we have proposed is based entirely on existing methodology; advances in the theory of integer programming and computer technology will surely lead to quite different algorithms.

General ILP and MILP algorithms cannot be expected to perform as well on highly structured problems as special purpose algorithms. A notable case is the edge matching algorithm of Section 3.7 that can solve very large problems more rapidly than one would expect to solve the corresponding LP.

Special purpose algorithms have also markedly outperformed general algorithms for more difficult combinatorial optimization problems associated with graphs. A good illustration is the traveling salesman problem, where branch and bound schemes based on assignment, matching, and spanning tree relaxations have solved large problems.

Structure in MILP's can be capitalized on, using the partitioning scheme of Section 4.9, since the decomposition of an MILP into a sequence of ILP's and LP's preserves the structure of the continuous part of the problem. For example, in fixed charge transportation problems, the network flow structure of the LP is preserved.

Special purpose algorithms for set covering and partitioning problems have been given in Chapter 8. Based on factors such as the efficient handling of binary data, the ease of finding basic feasible integer solutions for the covering problem, and the stringent requirements for feasibility in the partitioning problem, these algorithms have been somewhat more successful than general algorithms. However, the solution of many real covering and partitioning models is beyond our current capabilities because the problems involve a very large number of variables and/or constraints. The development of efficient special purpose algorithms for these problems awaits a better understanding of the properties of structured binary matrices.

Bibliography*

Abadie, J., ed. (1967), *Nonlinear Programming*, John Wiley & Sons.

———, ed. (1970), *Integer and Nonlinear Programming*, American Elsevier.

Agin, N. (1966), "Optimum Seeking with Branch-and-Bound," *Man. Sci. 13*, B176–B185.

Alcaly, R. E., and A. V. Klevorick (1966), "A Note on the Dual Prices of Integer Programs," *Econometrica 34*, 206–214.

Aldrich, D. W. (1969), "A Decomposition Approach to the Mixed Integer Programming Problem," Ph.D. Dissertation, School of Industrial Engineering, Purdue University.

Anderson, I. (1971), "Perfect Matchings in a Graph," *J. Comb. Theory 10*, 183–186.

Andrew, G., T. Hoffman, and C. Krabek (1968), "On the Generalized Set Covering Problem," Control Data Corporation.

Appelgren, L. H. (1971), "Integer Programming Methods for a Vessel Scheduling Problem," *Transp. Sci. 5*, 64–78.

Arabeyre, J. P., J. Fearnley, F. C. Steiger, and W. Teather (1969), "The Airline Crew Scheduling Problem: A Survey," *Trans. Sci. 3*, 140–163.

Armour, G. C., and E. S. Buffa (1965), "A Heuristic Algorithm and Simulation Approach to Relative Location of Facilities," *Man. Sci. 9*, 294–309.

* When a technical report has been superseded by a published paper, only the published paper is cited. Typically the date associated with a published paper will be one or two years later than the corresponding technical report.

Arnold, L. R., and M. Bellmore (1971a), "A Bounding Minimization Problem for Primal Integer Programming," Dept. of Operations Research, The Johns Hopkins University.

——— (1971b), "A Generated Cut for Primal Integer Programming," Dept. of Operations Research, The Johns Hopkins University.

——— (1971c), "Iteration Skipping in Primal Integer Programming," Dept. of Operations Research, The Johns Hopkins University.

Aronofsky, J., ed. (1969), *Progress in Operations Research*, Vol. 3: *Relationship between Operations Research and the Computer*, John Wiley & Sons.

Arthanari, T. S., and K. G. Ramamurthy (1970), "A Branch-and-Bound Algorithm for Sequencing *n*-Jobs on *m*-Parallel Processors," *Opsearch 7*, 147–156.

Ashour, S. (1970a), "A Branch-and-Bound Algorithm for Flow Shop Scheduling Problems," *AIIE Trans. II*, 172–176.

——— (1970b), "An Experimental Investigation and Comparative Evaluation of Flow Shop Scheduling Techniques," *Opns. Res. 18*, 541–548.

Ashour, S., and A. R. Char (1971), "Computational Experience on 0–1 Programming Approaches to Various Combinatorial Problems," *J. Op. Res. Soc. of Japan 13*, 78–107.

Avi-Itzhak, B., ed. (1971), *Developments in Operations Research*, Gordon and Breach.

Balas, E. (1965), "An Additive Algorithm for Solving Linear Programs with Zero-One Variables," *Opns. Res. 13*, 517–546.

——— (1967), "Discrete Programming by the Filter Method," *Opns. Res. 15*, 915–957.

——— (1968), "A Note on the Branch-and-Bound Principle," *Opns. Res. 16*, 442–445, errata *Opns. Res. 16*, 886.

——— (1969a), "Duality in Discrete Programming II: The Quadratic Case," *Man. Sci. 16*, 14–32.

——— (1969b), "Machine Sequencing via Disjunctive Graphs: An Implicit Enumeration Algorithm," *Opns. Res. 17*, 941–957.

——— (1970a), "Duality in Discrete Programming," 179–198, in Kuhn (1970).

——— (1970b), "Machine Sequencing, Disjunctive Graphs and Degree Constrained Subgraphs," *Nav. Res. Log. Quart. 17*, 1–10.

——— (1970c), "Minimax and Duality for Linear and Nonlinear Mixed-Integer Programming," 385–418, in Abadie (1970).

——— (1970d), "Alternative Strategies for Using Intersection Cuts in Integer Programming," Man. Sci. Res. Rep. No. 209, Carnegie-Mellon University.

——— (1970e), "Intersection Cuts from Maximal Convex Extensions of the Ball and Octahedron," Man. Sci. Res. Rep. No. 214, Carnegie-Mellon University.

——— (1971a), "Intersection Cuts—A New Type of Cutting Planes for Integer Programming," *Opns. Res. 19*, 19–39.

——— (1971b), "A Duality Theorem and an Algorithm for (Mixed) Integer Nonlinear Programming," *Lin. Alg. and Its Appl. 4*, 341–352.

——— (1971c), "Integer Programming and Convex Analysis," Man. Sci. Res. Rep. No. 246, Carnegie-Mellon University.

——— (1971d), "A Note on the Asymptotic Theory of Integer Programming and the 0–1 Case," Man. Sci. Res. Rep. No. 249, Carnegie-Mellon University.

——— (1971e), "Ranking the Facets of the Octahedron," Man. Sci. Rep. No. 252. Carnegie-Mellon University.

Balas, E., V. J. Bowman, F. Glover, and D. Sommer (1971), "An Intersection Cut from the Dual of the Unit Hypercube," *Opns. Res. 19*, 40–44.

Balas, E., and M. W. Padberg (1970), "On the Set Covering Problem," Man. Sci. Res. Rep. No. 197, Carnegie-Mellon University.

Balinski, M. L. (1961), "Fixed Cost Transportation Problems," *Nav. Res. Log. Quart. 11*, 41–54.

——— (1965), "Integer Programming: Methods, Uses, Computation," *Man. Sci. 12*, 253–313.

——— (1967), "Some General Methods in Integer Programming," 220–247, in Abadie (1967).

——— (1969a), "Labelling to Obtain a Maximum Matching," *Combinatorial Mathematics and its Applications*, R. C. Bose and T. A. Dowling (eds.), University of North Carolina Press, 585–602.

——— (1969b), "Establishing the Matching Polytope," City University of New York.

——— (1970a), "On Recent Developments in Integer Programming," 267–302, in Kuhn (1970).

——— (1970b), "On Maximum Matching, Minimum Covering and their Connections," 303–312, in Kuhn (1970).

Balinski, M. L., and R. E. Quandt (1964), "On an Integer Program for a Delivery Problem," *Opns. Res. 12*, 300–304.

Balinski, M. L., and K. Spielberg (1969), "Methods for Integer Programming: Algebraic, Combinatorial and Enumerative," 195–292, in Aronofsky (1969).

Baugh, C. R., T. Ibaraka, and S. Muroga (1971), "Results in Using Gomory's All-Integer Algorithm to Design Optimum Logic Networks," *Opns. Res. 19*, 1090–1096.

Baumol, W. J., and P. Wolfe (1958), "A Warehouse Location Problem," *Opns. Res. 6*, 252–263.

Beale, E. M. L. (1958), "A Method for Solving Linear Programming Problems When Some But Not All of the Variables Must Take Integral Values," Stat. Tech. Res. Group Tech. Rep. No. 19, Princeton University.

——— (1965), "Survey of Integer Programming," *Opnal. Res. Quart. 16*, 219–228.

Beale, E. M. L., and R. E. Small (1965), "Mixed Integer Programming by a Branch and Bound Technique," 450–451, in Kalenich (1965).

Beale, E. M. L., and J. A. Tomlin (1969), "Special Facilities in a General Mathematical Programming System for Non-convex Problems Using Ordered Sets of Variables," 8.21–8.28, in Lawrence (1969).

Bellman, R. (1957), *Dynamic Programming*, Princeton University Press.

Bellman, R. E., and S. E. Dreyfus (1962), *Applied Dynamic Programming*, Princeton University Press.

Bellman, R. E., and M. Hall, Jr., eds. (1960), *Combinatorial Analysis*, Proc. Tenth Symp. in Appl. Math. of the Am. Math. Soc.

Bellmore, M., G. Bennington, and S. Lubore (1970), "A Network Isolation Algorithm," *Nav. Res. Log. Quart. 17*, 461–470.

Bellmore, M., H. J. Greenberg, and J. J. Jarvis (1970), "Multi-Commodity Disconnecting Sets," *Man. Sci. 16*, B427–B433.

Bellmore, M., and J. C. Malone (1971), "Pathology of Traveling-Salesman Subtour Elimination Algorithms," *Opns. Res. 19*, 278–307.

Bellmore, M., and G. L. Nemhauser (1968), "The Traveling Salesman Problem: A Survey," *Opns. Res. 16*, 538–558.

Bellmore, M., and H. D. Ratliff (1971a), "Set Covering and Involutory Bases," *Man. Sci. 18*, 194–206.

——— (1971b), "Optimal Defense of Multi-Commodity Networks," *Man. Sci. 18*, B174–B185.

Benders, J. F. (1962), "Partitioning Procedures for Solving Mixed-Variables Programming Problems," *Numerische Mathematik 4*, 238–252.

Benichou, M., J. M. Gauthier, P. Girodet, G. Hentges, G. Ribiere, and O. Vincent

(1971), "Experiments in Mixed-Integer Linear Programming," *Math. Prog. 1*, 76–94.

Ben-Israel, A., and A. Charnes (1962), "On Some Problems in Diophantine Programming," *Cahiers Cent. d'Etudes Recherche Operationelle 4*, 215–280.

Berge, C. (1957), "Two Theorems in Graph Theory," *Proc. Nat. Acad. Sci. 43*, 842–844.

——— (1962), *The Theory of Graphs and Its Applications*, Methuen. Translated by A. Doig from the original French edition, Dunod, 1958.

Bertier, P., and B. Roy (1964), "Procédure de Résolution pour une Classe de Problems Pouvant Avoir un Caractère Combinatoire," *Cahiers Cent. d'Etudes Recherche Operationelle 6*, 202–208. Translated by W. Jewell (1967), ORC Rep. 17–34, University of California, Berkeley.

Bessiere, F. (1965), "Sur la Recherche du Nombre Chromatique d'un Graphe par un Programme Lineaire en Nombres Entiers," *Rev. Franc. Recherche Operationelle 9*, 143–148.

Birkhoff, G., and S. Maclane (1941), *A Survey of Modern Algebra*, Macmillan.

Blankenship, W. A. (1963), "A New Version of the Euclidean Algorithm," *Amer. Math. Monthly 70*, 742–745.

Bowman, E. H. (1959), "The Schedule-Sequencing Problem," *Opns. Res. 7*, 621–624.

——— (1960), "Assembly Line Balancing by Linear Programming," *Opns. Res. 8*. 385–389.

Bowman, V. J. (1969), "Determining Cuts in Integer Programming by Two Variable Diophantine Equations," Tech. Rep. No. 13, Dept. of Statistics, Oregon State University.

Bowman, V. J., and G. L. Nemhauser (1970), "A Finiteness Proof for Modified Dantzig Cuts in Integer Programming," *Nav. Res. Log. Quart. 17*, 309–313.

——— (1971), "Deep Cuts in Integer Programming," *Opsearch 8*, 89–111.

Bozoki, G. (1969), "Minimum Cost Multi-Commodity Networks Flows," Ph.D. Thesis, Purdue University.

Bozoki, G., and J. Richard (1970), "A Branch-and-Bound Algorithm for the Continuous Process Job-Shop Scheduling Problem," *AIIE Trans. II*, 246–252.

Bradley, G. H. (1969), "Equivalent Integer Programs," 455–463, in Lawrence (1969).

——— (1970a), "Algorithm and Bound for the Greatest Common Divisor of *n* Integers," *CACM 13*, 433–436.

——— (1970b), "Algorithms for Hermite and Smith Normal Matrices and Linear Diophantine Equations," Rep. No. 34, Dept. of Administrative Science, Yale University.

——— (1970c), "Equivalent Mixed Integer Programming Problems," Rep. No. 35, Dept. of Administrative Science, Yale University.

——— (1971a), "Equivalent Integer Programs and Canonical Problems," *Man. Sci. 17*, 354–366.

——— (1971b), "Heuristic Solution Methods and Transformed Integer Linear Programming Problems," Rep. No. 43, Dept. of Administrative Science, Yale University.

——— (1971c), "Transformation of Integer Programs to Knapsack Problems," *Discrete Math 1*, 29–45.

Bradley, G. H., and P. N. Wahi (1969), "An Algorithm for Integer Linear Programming: A Combined Algebraic and Enumeration Approach," Rep. No. 29, Dept. of Administrative Science, Yale University.

Brooks, R., and A. M. Geoffrion (1966), "Finding Everett's Lagrange Multipliers by Linear Programming," *Opns. Res. 14*, 1149–1153.

Burdet, C. (1970), "A Class of Cuts and Related Algorithms in Integer Programming," Man. Sci. Res. Rep. No. 220, Carnegie-Mellon University.

Busacker, R. G., and T. L. Saaty (1965), *Finite Graphs and Networks: An Introduction with Applications*, McGraw-Hill.

Cabot, V. A. (1970), "An Enumeration Algorithm for Knapsack Problems," *Opns. Res. 18*, 306–311.

Cabot, V. A., and A. P. Hurter (1968), "An Approach to 0–1 Integer Programming," *Opns. Res. 16*, 1206–1211.

Camion, P. (1965), "Characterization of Totally Unimodular Matrices," *Proc. Am. Math. Soc. 16*, 1068–1073.

Carlson, R. G., and G. L. Nemhauser (1966), "Scheduling to Minimize Interaction Cost," *Opns. Res. 14*, 52–58.

Chandrasekaran, R. (1969), "Total Unimodularity of Matrices," *SIAM J. 11*, 1032–1034.

Charlton, J. M., and C. C. Death (1970), "A Generalized Machine Scheduling Algorithm," *Opnal Res. Quart. 21*, 127–134.

Charnes, A., and W. W. Cooper (1961), *Management Models and Industrial Applications of Linear Programming*, Vols. 1 and 2, John Wiley & Sons.

Childress, J. P. (1969), "Five Petrochemical Industry Applications of Mixed Integer Programming," Bonner and Moore Assoc. Inc. Houston, Texas.

Clarke, G., and S. W. Wright (1964), "Scheduling of Vehicles from a Central Depot to a Number of Delivery Points," *Opns. Res. 12*, 568–581.

Cobham, A., R. Fridshal, and J. H. North (1961), "An Application of Linear Programming to the Minimization of Boolean Functions," Res. Rep. RC-472, IBM.

——— (1962), "A Statistical Study of the Minimization of Boolean Functions Using Integer Programming," Res. Rep. RC-756, IBM.

Cobham, A., and J. H. North (1963), "Extensions of the Integer Programming Approach to the Minimization of Boolean Functions," Res. Rep. RC-915, IBM.

Conway, R. W., W. L. Maxwell, and L. W. Miller (1967), *Theory of Scheduling*, Addison-Wesley.

Cooper, L., and C. Drebes (1967), "An Approximate Solution Method for the Fixed Charge Problem," *Nav. Res. Log. Quart. 14*, 101–113.

Cristofides, N., and S. Eilon (1969), "An Algorithm for the Vehicle Dispatching Problem," *Opnal. Res. Quart. 20*, 309–318.

Crowston, W. B. (1970), "Decision CPM: Network Reduction and Solution," *Opnal. Res. Quart. 21*, 435–452.

Cushing, B. E. (1970), "The Application Potential of Integer Programming," *J. of Bus. 43*, 457–467.

Dakin, R. J. (1961), "Application of Mathematical Programming Techniques to Cost Optimization in Power Generating Systems," Tech. Rep. 19, Basser Computing Dept., University of Sydney, Australia.

——— (1965), "A Tree-Search Algorithm for Mixed Integer Programming Problems," *Computer J. 8*, 250–255.

Dalton, R. E., and R. W. Llewellyn (1966), "An Extension of the Gomory Mixed-Integer Algorithm to Mixed-Discrete Variables," *Man. Sci. 12*, 569–575.

Dantzig, G. B. (1957), "Discrete Variable Extremum Problems," *Opns. Res. 5*, 266–277.

——— (1959), "Notes on Solving Linear Programs in Integers," *Nav. Res. Log. Quart. 6*, 75–76.

——— (1960a), "On the Significance of Solving Linear Programming Problems with Some Integer Variables," *Econometrica 28*, 30–44.

——— (1960b), "On the Shortest Route through a Network," *Man. Sci. 6*, 187–190.

——— (1963), *Linear Programming and Extensions*, Princeton University Press.

Dantzig, G. B., D. R. Fulkerson, and S. M. Johnson (1954), "Solution of a Large Scale Travelling Salesman Problem," *Opns. Res. 2*, 393–410.

——— (1959), "On a Linear Programming, Combinatorial Approach to the Travelling Salesman Problem," *Opns. Res. 7*, 58–66.

Dantzig, G. B., and J. H. Ramser (1959), "The Truck Dispatching Problem," *Man. Sci. 6*, 80–91.

Dantzig, G. B., and A. F. Veinott, Jr., eds. (1968), *Mathematics of the Decision Sciences*, Part I, Amer. Math. Soc.

Davis, E. W., and G. E. Heidorn (1971), "An Algorithm for Optimal Project Scheduling under Multiple Resource Constraints," *Man. Sci. 17*, B803–B816.

Davis, P. S., and T. L. Ray (1969), "A Branch-Bound Algorithm for the Capacitated Facilities Location Problem," *Nav. Res. Log. Quart. 16*, 331–344.

Davis, R. E. (1969), "A Simplex-Search Algorithm for Solving 0–1 Mixed Integer Programs," Tech. Rep. No. 5, Dept. of Operations Research, Stanford University.

Davis, R. E., D. A. Kendrick, and M. Weitzman (1971), "A Branch-and-Bound Algorithm for 0–1 Mixed Integer Programming Problems," *Opns. Res. 19*, 1036–1044.

Day, R. H. (1965), "On Optimal Extracting from a Multiple File Data Storage System: An Application of Integer Programming," *Opns. Res. 13*, 482–494.

Denzler, D. R. (1969), "An Approximative Algorithm for the Fixed Charge Problem," *Nav. Res. Log. Quart. 16*, 411–416.

Desler, J. F., and S. L. Hakimi (1969), "A Graph-Theoretic Approach to a Class of Integer Programming Problems," *Opns. Res. 17*, 1017–1033.

D'Esopo, D. A., and B. Lefkowitz (1964), "Note on an Integer Linear Programming Model for Determining a Minimum Embarkation Fleet," *Nav. Res. Log. Quart. 11*, 79–82.

Devine, M., and F. Glover (1969), "Generating the Nested Faces of the Gomory Polyhedron," Rep. R-69-1, School of Industrial Engineering, University of Oklahoma.

De Werra, D. (1970), "On Some Combinational Problems Arising in Scheduling," *CORS 8*, 165–175.

Dijkstra, E. W. (1959), "A Note on Two Problems in Connexion with Graphs," *Numer. Mathematik 1*, 269–271.

Dragan, I. (1968), "Un Algorithme Lexicographique pour la Résolution des Programmes Polynomiaux en Variables Entières," *RIRO2*, 81–90.

——— (1969), "Un Algorithme Lexicographique pour la Résolution des Programmes Lineaires en Variables Binaires," *Man. Sci. 16*, 246–252.

——— (1970), "An Improvement of the Lexicographical Algorithm for Solving Linear Discrete Programming Problems," CORE Discussion Paper No. 7029, Louvain, Belgium.

Dreyfus, S. E. (1969), "An Appraisal of Some Shortest-Path Algorithms," *Opns. Res. 17*, 395–412.

Driebeek, N. J. (1966), "An Algorithm for the Solution of Mixed Integer Programming Problems," *Man. Sci. 12*, 576–587.

Drysdale, J. K., and P. J. Sandiford (1969), "Heuristic Warehouse Location—A Case History Using a New Method," *CORS 7*, 45–61.

Eastman, W. L. (1958), "Linear Programming with Pattern Constraints," Ph.D. Dissertation, Harvard University.

Echols, R. E., and L. Cooper (1968), "Solution of Integer Linear Programming Problems by Direct Search," *JACM 15*, 75–84.

Edmonds, J. (1962), "Covers and Packings in a Family of Sets," *Bull. Am. Math. Soc. 68*, 494–499.

——— (1965a), "Paths, Trees and Flowers," *Can. J. Math. 17*, 449–467.

——— (1965b), "Maximum Matching and a Polyhedron with 0,1-Vertices," *J. Res. Nat. Bur. Stds. 69B*, 125–130.

——— (1967), "An Introduction to Matching," Lecture Notes, University of Michigan Summer Eng. Conf.

Edmonds, J., and D. R. Fulkerson (1970), "Bottleneck Extrema," *J. Comb. Theory 8*, 299–306.

Edmonds, J., and E. L. Johnson (1970), "Matching: A Well-Solved Class of Integer Linear Programs," *Proc. of the Calgary Int. Conf. on Comb. Structures and Their Appl.*, 89–92, Gordon and Breach.

Efroymson, M. A., and T. L. Ray (1966), "A Branch-and-Bound Algorithm for Plant Location," *Opns. Res. 14*, 361–368.

Eilon, S., and N. Cristofides (1971), "The Loading Problem," *Man. Sci. 17*, 259–268.

Elmaghraby, S. E. (1970), "The Theory of Networks and Management Science, Part I," *Man. Sci. 17*, 1–34.

Elmaghraby, S. E., and M. K. Wig (1970), "On the Treatment of Cutting Stock Problems as Diophantine Programs," Dept. of Industrial Engineering, North Carolina State University.

Everett, H. (1963), "Generalized Lagrange Multiplier Method for Solving Problems of Optimum Allocation of Resources," *Opns. Res. 11*, 399–417.

Faaland, B. (1970), "Generalized Equivalent Integer Programs and Canonical Problems," Tech. Rep. No. 21, Dept. of Operations Research, Stanford University.

——— (1971), "Solution of the Value Independent Knapsack Problem by Partitioning," Tech. Rep. No. 22, Dept. of Operations Research, Stanford University.

Feldman, E., F. A. Lehrer, and T. L. Ray (1966), "Warehouse Location under Continuous Economies of Scale," *Man. Sci. 12*, 670–684.

Finkelstein, J. J. (1970), "Estimation of the Number of Iterations for Gomory's All-Integer Algorithm," *Dokl. Akad. Nauk. SSSR. 193*, 988–992.

Fleischmann, B. (1967), "Computational Experience with the Algorithm of Balas," *Opns. Res. 15*, 153–155.

Florian, M., P. Trepant, and G. B. McMahon (1971), "An Implicit Enumeration Algorithm for the Machine Sequencing Problem," *Man. Sci. 17*, B782–B792.

Ford, L. R., Jr., and D. R. Fulkerson (1962), *Flows in Networks*, Princeton University Press.

Gale, D. (1960), *The Theory of Linear Economic Models*, McGraw-Hill.

Garfinkel, R. S. (1971), "An Improved Algorithm for the Bottleneck Assignment Problem," *Opns. Res. 19*, 1747–1751.

——— (1973), "On Partitioning the Feasible Set in a Branch-and-Bound Algorithm for the Asymmetric Traveling Salesman Problem," *Opns. Res.* (to appear).

Garfinkel, R. S., and G. L. Nemhauser (1969), "The Set Partitioning Problem: Set Covering with Equality Constraints," *Opns. Res. 17*, 848–856.

——— (1970), "Optimal Political Districting by Implicit Enumeration Techniques," *Man. Sci. 16*, B495–B508.

——— (1972), "Optimal Set Covering: A Survey," 164–183, in Geoffrion (1972).

Garfinkel, R. S., and M. R. Rao (1971), "The Bottleneck Transportation Problem," *Nav. Res. Log. Quart.* (to appear).

Gass, S. (1964), *Linear Programming Methods and Applications*, 2nd ed., McGraw-Hill.

Gately, D. (1970), "Investment Planning for the Electric Power Industry: An Integer Programming Approach," Res. Rep. 70, Dept. of Economics, University of West Ontario, London, Canada.

Gavett, J. W., and N. V. Plyter (1966), "The Optimal Assignment of Facilities to Locations by Branch and Bound," *Opns. Res. 14*, 210–232.

Geoffrion, A. M. (1967), "Integer Programming by Implicit Enumeration and Balas' Method," *SIAM Rev. 7*, 178–190.

——— (1969), "An Improved Implicit Enumeration Approach for Integer Programming," *Opns. Res. 17*, 437–454.

———, ed. (1972), *Perspectives on Optimization: A Collection of Expository Articles*, Addison-Wesley.

Geoffrion, A. M., and R. E. Marsten (1972), "Integer Programming: A Framework and State-of-the-Art Survey," *Man. Sci. 18*, 465–491.

Ghare, P. M., D. C. Montgomery, and W. C. Turner (1971), "Optimal Interdiction Policy for a Flow Network," *Nav. Res. Log. Quart. 18*, 37–46.

Giglio, R. J., and H. M. Wagner (1964), "Approximate Solutions to the Three-Machine Scheduling Problem," *Opns. Res. 12*, 306–324.

Gilmore, P. C. (1962), "Optimal and Sub-Optimal Algorithms for the Quadratic Assignment Problem," *SIAM J. 10*, 305–313.

Gilmore, P. C., and R. E. Gomory (1961), "A Linear Programming Approach to the Cutting Stock Problem," *Opns. Res. 9*, 849–859.

——— (1963), "A Linear Programming Approach to the Cutting Stock Problem, Part II," *Opns. Res. 11*, 863–888.

——— (1965), "Multistage Cutting Stock Problems of Two and More Dimensions," *Opns. Res. 13*, 94–120.

——— (1966), "The Theory and Computation of Knapsack Functions," *Opns. Res. 14*, 1045–1074.

Glover, F. (1965a), "A Bound Escalation Method for the Solution of Integer Linear Programs," *Cahiers Cent. d'Etudes Recherche Operationelle 6*, 131–168.

——— (1965b), "A Hybrid-Dual Integer Programming Algorithm," *Cahier Cent. d'Etudes Recherche Operationelle 7*, 5–23.

——— (1965c), "A Multiphase-Dual Algorithm for the Zero-One Integer Programming Problem," *Opns. Res. 13*, 879–919.

——— (1966), "Generalized Cuts in Diophantine Programming," *Man. Sci. 13*, 254–268.

——— (1967a), "A Pseudo Primal-Dual Integer Programming Algorithm," *J. Res. Nat. Bur. Stds. 71B*, 187–195.

——— (1967b), "Maximum Matching in a Convex Bipartite Graph," *Nav. Res. Log. Quart. 14*, 313–316.

——— (1967c), "Stronger Cuts in Integer Programming," *Opns. Res. 15*, 1174–1176.

——— (1968a), "A New Foundation for a Simplified Primal Integer Programming Algorithm," *Opns. Res. 16*, 727–740.

——— (1968b), "Surrogate Constraints," *Opns. Res. 16*, 741–749.

——— (1968c), "A Note on Linear Programming and Integer Feasibility," *Opns. Res. 16*, 1212–1216.

——— (1968d), "Faces of an Integer Polyhedron for an Additive Group," Pub. AMM-11, Graduate School of Business, University of Texas.

——— (1969a), "Integer Programming over a Finite Additive Group," *SIAM J. on Control 7*, 213–231.

——— (1969b), "Convexity Cuts," Graduate School of Business, University of Texas.

——— (1971a), "Faces of the Gomory Polyhedron for Cyclic Groups," *J. Math. Anal. and Appl. 35*, 195–208.

——— (1971b), "A Note on Extreme Point Solutions and a Paper by Lemke, Salkin and Spielberg," *Opns. Res. 19*, 1023–1026.

Glover, F., and L. Litzler (1969), "Extension of an Asymptotic Integer Programming Algorithm to the General Integer Programming Problem," Graduate School of Business, University of Texas.

Glover, F., and R. E. Woolsey (1970), "Aggregating Diophantine Equations," Rep. No. 70-4, University of Colorado.

Glover, F., and S. Zionts (1965), "A Note on the Additive Algorithm of Balas," *Opns. Res. 13*, 546–549.

Goldman, A. J., and G. L. Nemhauser (1967), "A Transport Improvement Problem Transformable to a Best-Path Problem," *Transp. Sci. 1*, 295–307.

Golomb, S. W., and L. O. Baumert (1965), "Backtrack Programming," *JACM 12*, 516–524.

Gomory, R. E. (1958), "Outline of an Algorithm for Integer Solutions to Linear Programs," *Bull. Amer. Math. Soc. 64*, 275–278.

——— (1960a), "Solving Linear Programming Problems in Integers," 211–216, in Bellman and Hall (1960).

——— (1960b), "An Algorithm for the Mixed Integer Problem," RM-2597, RAND Corp.

——— (1963a), "An Algorithm for Integer Solutions to Linear Programs," 269–302, in Graves and Wolfe (1963). (Originally appeared as Princeton-IBM Math. Res. Project Tech. Rep. No. 1, 1958.)

——— (1963b), "All-Integer Integer Programming Algorithm," 193–206, in Muth and Thompson (1963).

——— (1965), "On the Relation between Integer and Non-Integer Solutions to Linear Programs," *Proc. Nat. Acad. Sci. 53*, 260–265.

——— (1967), "Faces of an Integer Polyhedron," *Proc. Nat. Acad. Sci. 57*, 16–18.

——— (1969), "Some Polyhedra Related to Combinatorial Problems," *Lin. Alg. and Appl. 2*, 451–558.

Gomory, R. E., and W. J. Baumol (1960), "Integer Programming and Pricing," *Econometrica 28*, 521–550.

Gomory, R. E., and A. J. Hoffman (1963), "On the Convergence of an Integer Programming Process," *Nav. Res. Log. Quart. 10*, 121–123.

Gomory, R. E., and E. L. Johnson (1971), "Some Continuous Functions Related to Corner Polyhedra," RC-3311, IBM.

Hoffman, A. J., and I. Heller (1962), "On Unimodular Matrices," *Pac. J. Math. 12*, 1321–1327.

Hoffman, A. J., and J. B. Kruskal (1958), "Integral Boundary Points of Convex Polyhedra," 223–246, in Kuhn and Tucker (1958).

House, R. W., L. D. Nelson, and J. Rado (1966), "Computer Studies of a Certain Class of Linear Integer Problems," 241–280, in Lavi and Vogl (1966).

Hu, T. C. (1963), "Multi-Commodity Network Flows," *Opns. Res. 11*, 344–360.

——— (1969), *Integer Programming and Network Flows*, Addison-Wesley.

——— (1970), "On the Asymptotic Integer Algorithm," *Lin. Alg. and Appl. 3*, 279–294.

——— (1971), "Some Problems in Discrete Optimization," *Math. Prog. 1*, 102–112.

Huard, P. (1967), "Resolutions of Mathematical Programming with Nonlinear Constraints by the Method of Centers," 207–219, in Abadie (1967).

——— (1970), "Programmes Mathématiques Nonlineares à Variables Bivalentes," 313–322, in Kuhn (1970).

Ignall, E., and L. Schrage (1965), "Applications of the Branch-and-Bound Technique to Some Flow-Shop Scheduling Problems," *Opns. Res. 13*, 400–412.

Jensen, P. A. (1971), "Optimum Network Partitioning," *Opns. Res. 19*, 916–932.

Jeroslow, R. G. (1969), "On the Unlimited Number of Faces in Integer Hulls of Linear Problems with Two Constraints," Tech. Rep. No. 67, Dept. of Operations Research, Cornell University.

——— (1971), "Comments on Integer Hulls of Two Linear Constraints," *Opns. Res. 19*, 1061–1069.

——— (1972), "There Cannot Be Any Algorithm for Integer Programming with Quadratic Constraints," *Opns. Res.* (to appear).

Jeroslow, R. G., and K. O. Kortanek (1969), "Dense Sets of Two Variable Integer Programs Requiring Arbitrarily Many Cuts by Fractional Algorithms," Man. Sci. Res. Rep. No. 174, Carnegie-Mellon University.

——— (1971), "On an Algorithm of Gomory," *SIAM J. 21*, 55–60.

Jones, A. P., and R. M. Soland (1969), "A Branch-and-Bound Algorithm for Multilevel Fixed Charge Problems," *Man. Sci. 16*, 67–76.

Kalenich, W. A., ed. (1965), *Proc. IFIP Congress*, Vol. 2, Spartan Press, Washington, D.C.

Kalymon, B. A. (1971), "Note Regarding 'A New Approach to Discrete Mathematical Programming'," *Man. Sci. 17*, 777–778.

Kaplan, S. (1966), "Solution of the Lorie-Savage Problem and Similar Integer Programming Problems," *Opns. Res. 14*, 1130–1136.

Karp, R. M. (1972), "Reducibility Among Combinatorial Problems." Tech. Rep. 3, Computer Science, University of California, Berkeley.

Kianfar, F. (1971), "Stronger Inequalities for 0,1 Integer Programming Using Knapsack Functions," *Opns. Res. 19*, 1374–1392.

Kirby, M. J. L., and P. F. Scobey (1970), "Production Scheduling on *N* Identical Machines," *CORS 8*, 14–27.

Kolesar, P. J. (1967), "A Branch and Bound Algorithm for the Knapsack Problem," *Man. Sci. 13*, 723–735.

Kolner, T. N. (1966), "Some Highlights of a Scheduling Matrix Generator System," United Airlines.

Koopmans, T. C., and M. J. Beckmann (1957), "Assignment Problems and the Location of Economic Activities," *Econometrica 25*, 53–76.

Korte, B., W. Krelle, and W. Oberhoffer (1969), "Ein Lexikographischer Suchalgorithmus zur Losung Allgemeiner Ganzzahliger Programmierungsaufgaben, I, II," *Unternehmensforschung 13*, 73–98, 171–192.

——— (1970), "Ein Lexikographischer Suchalgorithmus zur Losung Allgemeiner Ganzzahliger Programmierungsaufgaben—Nachtrag," *Unternehmensforschung 14*, 228–234.

Krolak, P. (1969), "Computational Results of an Integer Programming Algorithm," *Opns. Res. 17*, 743–749.

Krolak, P., W. Felts, and G. Marble (1971), "A Man-Machine Approach toward Solving the Traveling Salesman Problem," *CACM 14*, 327–334.

Kruskal, J. B. (1956), "On the Shortest Spanning Subtree of a Graph and the Traveling Salesman Problem," *Proc. Amer. Math. Soc. 2*, 48–50.

Kuehn, A. A., and M. J. Hamburger (1963), "A Heuristic Program for Locating Warehouses," *Man. Sci. 9*, 643–666.

Kuhn, H. W., ed. (1970), *Proceedings of the Princeton Symposium on Mathematical Programming*, Princeton University Press.

Kuhn, H. W., and W. J. Baumol (1962), "An Approximate Algorithm for the Fixed Charge Transportation Problem," *Nav. Res. Log. Quart. 9*, 1–15.

Kuhn, H. W., and A. W. Tucker, eds. (1958), *Linear Inequalities and Related Systems*, Princeton University Press.

Kunzi, H. P., and W. Oettli (1963), "Integer Quadratic Programming," 303–308, in Graves and Wolfe (1963).

Land, A. H., and A. G. Doig (1960), "An Automatic Method for Solving Discrete Programming Problems," *Econometrica 28*, 497–520.

——— (1965), "A Problem of Assignment with Interrelated Costs," *Opnal. Res. Quart. 14*, 185–199.

Laughhunn, D. J. (1970), "Quadratic Binary Programming with Applications to Capital-Budgeting Problems," *Opns. Res. 18*, 454–461.

Lavi, A., and T. Vogl (1966), *Recent Advances in Optimization Techniques*, John Wiley & Sons.

Lawler, E. L. (1963), "The Quadratic Assignment Problem," *Man. Sci. 9*, 586–599.

——— (1966), "Covering Problem: Duality Relations and a New Method of Solution," *SIAM J. 14*, 1115–1132.

Lawler, E. L., and M. D. Bell (1966), "A Method for Solving Discrete Optimization Problems," *Opns. Res. 14*, 1098–1112.

Lawler, E. L., and D. E. Wood (1966), "Branch-and-Bound Methods: A Survey," *Opns. Res. 14*, 699–719.

Lawrence, J., ed. (1969), *Proc. Fifth Int. Conf. of Operational Res.*, Venice, Tavestock Pub., London.

Lemke, C. E., H. M. Salkin, and K. Spielberg (1971), "Set Covering by Single Branch Enumeration with Linear Programming Subproblems," *Opns. Res. 19*, 998–1022.

Lemke, C. E., and K. Spielberg (1967), "Direct Search Zero-One and Mixed Integer Programming," *Opns. Res. 15*, 892–914.

Lin, S. (1965), "Computer Solutions of the Traveling Salesman Problem," *Bell System Tech. J. 44*, 2245–2269.

Little, J. D. C. (1966), "The Synchronization of Traffic Signals by Mixed Integer Linear Programming," *Opns. Res. 14*, 568–594.

Little, J. D. C., K. G. Murty, D. W. Sweeney, and C. Karel (1963), "An Algorithm for the Traveling Salesman Problem," *Opns. Res. 11*, 979–989.

Lomnicki, Z. (1965), "A Branch-and-Bound Algorithm for the Exact Solution of the Three-Machine Scheduling Problem," *Opnal. Res. Quart. 16*, 89–100.

Lorie, J., and L. J. Savage (1955), "Three Problems in Capital Rationing," *J. of Bus. 28*, 229–239.

MacDuffee, C. C. (1940), *An Introduction to Abstract Algebra*, John Wiley & Sons.

Manne, A. S. (1960), "On the Job-Shop Scheduling Problem," *Opns. Res. 8*, 219–223.

——— (1964), "Plant Location under Economies of Scale-Decentralization and Computation," *Man. Sci. 11*, 213–235.

——— (1971), "A Mixed Integer Algorithm for Project Evaluation," Int. Bank. in Reconstruction and Development, Washington, D.C.

Mao, J. C. T., and B. A. Wallingford (1968), "An Extension of Lawler and Bell's Method of Discrete Optimization with Examples from Capital Budgeting," *Man. Sci. 15*, B51–B60.

Markowitz, H. M., and A. S. Manne (1957), "On the Solution of Discrete Programming Problems," *Econometrica 25*, 84–110.

Marsten, R. E. (1971), "An Implicit Enumeration Algorithm for the Set Partitioning Problem with Side Constraints," Ph.D. Dissertation, University of California, Los Angeles.

Martin, G. T. (1963), "An Accelerated Euclidean Algorithm for Integer Linear Programming," 311–318, in Graves and Wolfe (1963).

——— (1966), "Solving the Traveling Salesman Problem by Integer Linear Programming," CEIR, New York.

Martin-Lof, A. (1970), "A Branch-and-Bound Algorithm for Determining the Minimal Fleet Size of a Transportation System," *Transp. Sci. 4*, 159–163.

Mason, A. T., and C. L. Moodie (1971), "A Branch-and-Bound Algorithm for Minimizing Cost in Project Scheduling," *Man. Sci. 18*, B158–B173.

Mathis, S. J., Jr. (1971), "A Counterexample to the Rudimentary Primal Integer Programming Algorithm," *Opns. Res. 19*, 1518–1522.

McMahon, G. B., and P. G. Burton (1967), "Flow Shop Scheduling with the Branch-and-Bound Method," *Opns. Res. 15*, 473–481.

Miercort, R. A., and R. M. Soland (1971), "Optimal Allocation of Missiles against Area and Point Defenses," *Opns. Res. 11*, 605–617.

Miller, B. L. (1971), "On Minimizing Non-Separable Functions Defined on the Integers with an Inventory Application," *SIAM J. 21*, 166–185.

Miller, C. E., A. W. Tucker, and R. A. Zemlin (1960), "Integer Programming Formulation of Travelling Salesmen Problems," *JACM 7*, 326–329.

Mitra, G., B. Richards, and K. Wolfenden (1970), "An Improved Algorithm for the Solution of Linear Programs by the Solution of Associated Diophantine Equations," *R.I.R.O-R.1*, 47–66.

Mitten, L. G. (1970), "Branch-and-Bound Methods: General Formulation and Properties," *Opns. Res. 18*, 24–34.

Mizukami, K. (1968), "Optimum Redundancy for Maximum System Reliability by the Method of Convex and Integer Programming," *Opns. Res. 16*, 392–406.

Morrison, D. R. (1969), "Matching Algorithms," *J. Comb. Theory 6*, 20–32.

Mueller-Merbach, H. (1970), "Approximation Methods for Integer Programming," Johannes Gutenberg University, Mainz, Germany.

——— (1971), "Heuristic Methods: Structures, Applications, Computational Experience," Johannes Gutenberg University, Mainz, Germany.

Murty, K. G. (1968), "Solving the Fixed Charge Problem by Ranking the Extreme Points," *Opns. Res. 16*, 268–279.

Muth, J. F., and G. L. Thompson, eds. (1963), *Industrial Scheduling*, Prentice-Hall.

Nemhauser, G. L. (1966), *Introduction to Dynamic Programming*, John Wiley & Sons.

——— (1972), "A Generalized Permanent Label Setting Algorithm for the Shortest Path between Specified Nodes," *J. Math. Anal. and Appl.* (to appear).

Nemhauser, G. L., and Z. Ullman (1968), "A Note on the Generalized Lagrange Multiplier Solution to an Integer Programming Problem," *Opns. Res. 16*, 450–452.

——— (1969), "Discrete Dynamic Programming and Capital Allocation," *Man. Sci. 15*, 494–505.

Nemhauser, G. L., and W. B. Widhelm (1971), "A Modified Linear Program for Columnar Methods in Mathematical Programming," *Opns. Res. 19*, 1051–1060.

Norman, R. Z., and M. O. Rabin (1959), "An Algorithm for the Minimum Cover of a Graph," *Proc. Amer. Math. Soc. 10*, 315–319.

Orchard-Hays, W. (1968), *Advanced Linear Programming Computing Techniques*, McGraw-Hill.

Padberg, M. (1970), "Equivalent Knapsack-type Formulations of Bounded Integer Linear Programs," Man. Sci. Res. Rep. No. 227, Carnegie-Mellon University.

——— (1971), "Simple Zero-One Problems: Set Covering, Matchings and Coverings in Graphs," Man. Sci. Res. Rep. No. 235, Carnegie-Mellon University.

Palmer, D. S. (1965), "Sequencing Jobs through A Multi-stage Process in the Minimum Total Time," *Opnal. Res. Quart. 16*, 101–107.

Pandit, S. N. N. (1962), "The Loading Problem," *Opns. Res. 10*, 639–646.

Paul, M. C., and S. H. Unger (1959), "Minimizing the Number of States in Incompletely Specified Sequential Functions," *IRE Trans. Electronic Computers EC-8*, 356–367.

Petersen, C. C. (1967), "Computational Experience with Variants of the Balas Algorithm Applied to the Selection of R and D Projects," *Man. Sci. 13*, 736–750.

Pierce, J. F. (1968), "Application of Combinatorial Programming to a Class of All-Zero-One Integer Programming Problems," *Man. Sci. 15*, 191–209.

Pierce, J. F., and W. B. Crowston (1971), "Tree Search Algorithms in Quadratic Assignment Problems," *Nav. Res. Log. Quart. 18*, 1–36.

Pierce, J. F., and J. S. Lasky (1970), "Improved Combinatorial Programming Algorithms for a Class of All-Zero-One Integer Programming Problems," IBM Cambridge Sci. Cent. Rep.

Plane, D. R., and C. McMillan (1971), *Discrete Optimization: Integer Programming and Network Analysis for Management Decisions*, Prentice-Hall.

Pnueli, A. (1971), "An Improved Starting Point for Integer Linear Programming Algorithms," 83–93, in Avi-Itzhak (1971).

Pollatschek, M. A., and B. Avi-Itzhak (1971), "A Class of Deep-Cut Procedures for Integer or Mixed Linear Programs," Rep. No. 85, Faculty of Industrial and Management Engineering, Technion, Israel.

Polya, G. (1957), *How To Solve It*, Doubleday.

Pritsker, A. A., L. J. Watters, and P. M. Wolfe (1969). "Multiproject Scheduling with Limited Resources: A 0–1 Programming Approach," *Man. Sci. 16*, 93–108.

Pyne, I. B., and E. J. McCluskey, Jr. (1961), "An Essay on Prime Implicant Tables," *SIAM J. 9*, 604–631.

Quine, W. V. (1955), "A Way to Simplify Truth Functions," *Am. Math. Mon. 62*, 627–631.

Raghavachari, M. (1969), "On Connections between 0–1 Integer Programming and Concave Programming under Linear Constraints," *Opns. Res. 17*, 680–684.

Rebelein, P. R. (1968), "An Extension of the Algorithm of Driebeek for Solving Mixed Integer Programming Problems," *Opns. Res. 16*, 193–197.

Rebman, K. R. (1969), "Non-Unimodular Network Programming," Ph.D. Dissertation, University of Michigan.

Reiter, S., and D. B. Rice (1966), "Discrete Optimizing Solution Procedures for Linear and Nonlinear Integer Programming Problems," *Man. Sci. 12*, 829–850.

Reiter, S., and G. Sherman (1965), "Discrete Optimizing," *SIAM J. 13*, 864–899.

Revelle, C., D. Marks, and J. C. Liebman (1970), "An Analysis of Private and Public Sector Location Models," *Man. Sci. 11*, 692–707.

Roberts, S. D., and C. D. Villa (1970), "On A Multiproduct Assembly Line Balancing Problem," *AIIE Trans. II*, 361–364.

Robillard, P. (1971), "(0,1) Hyperbolic Programming," *Nav. Res. Log. Quart. 18*, 47–58.

Roth, R. H. (1969), "Computer Solutions to Minimum Cover Problems," *Opns. Res. 17*, 455–466.

——— (1970), "An Approach to Solving Linear Discrete Optimization Problems," *JACM 17*, 303–313.

Rothschild, B., and A. B. Whinston (1966a), "On Two-Commodity Network Flows," *Opns. Res. 14*, 377–387.

——— (1966b), "Feasibility of Two-Commodity Network Flows," *Opns. Res. 14*, 1121–1129.

Roveda, C., and R. Schmid (1971), "Two Algorithms for a Plant Storehouse Location Problem," *Unternehmensforschung 15*, 30–44.

Roy, B., R. Benayoun, and J. Tergny (1970), "From S. E. P. Procedure to the Mixed Ophelie Program," 419–436, in Abadie (1970).

Rubin, A. A., and P. L. Hammer (1969), "Quadratic Programming with 0–1 Variables," Rep. No. 53, Faculty of Industrial and Management Engineering, Technion, Israel.

Rubin, D. S. (1970), "On the Unlimited Number of Faces in Integer Hulls of Linear Programs with a Single Constraint," *Opns. Res. 18*, 940–946.

——— (1972), "Redundant Constraints and Extraneous Variables in Integer Programs," *Man. Sci. 18*, 423–427.

Rubin, D. S. and R. L. Graves (1972), "Strengthened Dantzig Cuts for Integer Programming." *Opns. Res. 20*, 173–177.

Rudeanu, S. (1969), "Irredundant Optimization of a Pseudo-Boolean Function," *J.O.T.A. 4*, 253–259.

Rutledge, R. W. (1967), "A Simplex Method for 0–1 Mixed Integer Linear Programs," *J. Math. Anal. and Appl. 18*, 377–390.

Saaty, T. L. (1970), *Optimization in Integers and Related Extremal Problems*, McGraw-Hill.

Salkin, H. M. (1970), "On the Merit of the Generalized Origin and Restarts in Implicit Enumeration," *Opns. Res. 18*, 549–554.

——— (1971), "A Note on Gomory Fractional Cuts," *Opns. Res. 19*, 1538–1541.

Salkin, H. M., and P. Breining (1971), "Integer Points on the Gomory Fractional Cut," Dept. of Operations Research, Case Western Reserve University.

Salveson, M. E. (1955), "The Assembly Line Balancing Problem," *J. of Ind. Eng. 6*, 18–25.

Schrage, L. (1970), "Solving Resource-Constrained Network Problems by Implicit Enumeration—Nonpreemptive Case," *Opns. Res. 18*, 263–278.

Senju, S., and Y. Toyoda (1968), "An Approach to Linear Programming with 0–1 Variables," *Man. Sci. 15*, B196–B207.

Shaftel, T. (1971), "An Integer Approach to Modular Design," *Opns. Res. 19*, 130–134.

Shannon, R. E., and J. P. Ignizio (1970), "A Heuristic Programming Algorithm for Warehouse Location," *AIIE Trans. II*, 361–364.

Shapiro, D. (1966), "Algorithms for the Solution of the Optimal Cost Traveling Salesman Problem," Sc.D. Thesis, Washington University, St. Louis.

Shapiro, J. F. (1968a), "Dynamic Programming Algorithms for the Integer Programming Problem-I: The Integer Programming Problem Viewed as a Knapsack Type Problem," *Opns. Res. 16*, 103–121.

——— (1968b), "Group Theoretic Algorithms for the Integer Programming Problem-II: Extension to a General Algorithm," *Opns. Res. 16*, 928–947.

——— (1968c), "Shortest Route Methods for Finite State Space Deterministic Dynamic Programming Problems," *SIAM J. 16*, 1232–1250.

——— (1970), "Turnpike Theorems for Integer Programming Problems," *Opns. Res. 18*, 432–440.

——— (1971), "Generalized Lagrange Multipliers in Integer Programming," *Opns. Res. 19*, 68–76.

Shapiro, J. F., and Wagner, H. M. (1967), "A Finite Renewal Algorithm for the Knapsack and Turnpike Models," *Opns. Res. 15*, 319–341.

Shareshian, R. (1966), "A Modification of the Mixed Integer Algorithm of N. Driebeek," Rep. No. 939007, IBM.

Shareshian, R., and K. Spielberg (1966), "The Mixed Integer Algorithm of N. Driebeek," Rep. No. 939013, IBM.

Simonnard, M. (1966), *Linear Programming*, Prentice-Hall. Translated from the original French edition by W. S. Jewell.

Spielberg, K. (1969a), "Algorithms for the Simple Plant-Location Problem with Some Side Conditions," *Opns. Res. 17*, 85–111.

——— (1969b), "Plant Location with Generalized Search Origin," *Man. Sci. 16*, 165–178.

Srinivasan, A. V. (1965), "An Investigation of Some Computational Aspects of Integer Programming," *JACM 12*, 525–535.

Steinberg, D. I. (1970), "The Fixed Charge Problem," *Nav. Res. Log. Quart. 17*, 217–236.

Steinmann, H., and R. Schwinn (1969), "Computational Experience with a 0–1 Programming Problem," *Opns. Res. 17*, 917–920.

Swarc, W. (1971), "Some Remarks on the Time Transportation Problem," *Nav. Res. Log. Quart.* (to appear).

Taha, H. A. (1971), "A Class of Convexity Cuts for 0–1 Linear Programs with Application to the Minimization of Certain Concave Problems," Rep. No. 71–3, Dept. of Industrial Engineering, University of Arkansas.

Teruo, S. (1970), "A Method for the Optimal Scheduling of a Project with Resource Constraints," *J. Opns. Res. Soc. of Jap. 12*, 52–64.

Thangavelu, S. R., and C. M. Shetty (1971), "Assembly Line Balancing by 0–1 Integer Programming," *AIIE Trans. III*, 64–69.

Thiriez, H. (1969), "Airline Crew Scheduling: A Group Theoretic Approach," Rep. R-67, Flight Transportation Laboratory, Massachusetts Institute of Technology.

Thompson, G. L. (1964), "The Stopped Simplex Method: Basic Theory for Mixed Integer Programming," *Rev. Franc. Recherche Operationelle 8*, 159–182.

Tomlin, J. A. (1970), "Branch and Bound Methods for Integer and Non-Convex Programming," 437–450, in Abadie (1970).

——— (1971), "An Improved Branch-and-Bound Method for Integer Programming," *Opns. Res. 19*, 1070–1074.

Toregas, C., R. Swain, C. Revelle, and L. Bergman (1971), "The Location of Emergency Service Facilities," *Opns. Res. 19*, 1363–1373.

Trauth, C. A., and R. E. Woolsey (1968), "MESA; A Heuristic Integer Linear Programming Technique," Res. Rep. SC-RR-68-299, Sandia Labs, Albuquerque, New Mexico.

——— (1969), "Integer Linear Programming: A Study in Computational Efficiency," *Man. Sci. 15*, 481–493.

Trotter, L. E., Jr., and C. M. Shetty (1971), "An Algorithm for the Bounded Variable Integer Programming Problem," Dept. of Industrial Engineering, Georgia Institute of Technology.

Tuan, N. Ph. (1971), "A Flexible Tree Search Method for Integer Programming Problems," *Opns. Res. 19*, 115–119.

Tutte, W. T. (1947), "The Factorisations of a Graph," *J. Lond. Math. Soc. 22*, 107–111.

Unger, V. E. (1970), "Capital Budgeting and Mixed 0–1 Integer Programming," *AIIE Trans. II*, 28–36.

Veinott, A. F., Jr. (1968), "Extreme Points of Leontief Substitution Systems," *Lin. Alg. Appl. 1*, 181–194.

Veinott, A. F., Jr., and G. B. Dantzig (1968), "Integral Extreme Points," *SIAM Rev. 10*, 371–372.

Von Lanzenauer, C. H. (1970), "A Linear Programming Solution to the General Sequencing Problem," *CORS 8*, 129–134.

Wagner, H. M. (1959), "An Integer Linear Programming Model for Machine Scheduling," *Nav. Res. Log. Quart. 6*, 131–140.

Wagner, H. M., R. J. Giglio, and R. G. Glaser (1964), "Preventive Maintenance Scheduling by Mathematical Programming," *Man. Sci. 10*, 316–334.

Wagner, W. H. (1968), "An Application of Integer Programming to Legislative Redistricting," presented at the 34th Nat. Meeting of ORSA.

Wahi, P. N., and G. H. Bradley (1969), "Integer Programming Test Problems," Rep. No. 28, Dept. of Administrative Science, Yale University.

Walker, R. J. (1960), "An Enumerative Technique for a Class of Combinational Problems," 90–94, in Bellman and Hall (1960).

Webb, M. H. J. (1971), "Some Methods of Producing Approximate Solutions to Travelling Salesmen Problems with Hundreds or Thousands of Cities," *Opnal. Res. Quart. 22*, 49–66.

Weingartner, H. M. (1963), *Mathematical Programming and the Analysis of Capital Budgeting Problems*, Prentice-Hall.

——— (1966), "Capital Budgeting of Interrelated Projects: Survey and Synthesis," *Man. Sci. 12*, 485–516.

Weingartner, H. M., and D. N. Ness (1967), "Methods for the Solution of Multi-Dimensional 0/1 Knapsack Problems," *Opns. Res. 15*, 83–103.

White, J. L., and R. L. Francis (1971), "Solving a Segregated Storage Problem Using Branch-and-Bound and Extreme Point Ranking Methods," *AIIE Trans. III*, 37–44.

White, L. J. (1971), "Minimum Covers of Fixed Cardinality in Weighted Graphs," *SIAM J. 21*, 104–113.

White, W. W. (1961), "On Gomory's Mixed Integer Algorithm," Senior Thesis, Princeton University.

——— (1966), "On a Group Theoretic Approach to Linear Integer Programming," ORC 66–27, Operations Research Center, University of California, Berkeley.

Wilson, R. B. (1967), "Stronger Cuts in Gomory's All-Integer Programming Algorithm," *Opns. Res. 15*, 155–157.

——— (1970), "Integer Programming via Modular Representations," *Man. Sci. 16*, 289–294.

Witzgall, C. (1963), "An All-Integer Programming Algorithm with Parabolic Constraints," *SIAM J. 11*, 855–871.

Witzgall, C., and C. T. Zahn, Jr. (1965), "Modification of Edmonds' Maximum Matching Algorithm," *J. Res. Nat. Bur. Stds. 69B*, 91–98.

Wolsey, L. A. (1970), "Mixed Integer Programming: Discretization and the Group-Theoretic Approach," Ph.D. Dissertation, Massachusetts Institute of Technology.

——— (1971a), "Extensions of the Group Theoretic Approach in Integer Programming," *Man. Sci. 18*, 74–83.

——— (1971b), "Group-Theoretic Results in Mixed Integer Programming," *Opns. Res. 19*, 1691–1697.

Woolsey, R. E. (1971), "Comment on Briskin's Note," *Man. Sci. 17*, 500–501.

——— (1972), "A Candle to Saint Jude, or Four Real World Applications of Integer Programming," *Interfaces 2*, (Bulletin of TIMS) 20–27.

Young, R. D. (1965), "A Primal (All-Integer) Integer Programming Algorithm," *J. Res. Nat. Bur. Stds. 69B*, 213–250.

——— (1968a), "A Simplified Primal (All-Integer) Integer Programming Algorithm," *Opns. Res. 16*, 750–782.

——— (1968b), "New Cuts for a Special Class of 0–1 Integer Programs," Rice University.

——— (1971), "Hypercylindrically Deduced Cuts in 0–1 Integer Programming," *Opns. Res. 19*, 1393–1405.

Zionts, S. (1968), "On an Algorithm for the Solution of Mixed Integer Programming Problems," *Man. Sci. 15*, 113–116.

——— (1969), "Toward a Unifying Theory of Integer Linear Programming," *Opns. Res. 17*, 359–367.

Zoutendijk, G. (1970), "Enumeration Algorithms for the Pure and Mixed Integer Programming Problem," 323–338, in Kuhn (1970).

Author Index

Subject Index